A Commemorative Issue in Honor of Professor Nick Hadjiliadis

A Commemorative Issue in Honor of Professor Nick Hadjiliadis

Metal Complex Interactions with Nucleic Acids and/or DNA

Special Issue Editors

Sotiris K. Hadjikakou
Christina N. Banti

MDPI • Basel • Beijing • Wuhan • Barcelona • Belgrade

MDPI

Special Issue Editors
Sotiris K. Hadjikakou
University of Ioannina
Greece

Christina N. Banti
University of Ioannina
Greece

Editorial Office
MDPI
St. Alban-Anlage 66
4052 Basel, Switzerland

This is a reprint of articles from the Special Issue published online in the open access journal *International Journal of Molecular Sciences* (ISSN 1422-0067) in 2018 (available at: https://www.mdpi.com/journal/ijms/special_issues/metal_nucleic_acids)

For citation purposes, cite each article independently as indicated on the article page online and as indicated below:

LastName, A.A.; LastName, B.B.; LastName, C.C. Article Title. *Journal Name* **Year**, *Article Number*, Page Range.

ISBN 978-3-03897-511-3 (Pbk)
ISBN 978-3-03897-512-0 (PDF)

Cover image courtesy of shutterstock.com.

Contents

About the Special Issue Editors

Sotiris K. Hadjikakou was born in Famagusta, Cyprus. He graduated from the Chemistry Department of the Aristotle University of Thessaloniki, Greece. He was awarded a Ph.D. in chemistry in 1992. He was appointed to the position of Lecturer in Inorganic Chemistry in the University of Ioannina, Greece, where he is serving as Professor (since 2013). He has worked as a visiting researcher at the University of Dortmund, Germany and at the University of Essex in Colchester, UK. His research interests in the field of inorganic biochemistry mainly focus on drug design and development from the modification of NSAIDs, antibiotics, or ingredient of natural products, for targeted chemotherapy. He currently coordinates the International Graduate Program "Biological Inorganic Chemistry" and he is a member of the Editorial Board of the Section of Bioinorganic Chemistry in *Int. J. Mol. Sci.* (*IJMS*).

Christina N. Banti received her bachelor's degree in Biological Applications and Technologies, University of Ioannina, Greece in 2009. She was awarded a M.Sc. in Bioinorganic Chemistry, University of Ioannina, and obtained her Ph.D. from the Department of Physiology, School of Medicine, University of Ioannina, Greece. She is now a Postdoctoral Researcher and Adjunct Lecturer of the Department of Chemistry, University of Ioannina. Her research interests are mainly focused on bioinorganic chemistry and particularly on the evaluation of new anticancer or antibacterial metallodrugs and their study of their interaction with DNA or enzymes.

Preface to "A Commemorative Issue in Honor of Professor Nick Hadjiliadis"

This Special Issue of the International Journal of Molecular Science comprises a comprehensive study on "Metal Complex Interactions with Nucleic Acids and/or DNA". This Special Issue has been inspired by the important contribution of Prof. Nick Hadjiliadis to the field of palladium or/and platinum/nucleic acid interactions. It covers a selection of recent research and review articles in the field of metal complex interactions with nucleic acids and/or DNA.

Metal complexes have long been recognized as critically important components of nucleic acid chemistry, both in the regulation of gene expression and as promising therapeutic agents. The ability to recognize, to understand at the molecular level of how metal complexes interact with DNA has become an active research area at the interface between biological inorganic chemistry, molecular biology, and medicine.

Arguably, the most prominent drug which contains a metal is cisplatin, the most widely used anti-cancer drug. The success of cisplatin in chemotherapy and the clarification of its mechanism of action through interaction with DNA has motivated a large number of studies on metal complexes interaction with nucleic acids or/and DNA. From the foregoing that the reader of this Special Issue will gain an appreciation of the real role of the interactions of metal complexes with nucleic acids or/and DNA in modern medicine.

This Special Issue on the "Metal complexes interactions with Nucleic acids and/or DNA" aims to provide an overview of this increasingly diverse field, presenting recent developments, and the latest research with particular emphasis on metal-based drugs and metal ion toxicity.

Sotiris K. Hadjikakou, Christina N. Banti

Special Issue Editors

International Journal of
Molecular Sciences

MDPI

Editorial

Preface to "A Commemorative Issue in Honour of Professor Nick Hadjiliadis: Metal Complex Interactions with Nucleic Acids and/or DNA"

Christina N. Banti and Sotiris K. Hadjikakou *

Section of Inorganic and Analytical Chemistry, Department of Chemistry, University of Ioannina, 45110 Ioannina, Greece; cbanti@cc.uoi.gr
* Correspondence: shadjika@uoi.gr; Tel.: +30-2651-008374

Received: 27 November 2018; Accepted: 30 November 2018; Published: 30 November 2018

This Special Issue of the *International Journal of Molecular Science* comprises a comprehensive study on "Metal Complex Interactions with Nucleic Acids and/or DNA". This Special Issue has been inspired by the important contribution of Prof. Nick Hadjiliadis in the field of palladium or/and platinum/nucleic acid interactions. It covers a selection of recent research and review articles in the field of metal complex interactions with nucleic acids and/or DNA.

Metal complexes have long been recognized as critically important components of nucleic acid chemistry, both in the regulation of gene expression and as promising therapeutic agents. The ability to recognize and understand how metal complexes interact with DNA at the molecular level has become an active research area at the interface between biological inorganic chemistry, molecular biology, and medicine. Arguably the most prominent drug which contains a metal is cisplatin, the most widely used anti-cancer drug. The success of cisplatin in chemotherapy and the clarification of its mechanism of action through its interaction with DNA has motivated a large number of studies on metal complex interactions with nucleic acids or/and DNA. Thus, the reader of this Special Issue will gain an appreciation of the real role of the interactions of metal complexes with nucleic acids or/and DNA in modern medicine.

This Special Issue on "Metal complex Interactions with Nucleic Acids and/or DNA" provides an overview of this increasingly diverse field, presenting recent developments and the latest research with particular emphasis on metal-based drugs and metal ion toxicity.

Inorganic biochemistry or bioinorganic chemistry is a multidiscipline field which involves inorganic chemistry, biochemistry, spectroscopy, material science, biology, and medicine. The introduction of metal ions or metal ion binding components into a biological system for the treatment of diseases is one of the main subdivisions in the field of inorganic biochemistry. Nowadays, at the forefront of the field is the development of new metallodrugs for diagnostics (radiopharmaceuticals drugs), medicines (anticancer, antimicrobial/antiparasitic, therapeutic radiopharmaceuticals, photochemotherapeutic metallodrugs, antiarthritic, antidiabetes, antiviral, metallodrugs addressing deficiencies syndromes), and tools for chemical biology, biocatalysis, and bioelectronics, as well as the characterization of metalloproteins, enzymes, and their model complexes. Recently, biomaterials have been applied in many cases such as cardiovascular medical devices, orthopedic and dental applications, ophthalmologic applications, bioelectrodes and biosensors, burn dressings and skin substitutes, sutures, and drug delivery systems. For the coming years, the well-defined bioavailability, absorption, distribution, metabolism, and excretion of metal-based drugs will be the main target of new metallodrugs. The use of nanoparticles, micelle emulsions, and liposomal formulations can open new opportunities for improved delivery, cell uptake, and targeting.

Emeritus Professor Nick Hadjiliadis is one of the pioneers in the field of bioinorganic chemistry. He graduated from the University of Athens, and subsequently earned a Masters in Science from the University of Montreal and a Doctor of Chemistry, PhD from the University of Montreal in 1975. He served as a Professor of Inorganic and General Chemistry at the Department of Chemistry of the University of Ioannina, Greece, since 1980. He worked on the interaction of metal ions with nucleobases and nucleosides, as well as with DNA and RNA. This pioneering work led to the first conclusions elucidating the mechanism of the antitumoral action of cisplatin. His research interests also included: (i) the synthesis, characterization, and study of the antitumor properties of new metal complexes (e.g., Pt, Pd, Sn, Sb, etc.); (ii) the study of metalloenzyme models and the clarification of their mechanism of action. His research on thiamine enzymes in the presence of divalent metal ions, for which he proposed a mechanism, was also a pioneering work in the field of enzyme/metal ion interactions. He also studied (iii) the mechanism of action of superoxide dismutase, Cu(II), and Zn(II); (iv) peptide interactions with metal ions, as models of enzymes or other biological systems; (v) the involvement of metal ions such as Ni(II) and Cu(II) in carcinogenesis; (vi) the biocatalysis of new materials; (vii) the mechanism of action of anti-thyroid drugs; (viii) organometallic chemistry, etc. He has numerous publications (up to 260), which have been cited more than 7000 times with an h-index of 44. He has been invited to 70 international conferences (as a Lead Speaker). He has coordinated many research and teaching programs. Important international conferences dedicated to the field of inorganic biochemistry—such as 5-ISABC, HALCHEM-III, 12-EURASIA, etc.—have been organized and chaired by him. Furthermore, he mentored many scientists, initiating their careers as academicians worldwide. Special attention should be paid to his contribution to the foundation and operation of the Interdisciplinary Program of Postgraduate Studies in Bioinorganic Chemistry, which he led for a decade. His contribution to the field of Inorganic-Bioinorganic Chemistry is undoubtedly superior in quality and unique for a Greek scientist. Overall, he contributed to the progress of inorganic chemistry not only at the University of Ioannina, Greece, but in the whole nation.

Recognizing this contribution of Emeritus Professor Nick Hadjiliadis to the field of inorganic biochemistry and especially to the field of palladium or/and platinum/nucleic acid interactions, it is our honour to dedicate the prologue of this commemorative issue of the *International Journal of Molecular Sciences* to him.

This Special Issue is composed of 14 articles, which are briefly reviewed below.

Yu-Wen Chen et al. showed that the i-motif DNA sequence may transition to a base-extruded duplex structure with a GGCC tetranucleotide tract when it is bound to the (Co^{II})-mediated dimer of chromomycin A3 [1]. G. Momekov et al. investigated two paramagnetic palladium(III) complexes of hematoporphyrin IX for their ability to process DNA adducts as well as for their antineoplastic and apoptogenic activities [2]. E. Makkonen et al. reported a combined quantum mechanics/molecular mechanics molecular dynamics and time-dependent density functional study of silver-mediated deoxyribonucleic acid nanostructures [3]. S. Liu et al. investigated the interactions between ruthenium(II) complexes and 15-mer single- and double-stranded oligodeoxynucleotides and they tested the thermodynamic base and sequence selectivity [4]. G.K. Latsis et al. investigated two polyorganotic acetate complexes against DNA with possible implementation towards breast cancer cells [5]. S. Savino et al. tested the ability of platinum prodrugs of kiteplatin with a-lipoic acid in the axial position to target in mitochondria [6]. Q.Y. Yang et al. tested the molecular mechanism of two transition metal complexes with 2-((2-(pyridin-2-yl)hydrazono)methyl)quinolin-8-ol against tumor cells [7]. M. Hande et al. described the synthesis and hybridization properties of short oligonucleotides incorporating cyclopalladated benzylamine "warheads" at their 5′-termini [8]. M.F. AlAjmi et al. evaluated the benzimidazole-derived biocompatible copper(II) and zinc(II) complexes as anticancer chemotherapeutics [9]. T. Qin et al. examined the binding of zinc cationic porphyrin towards B-DNA and Z-DNA [10] A.B. Olejniczak et al. presented an overview of the methods for incorporating metal centers into nucleic acids based on metal–boron cluster complexes (metallacarboranes) as the metal carriers [11]. Y.H. Lai et al. reviewed the mechanisms by which the regulation of copper homeostasis modulates the chemosensitivity of tumors to platinum drugs [12]. V. Murray et al. reviewed the abilities of bleomycin to interact with DNA [13]. N.C. Sabharwal et al. investigated the interactions between spermine-derivitized tentacle 2 porphyrins and the human telomeric DNA G quadruplex [14].

Conflicts of Interest: The authors declare no conflict of interest.

References

1. Chen, Y.-W.; Satange, R.; Wu, P.-C.; Jhan, C.-R.; Chang, C.-K.; Chung, K.-R.; Waring, M.J.; Lin, S.-W.; Hsieh, L.-C.; Hou, M.-H. Co^{II}(Chromomycin)$_2$ Complex Induces a Conformational Change of CCG Repeats from i-Motif to Base-Extruded DNA Duplex. *Int. J. Mol. Sci.* **2018**, *19*, 2796. [CrossRef] [PubMed]
2. Momekov, G.; Ugrinova, I.; Pasheva, E.; Tsekova, D.; Gencheva, G. Cellular Pharmacology of Palladinum(III) Hematoporphyrin IX Complexes: Solution Stability, Antineoplastic and Apoptogenic Activity, DNA Binding, and Processing of DNA-Adducts. *Int. J. Mol. Sci.* **2018**, *19*, 2451. [CrossRef] [PubMed]
3. Makkonen, E.; Rinke, P.; Lopez-Acevedo, O.; Chen, X. Optical Properties of Silver-Mediated DNA from Molecular Dynamics and Time Dependent Density Functional Theory. *Int. J. Mol. Sci.* **2018**, *19*, 2346. [CrossRef] [PubMed]
4. Liu, S.; Liang, A.; Wu, K.; Zeng, W.; Luo, Q.; Wang, F. Binding of Organometallic Ruthenium Anticancer Complexes to DNA: Thermodynamic Base and Sequence Selectivity. *Int. J. Mol. Sci.* **2018**, *19*, 2137. [CrossRef] [PubMed]
5. Latsis, G.K.; Banti, C.N.; Kourkoumelis, N.; Papatriantafyllopoulou, C.; Panagiotou, N.; Tasiopoulos, A.; Douvalis, A.; Kalampounias, A.G.; Bakas, T.; Hadjikakou, S.K. Poly Organotin Acetates against DNA with Possible Implementation on Human Breast Cancer. *Int. J. Mol. Sci.* **2018**, *19*, 2055. [CrossRef] [PubMed]
6. Savino, S.; Marzano, C.; Gandin, V.; Hoeschele, J.D.; Natile, G.; Margiotta, N. Multi-Acting Mitochondria-Targeted Platinum(IV) Prodrugs of Kiteplatin with α-Lipoic Acid in the Axial Positions. *Int. J. Mol. Sci.* **2018**, *19*, 2050. [CrossRef] [PubMed]
7. Yang, Q.-Y.; Cao, Q.-Q.; Qin, Q.-P.; Deng, C.-X.; Liang, H.; Chen, Z.-F. Syntheses, Crystal Structures, and Antitumor Activities of Copper(II) and Nickel(II) Complexes with 2-((2-(Pyridin-2-yl)hydrazono)methyl)quinolin-8-ol. *Int. J. Mol. Sci.* **2018**, *19*, 1874. [CrossRef] [PubMed]
8. Hande, M.; Maity, S.; Lönnberg, T. Palladacyclic Conjugate Group Promotes Hybridization of Short Oligonucleotides. *Int. J. Mol. Sci.* **2018**, *19*, 1588. [CrossRef] [PubMed]
9. AlAjmi, M.F.; Hussain, A.; Rehman, M.T.; Khan, A.A.; Shaikh, P.A.; Khan, R.A. Design, Synthesis, and Biological Evaluation of Benzimidazole-Derived Biocompatible Copper(II) and Zinc(II) Complexes as Anticancer Chemotherapeutics. *Int. J. Mol. Sci.* **2018**, *19*, 1492. [CrossRef] [PubMed]
10. Qin, T.; Liu, K.; Song, D.; Yang, C.; Zhao, H.; Su, H. Binding Interactions of Zinc Cationic Porphyrin with Duplex DNA: From B-DNA to Z-DNA. *Int. J. Mol. Sci.* **2018**, *19*, 1071. [CrossRef] [PubMed]
11. Olejniczak, A.B.; Nawrot, B.; Leśnikowski, Z.J. DNA Modified with Boron–Metal Cluster Complexes [M(C$_2$B$_9$H$_{11}$)$_2$]—Synthesis, Properties, and Applications. *Int. J. Mol. Sci.* **2018**, *19*, 3501. [CrossRef] [PubMed]
12. Lai, Y.-H.; Kuo, C.; Kuo, M.T.; Chen, H.H.W. Modulating Chemosensitivity of Tumors to Platinum-Based Antitumor Drugs by Transcriptional Regulation of Copper Homeostasis. *Int. J. Mol. Sci.* **2018**, *19*, 1486. [CrossRef] [PubMed]
13. Murray, V.; Chen, J.K.; Chung, L.H. The Interaction of the Metallo-Glycopeptide Anti-Tumour Drug Bleomycin with DNA. *Int. J. Mol. Sci.* **2018**, *19*, 1372. [CrossRef] [PubMed]
14. Sabharwal, N.C.; Chen, J.; Lee, J.H.J.; Gangemi, C.M.A.; D'Urso, A.; Yatsunyk, L.A. Interactions Between Spermine-Derivatized Tentacle Porphyrins and The Human Telomeric DNA G-Quadruplex. *Int. J. Mol. Sci.* **2018**, *19*, 3686. [CrossRef] [PubMed]

International Journal of
Molecular Sciences

MDPI

Article

Interactions Between Spermine-Derivatized Tentacle Porphyrins and The Human Telomeric DNA G-Quadruplex

Navin C. Sabharwal [1,2], Jessica Chen [1,3], Joo Hyun (June) Lee [1,4], Chiara M. A. Gangemi [5], Alessandro D'Urso [5,*] and Liliya A. Yatsunyk [1,*]

[1] Department of Chemistry and Biochemistry, Swarthmore College, Swarthmore, PA 19081, USA; navin.sabharwal.424@gmail.com (N.C.S.); ymchen.017@gmail.com (J.C.); jlee2143@gmail.com (J.H.(J.)L.)

[2] Lerner College of Medicine, Cleveland Clinic, Cleveland, OH 44195, USA

[3] School of Dental Medicine, University of Pennsylvania, Philadelphia, PA 19104, USA

[4] College of Dentistry, New York University, New York, NY 10010, USA

[5] Department of Chemical Science, University of Catania, 95125 Catania, Italy; gangemichiara@unict.it

* Correspondence: adurso@unict.it (A.D.); lyatsun1@swarthmore.edu (L.A.Y.);
  Tel.: +39-095-738-5095 (A.D.); +1-610-328-8558 (L.A.Y.)

Received: 16 October 2018; Accepted: 17 November 2018; Published: 21 November 2018

Abstract: G-rich DNA sequences have the potential to fold into non-canonical G-Quadruplex (GQ) structures implicated in aging and human diseases, notably cancers. Because stabilization of GQs at telomeres and oncogene promoters may prevent cancer, there is an interest in developing small molecules that selectively target GQs. Herein, we investigate the interactions of *meso*-tetrakis-(4-carboxysperminephenyl)porphyrin (TCPPSpm4) and its Zn(II) derivative (ZnTCPPSpm4) with human telomeric DNA (Tel22) via UV-Vis, circular dichroism (CD), and fluorescence spectroscopies, resonance light scattering (RLS), and fluorescence resonance energy transfer (FRET) assays. UV-Vis titrations reveal binding constants of 4.7×10^6 and 1.4×10^7 M^{-1} and binding stoichiometry of 2–4:1 and 10–12:1 for TCPPSpm4 and ZnTCPPSpm4, respectively. High stoichiometry is supported by the Job plot data, CD titrations, and RLS data. FRET melting indicates that TCPPSpm4 stabilizes Tel22 by 36 ± 2 °C at 7.5 eq., and that ZnTCPPSpm4 stabilizes Tel22 by 33 ± 2 °C at ~20 eq.; at least 8 eq. of ZnTCPPSpm4 are required to achieve significant stabilization of Tel22, in agreement with its high binding stoichiometry. FRET competition studies show that both porphyrins are mildly selective for human telomeric GQ vs duplex DNA. Spectroscopic studies, combined, point to end-stacking and porphyrin self-association as major binding modes. This work advances our understanding of ligand interactions with GQ DNA.

Keywords: G-quadruplex; tentacle porphyrins; Zn(II) porphyrin; anti-cancer therapy; end-stacking

1. Introduction

DNA can exist in a variety of secondary structures [1] in addition to the right-handed double-stranded (dsDNA) form first proposed by Watson and Crick in 1953. One example is G-Quadruplex (GQ) DNA, a non-canonical DNA structure formed by guanine rich sequences [2]. The primary structural unit of GQ DNA is a G-tetrad which consists of four guanines associated through Hoogsteen hydrogen bonding (Figure 1A). G-tetrads interact with each other via π-π stacking, and are linked by the phosphate sugar backbone, forming GQs. The stability of the GQ is further enhanced by coordinating cations [3,4]. In fact, biological GQs with 2–4 G-tetrads would not fold without a cation due to a strong repulsion of guanine carbonyls in the center of each tetrad (Figure 1A). Unlike dsDNA, GQs exhibit high structural diversity, adopting parallel, mixed-hybrid, and antiparallel topologies (Figure 1B). Bioinformatics studies suggest that sequences with GQ-forming potential are prevalent in highly-conserved functional regions of the human genome including telomeres, oncogene promoters, immunoglobulin switch regions, and ribosomal DNA [5–8], and may regulate numerous biological processes. Evidence for GQ formation inside the cell was recently presented [9–12], and studies are underway to better assess their in vivo roles [2].

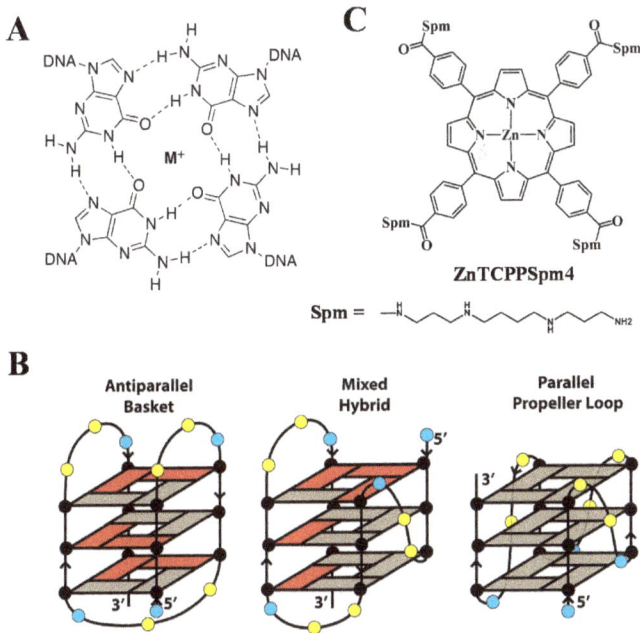

Figure 1. (**A**) Four guanines associate via Hoogsteen hydrogen bonding to form a G-tetrad. M^+ represents a central coordinating cation, such as Na^+, K^+, or NH_4^+. (**B**) Schematics of the physiologically-relevant structures of human telomeric DNA, dAGGG(TTAGGG)$_3$. Grey and red rectangles represent guanines in *anti* and *syn* conformations. Adenines and thymines are represented as blue and yellow circles, respectively. Strand orientations are depicted with arrows. Mixed-hybrid conformation is that of Form 2. (**C**) Structure of ZnTCPPSpm4; the fifth axial water ligand attached to Zn(II) is not depicted for clarity of the image.

Telomeres protect the ends of eukaryotic chromosomes from degradation and fusion and contain tandem repeats of dTTAGGG [13]. The 22-mer human telomeric DNA sequence dAGGG(TTAGGG)$_3$ (Tel22) is well-studied and has been shown to form diverse GQ structures in vitro [14–16], see Figure 1B. The topology, stability, and homogeneity of the human telomeric DNA depends on the DNA length and the identity of the nucleotides at 5' and 3' ends. In addition, the nature of the central stabilizing cation, the presence of small molecules, annealing temperature and rate, and molecular crowding reagents impact the resulting secondary structure. In K^+, Tel22 forms a parallel GQ with three G-tetrads and three TTA propeller loops, but only in the presence of molecular crowding conditions [17,18], some small molecules (e.g., N-methylmesoporphyrin IX, NMM) [19,20], under crystallization conditions [21], or at high DNA concentration [22]. In Na^+, Tel22 adopts an antiparallel topology with three G-tetrads connected by two lateral loops and one central diagonal loop [23]. In the dilute K^+ solutions favored in this work, Tel22 adopts at least two (3 + 1) mixed-hybrid structures called Form 1 and Form 2 [24–28]. The two forms have one propeller loop and two lateral loops, but differ by loop orders; three G-rich strands run in the same direction and opposite from that of the fourth strand, hence the name (3 + 1). Other GQ topologies exist under these conditions (e.g., an antiparallel GQ with two G-tetrads) [29], but at low abundance. It has been proposed that formation of GQ structures at telomeres inhibits the activity of telomerase, the enzyme responsible for maintenance of telomeres integrity, leading to cell immortality. Because telomerase is upregulated in 85–90% of cancers [30], stabilization of GQs by small molecule ligands has emerged as a novel, selective, anti-cancer therapeutic strategy [31,32].

Porphyrins are one of the earliest classes of DNA ligands. Their interactions with GQ DNA were first studied in 1998 [33], and with dsDNA as far back as 1979 [34], and are still of great interest [35]. Porphyrins are aromatic, planar, and the size of their macrocycle (~10 Å) matches that of a G-tetrad (~11 Å), leading to an efficient π-π stacking. Cellular uptake and localization studies demonstrate that porphyrins accumulate rapidly in nuclei of normal and tumor cells [36,37] at levels sufficient for tumor growth arrest; yet they are non-toxic to somatic cells [38]. Porphyrins can be readily functionalized to optimize their GQ-stabilizing ability and selectivity, solubility, and cell permeability. Our laboratory and others have characterized binding of numerous porphyrins, including NMM [19,20,39], *meso*-tetrakis-(*N*-methyl-4-pyridyl) porphyrin (TMPyP4) [38,40], and its various derivatives [41–44] to human telomeric DNA. Porphyrins can bind to GQ DNA via end-stacking, which has been characterized spectroscopically [45,46], and observed in structural studies [20,47]. Intercalation has been suggested [46,48–50], but is considered energetically unfavorable for short GQs with 2-4 G-tetrads. Porphyrins can also interact with the grooves [51] and loops [52] of GQs. Porphyrin metallation is expected to enhance its GQ binding due to the electron-withdrawing property of the metal, which reduces the electron density on the porphyrin, improving its π-π stacking ability. The enhancement of porphyrin's binding to GQ is especially strong when the metal is positioned above the ion channel of the GQ.

In this work, we focus on two novel tentacle porphyrins, *meso*-tetrakis-(4-carboxyspermine-phenyl)porphyrin, TCPPSpm4 and its Zn(II)-derivative, ZnTCPPSpm4, Figure 1C. Binding of tentacle porphyrins to dsDNA is well studied [53–56], but their interactions with GQ DNA remain poorly characterized. We introduced spermine groups to enhance the GQ-binding potential, solubility, and biocompatibility of the porphyrins. Polyamines have been reported to interact with DNA by both electrostatic forces and via site-specific interactions with the phosphate backbone and DNA bases [57–59]. In some cases polyamines induced conformational modifications [60]. Spermine was shown to preferentially bind to the major groove of dsDNA [59]. A variety amines (e.g., pyrrolidine, piperidine, morpholine, 1-ethylpiperazine, *N,N*-diethylethylenediamine, and guanidine) have been incorporated into GQ ligands, leading to improvements in their GQ binding affinities and water solubility [61–65]. Of equally strong importance, spermine is essential for cellular growth, differentiation [66], and protection against double-strand breaks. Polyamines are currently being exploited as a transport system for cancer drugs due to their well-known ability to accumulate in neoplastic tissues [67–71]. Therefore, we added spermine to *meso*-tetrakis-(4-carboxyphenyl)porphyrin not only to improve its GQ-binding, but also to facilitate its delivery to cancer cells in future biological studies.

We characterized the interactions between human telomeric DNA and TCPPSpm4 or ZnTCPPSpm4 in a K^+ buffer through UV-Vis, fluorescence, and circular dichroism (CD) spectroscopies, resonance light scattering (RLS), and fluorescence resonance energy transfer (FRET) assays. We demonstrate that both porphyrins bind tightly to Tel22 GQ with a high binding stoichiometries (2–4:1 for TCPPSpm4 and 10–12:1 for ZnTCPPSpm4) and stabilize it strongly with mild selectivity over dsDNA. Our data are consistent with end-stacking binding mode and DNA-assisted porphyrin self-stacking.

2. Results and Discussion

In this work, we focus on two tentacle porphyrins, *meso*-tetrakis(4-carboxysperminephenyl) porphyrin, TCPPSpm4, and its Zn(II) derivative, ZnTCPPSpm4. Both porphyrins are modified with four spermine arms, see Figure 1C. The *pKa* of the spermine amine groups in TCPPSpm4 was measured to be ~5.8 for the first protonation and ~8 for the second protonation [72]. Therefore, this porphyrin is expected to be at least tetracationic at pH 7.2 used in this work. Zn(II) was introduced into TCPPSpm4 to improve its GQ binding due to electron-poor nature of the metal. In addition, Zn(II) is coordinated to an axial water, which is expected to prevent its intercalation into dsDNA, and thus, to improve its selectivity. Binding of TCPPSpm4 to the GQ aptamer (dTGGGAG)$_4$ was recently characterized [73], whereas binding of ZnTCPPSpm4 to any of the GQs has not yet been tested. Here, we explore in detail how both porphyrins interact with human telomeric GQ DNA, Tel22.

2.1. UV-Vis Spectroscopy Demonstrates that TCPPSpm4 and ZnTCPPSpm4 Bind Tightly to Tel22

Due to the excellent chromophoric properties of both porphyrins, their binding to Tel22 was monitored using Soret band of 415 nm for TCPPSpm4 and 424 nm for ZnTCPPSpm4. We first performed a dilution study which indicated that the porphyrins maintain their aggregation state, assumed to be monomeric, in the concentration range of 1–40 μM (Figure S1). Subsequently, both porphyrins were titrated with Tel22; representative UV-Vis titrations are shown in Figure 2. The extinction coefficient for the TCPPSpm4-Tel22 complex was determined to be $(1.2 \pm 0.2) \times 10^5$ M^{-1}cm^{-1} at 429 nm and $(0.54 \pm 0.04) \times 10^5$ M^{-1}cm^{-1} for ZnTCPPSpm4-Tel22 at 435 nm. The Soret band of TCPPSpm4 displayed a pronounced red shift ($\Delta\lambda$) of 13.5 ± 0.5 nm and hypochromicity (%H) of 58 ± 6 % upon addition of Tel22. The corresponding values for ZnTCPPSpm4 are similar with $\Delta\lambda$ of 11.3 ± 0.6 and % H of 58 ± 5%. Red shift of ~15 nm and %H of ~50% were obtained for TCPPSpm4 binding to another GQ structure formed by (dTGGGAG)$_4$ aptamer [73]. High values of $\Delta\lambda$ and %H indicate strong interactions between the π-systems of porphyrins and GQ, characteristic of either end-stacking or intercalation. Pasternack et al. found that intercalation of a porphyrin into dsDNA can be identified by %H > 40% and $\Delta\lambda \geq 15$ nm [74]. Although supported by molecular dynamics stimulation studies [50], this mode of binding has not yet been detected in structural studies. On the other hand, both end-stacking [20,47] and loop binding [52] have been observed in X-ray structures of porphyrin-GQ complexes.

To extract binding constants, we employed the Direct Fit method, which is the simplest way of treating the titration data, as it assumes equivalent and independent binding sites. Such data treatment is justified by the presence of the isosbestic points, yet it is an oversimplification in view of high stoichiometric ratios obtained (see below) and the presence of detectable shoulders, especially in final samples. Data analysis yielded a binding constant, *Ka*, of $(4.7 \pm 0.7) \times 10^6$ M^{-1} for TCPPSpm4 assuming a binding stoichiometry of 4:1; and *Ka* of $(1.4 \pm 0.7) \times 10^7$ M^{-1} for ZnTCPPSpm4 assuming a binding stoichiometry of 12:1. The high *Ka* values indicate strong binding between Tel22 and the porphyrins and correlate well with the high values of $\Delta\lambda$ and %H. ZnTCPPSpm4 binds three times tighter than its free-base analogue, possibly due to the presence of electron withdrawing metal. This binding is likely further enhanced by electrostatic attractions due to high charges on the porphyrins and by interactions of four spermine arms with the grooves of Tel22 GQ.

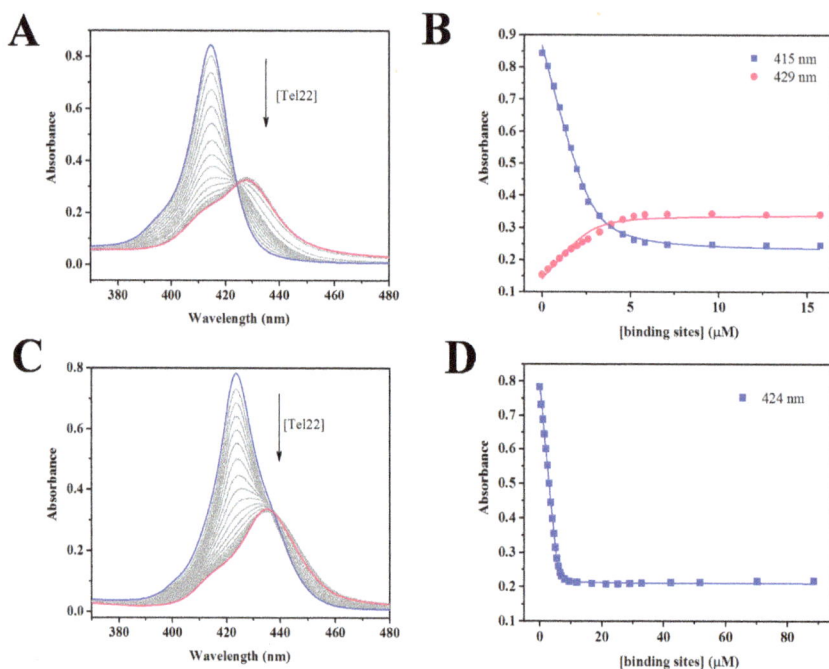

Figure 2. Interactions between porphyrins and Tel22 GQ probed by UV-Vis spectroscopy. (**A**) A representative UV-Vis titration of 2.8 µM TCPPSpm4 with 82.6 µM Tel22. Clear isosbestic point is observed at 424 nm. (**B**) Best fit (solid line) to the titration data monitored at 415 nm (squares) and 429 nm (circles). (**C**) A representative UV-Vis titration of 5.8 µM ZnTCPPSpm4 with 46.3 (followed by 185) µM Tel22. Clear isosbestic point is observed at 442 nm. (**D**) Best fit (solid line) to the titration data monitored at 424 nm (squares). Concentration of binding sites is defined as the concentration of Tel22 multiplied by the binding stoichiometry (4:1 for TCPPSpm4 and 12:1 for ZnTCPPSpm4). Blue lines and points correspond to porphyrins alone and pink corresponds to porphyrin-Tel22 complex.

To independently verify the stoichiometry for porphyrin-Tel22 binding, we used Job's method, also known as the method of continuous variation [75]. In this method, the mole fraction of DNA and porphyrin is varied while their total concentration is kept constant. The mole fraction at the maximum or minimum on the plot of absorbance vs mole fraction corresponds to the binding stoichiometry between the two binding partners [76]. Representative Job plots are depicted in Figure 3. Job plot experiments for TCPPSpm4-Tel22 system yielded an average mole fraction of 0.70 ± 0.04, which corresponds to the binding of 2–3 porphyrins to one Tel22. For the ZnTCPPSpm4-Tel22 system, Job plot yielded a mole fraction value of ~0.9, which corresponds to the binding of nine porphyrin molecules to one Tel22 GQ. In both cases, binding stoichiometries are somewhat lower than those obtained via fitting of the UV-vis titration data. Similar discrepancy was also observed in our previous work where we investigated binding of four different cationic porphyrins to two parallel GQs [77]. Job plot stoichiometry is lower because it represents only the major binding event, while stoichiometry obtained via fitting of UV-vis titration data encompasses strong, weak, and non-specific binding. It is also important to remember that binding stoichiometries of 1:1 and 2:1 can be clearly differentiated via Job's method, but higher binding stoichiometries are difficult to determine precisely. For example, binding ratios of 4:1 and 5:1 correspond to molar fractions of 0.8 and 0.83, respectively, which would likely be impossible to distinguish, given the expected level of data accuracy. The unusually high binding stoichiometry supports the involvement of multiple binding modes such as end-stacking, electrostatic interactions, and groove binding, the latter two resulting from the presence of spermine arms. It also suggests the possibility of porphyrin self-association on the DNA backbone. The much higher binding stoichiometry for ZnTCPPSpm4 is puzzling, especially in light of ZnTCPPSpm4's axial water molecule, which is expected to inhibit some binding modes, such as porphyrin self-association. However, slipped self-stacking is still possible.

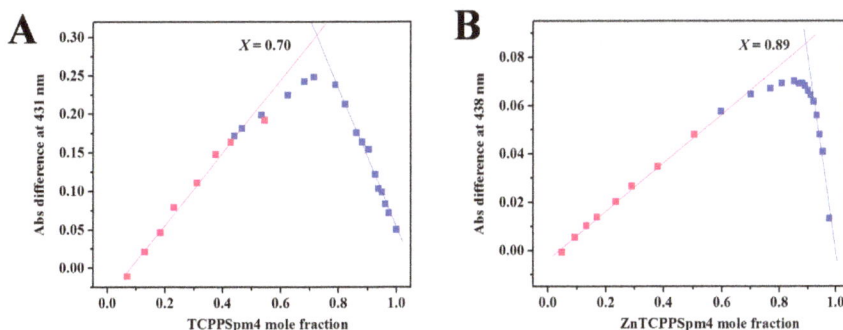

Figure 3. Representative Job plots for (**A**) 3.1 μM TCPPSpm4 and (**B**) 2.9 μM ZnTCPPSpm4 in complex with Tel22 at 25 °C. Porphyrins and Tel22 GQ DNA concentrations were maintained equal within 20%. The Job plots were constructed by plotting the difference in the absorbance values at a specified wavelength vs mole fraction of the porphyrin, *X*. Pink squares represent data collected by titrating porphyrins into DNA; blue squares represent data collected by titrating DNA into porphyrins.

2.2. RLS Indicates the Formation of Discrete Stoichiometric Porprhyrin-Tel22 Complexes

Because UV-vis titrations yielded high stoichiometry for porphyrin-Tel22 complexes, we employed the RLS method [78] to check for possible aggregation. In RLS, porphyrin solution is excited close to its Soret maximum and the scattering is measured at the same wavelength. If aggregated (alone or on a substrate), porphyrins display enhanced Rayleigh scattering originating from electronic coupling between the individual molecules in the assembly. To detect communication between porphyrins, RLS experiments are performed under porphyrin excess, unlike UV-vis titrations, where DNA excess is used.

The RLS intensity of TCPPSpm4 alone is low (Figure 4A), indicating an absence of aggregation in agreement with UV-vis dilution studies (Figure S1). The addition of Tel22 does not change the RLS signal in the [TCPPSpm4]/[Tel22] range of 40–8. Below this ratio, however, RLS signal starts to increase and reaches a maximum at [TCPPSpm4]/[Tel22] = 2, suggesting the formation of an assembly with strong electronic communication between porphyrins. Further addition of Tel22 does not change RLS, indicating that the TCPPSpm4-Tel22 complex is stable. Adding more Tel22 to this solution eventually leads to drastic decline in RLS signal, owing to the precipitation of the complex (data not shown).

ZnTCPPSpm4 does not aggregate alone or in the [ZnTCPPSpm4]/[Tel22] range of 40–14 (Figure 4B). When more Tel22 is added, however, stable aggregates are formed at [ZnTCPPSpm4]/[Tel22] ~13, in line with the stoichiometry determined in UV-vis experiments. Subsequent addition of Tel22 does not change RLS until [ZnTCPPSpm4]/[Tel22] ~2, at which point the RLS signal rises up to 1:1 ratio then starts to decrease, although the observed changes are small.

Taken together, the RLS data allow us to (i) exclude porphyrin aggregation in the absence of DNA; (ii) confirm formation of discrete porphrin-Tel22 complexes with a stoichiometry consistent with that measured in UV-vis; and (iii) exclude existence of large, non-stoichiometric porphryin-Tel22 aggregates. Overall, RLS and UV-vis data support our hypothesis of DNA-assisted porphyrin self-aggregation on Tel22 which leads to strong electronic communication between individual porphyrins in the assembly.

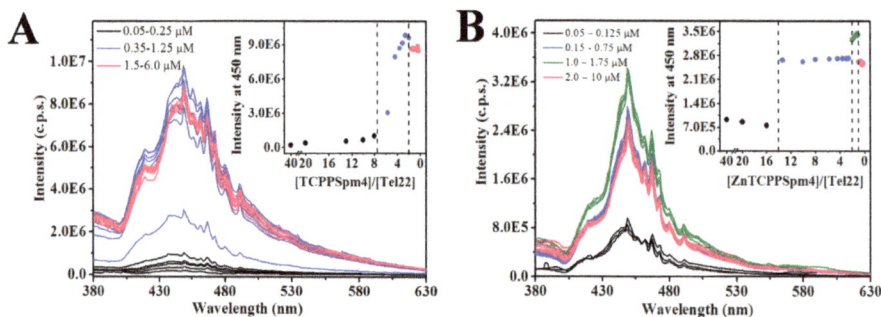

Figure 4. Representative RLS titration of 2.0 μM (**A**) TCPPSpm4 and (**B**) ZnTCPPSpm4 with 500 μM Tel22 at 25 °C. The amounts of Tel22 added are specified in the legend. Inset reports RLS intensity at 450 nm vs [porphyrin]/[Tel22] ratio. Note, the scale in the inset is inverted to follow the progress of the titration which starts with the solution of porphyrin and proceeds toward lower [porphyrin]/[Tel22] ratios.

2.3. Fluorescence of TCPPSpm4 and ZnTCPPSpm4 Decreases in the Presence of Tel22 Suggesting DNA-Assisted Porphyrin Self-Association

The steady-state fluorescence emission spectrum of a porphyrin is produced by the first excited state, S_1, and the charge-transfer state between the porphyrin ring and its peripheral substituents (in this case carboxysperminephenyl groups). The coupling between these two states leads to quenching of the fluorescence signal, which occurs in polar solvents or when the rotation of peripheral substituents is unrestricted. TCPPSpm4 fluoresces in aqueous solution, producing a peak at 643 nm and a shoulder at 702 nm, as has been previously observed [72]. At the same time, ZnTCPPSpm4 produces a split peak at 607 and 657 nm, Figure 5, suggesting that the rotation of its side-chains is more restricted.

Position and intensity of the fluorescence peak of a porphyrin is strongly sensitive to its environment and, thus, can report on porphyrin binding to GQ DNA [79]. Addition of Tel22 GQ to TCPPSpm4 leads to a dramatic decrease in fluorescence intensity and a red shift of 10 and 15 nm for the 643 and 702 nm peaks, respectively. The spectra at saturating amount of Tel22 are sharper and better resolved, Figure 5A, suggesting restriction in rotation of the peripheral groups upon GQ binding. Similarly, the fluorescence intensity of ZnTCPPSPm4 decreased dramatically upon addition of Tel22, but the red shift observed was significantly smaller, i.e., 5 and 3 nm for the 607 and 657 nm bands, respectively. In both cases, the original dramatic decrease in signal intensity is followed by a small increase in the signal at high [Tel22]/[porphyrin] ratios (see Figure S2) suggesting a change in a mechanism of ligand interactions with Tel22 or with each other. The strong decrease in fluorescence could be explained by close interactions between porphyrins and Tel22 as well as by self-association of porphyrins assisted by the DNA backbone. Such interpretation is consistent with reported high binding stoichiometry, especially for ZnTCPPSpm4. Similar to our case, the steady-state fluorescence of the Zn(II) derivative of a widely-studied porphyrin, TMPyP4, decreased upon addition of tetrastranded parallel GQs [77] and poly(dG-dC) dsDNA [80], although in both cases the decrease was not as dramatic as in the present case.

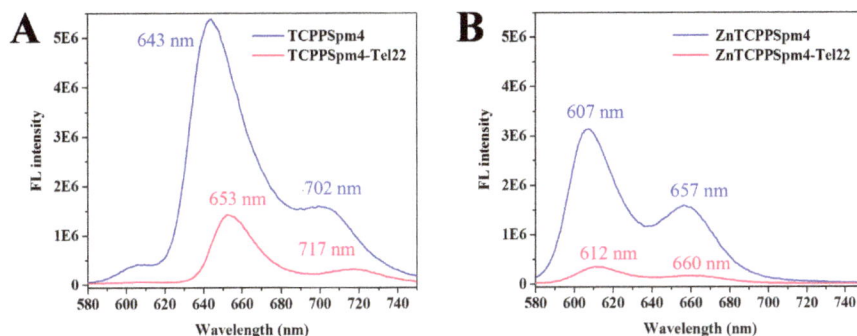

Figure 5. Steady-state fluorescence emission spectra for (**A**) 0.33 µM TCPPSpm4 alone and in the presence of 19.5 fold excess of Tel22 and (**B**) 0.47 µM ZnTCPPSpm4 alone and in the presence of 9.1 fold excess of Tel22. Note, for the ease of comparison, the data were scaled to 1 µM porphyrin.

2.4. FRET Studies Indicate that Both Porphyrins Have Exceptional Stabilizing Ability and Modest Selectivity toward Tel22 GQ

FRET is a benchmark technique in the quadruplex field enabling facile and reliable measurement of ligands' stabilizing ability and selectivity for GQ DNA [81]. We used F21D, a 21-nt sequence of the human telomeric DNA labeled with 6-FAM fluorescent dye at the 5' end and a quencher, Dabcyl, at the 3' end (5'-6-FAM-GGG(TTAGGG)$_3$-Dabcyl-3'). We have thoroughly characterized the fold and stability of this sequence in our earlier work [19]. The addition of up to 7.5 eq. of TCPPSpm4 and up to 20 eq. of ZnTCPPSpm4 to F21D resulted in a concentration-dependent increase in Tm of F21D by 36 ± 2 °C and 33 ± 2 °C, respectively (Figure 6A; raw data are shown in Figure S3). Our data shows that both porphyrins stabilize Tel22 GQ to a great extent, but the stabilization curve for ZnTCPPSpm4 is sigmoidal, and only weak stabilization is observed up to 1.6 µM (8 eq.) of the porphyrin. This data is in agreement with high stoichiometry of the ZnTCPPSpm4-Tel22 complex determined in UV-vis and Job plot studies.

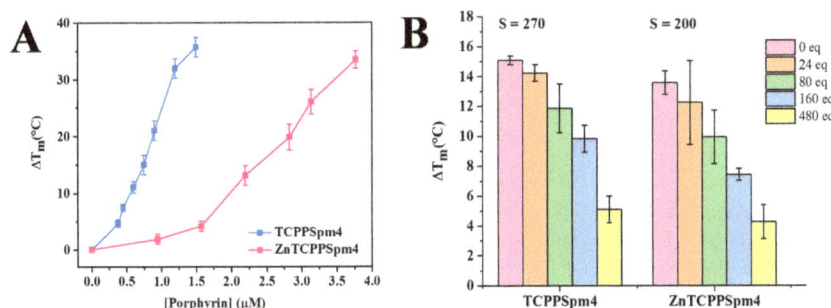

Figure 6. Stabilizing ability and selectivity of TCPPSpm4 and ZnTCPPSpm4 toward human telomeric DNA investigated via FRET. (**A**) Dose dependent stabilization, ΔTm, of 0.2 µM F21D as a function of porphyrin concentration. (**B**) Stabilization of 0.2 µM F21D with 0.75 µM TCPPSpm4 or 2.2 µM ZnTCPPSpm4 in the presence of increasing amount of CT DNA (equivalents relative to F21D are specified in the legend). Concentration of porphyrins was chosen in order to achieve similar starting Tm for the first sample before any CT DNA was added in order to facilitate the comparison. The concentration of F21D is expressed per strand, while the concentration of CT DNA is expressed per base pair. Note, all raw data are presented in Figure S3.

Selectivity is an essential characteristic of an ideal anticancer GQ ligand, because a drug that binds readily to dsDNA will require a greater concentration to achieve its therapeutic effect, or even cause cytotoxicity. Thus, we conducted FRET competition studies in the presence of large excess of CT DNA and a fixed ligand concentration (Figure 6B). The selectivity ratio, defined as the fold of competitor necessary to reduce ΔTm by 50%, was calculated to be 270 for TCPPSpm4 and 200 for ZnTCPPSpm4. While the porphyrins prefer GQ to dsDNA, the observed selectivity ratios are rather modest. Such modest selectivity is likely due to strong electrostatic interactions between the positively charged porphyrins and negatively charged DNA (GQ, dsDNA, etc). This hypothesis is supported by our earlier work showing that reducing the charge on a porphyrin increases its selectivity for GQ DNA [44]. Our laboratory previously demonstrated that another Zn(II)-metallated porphyrin, ZnTMPyP4, displays selectivity ratio of 100 toward F21D vs CT DNA, while its free-base analogue displays a selectivity ratio of 300 [42]. These values are on the same scale and display the same trend as the one obtained in this work. Overall, FRET studies suggest that both porphyrins are robust stabilizers of human telomeric DNA, with TCPPSpm4 displaying both superior selectivity and stabilizing ability.

2.5. Circular Dichroism (CD) Signal Decreases upon Addition of Porphyrins Signifying Interaction between Porphyrins and Tel22

To determine if porphyrin binding alters the topology of the Tel22 GQ, we performed CD annealing and titration studies. CD is an excellent method to report on the type of GQ fold and its alteration upon ligand binding. The CD signature of Tel22 in potassium buffer (5 mM KCl) is well characterized in our previous works [19] and that of others [22], and contains a peak at 295 nm and a shoulder at ~250 nm. Titration of TCPPSpm4 under kinetic conditions (with short 12 min equilibration) did not alter the conformation of Tel22, but lead to dramatic decrease in the intensity of 295 nm peak (Figure 7A). Under similar conditions, ZnTCPPSpm4 caused only a mild decrease of CD signal intensity (Figure 7B). To investigate the system under thermodynamic equilibrium, Tel22 samples were annealed with ~2 eq. of porphyrins and equilibrated overnight. The CD signals displayed stronger decrease (Figure 7C,D), in part caused by minor precipitation. Decrease in CD signal intensity was also observed upon interaction of TCPPSpm4 with (dTGGGAG)$_4$ GQ aptamer [73]. Other metallated porphyrins, such as PtTMPyP4 [43], CuTMPyP4, and NiTMPyP4 [82] caused decrease in the intensity of Tel22 CD signal in potassium buffer, while CoTMPyP4 and ZnTMPyP4 did not [82].

The porphyrin-induced decrease in CD signal intensity could be explained, in part, by DNA precipitation, most likely caused by highly charged spermine arms of the porphyrin ligands. The precipitation was minor and was only observed at high porphyrin and DNA concentrations (above 10 µM DNA). In addition, the observed behavior in CD titrations could be explained by preferential binding of porphyrins to single-stranded (ssDNA), which disfavors GQ in the GQ DNA $\leftrightarrow$ ssDNA equilibrium. This mode of binding was observed for TMPyP4 [83], triarylpyridines [84], and anthrathiophenedione [85]. However, such data interpretation seems to contradict the observed stabilization of human telomeric DNA in our FRET studies (Figure 6A). Alternatively, we can explain the observed decrease in CD signal intensity by proposing that porphyrins bind to GQ DNA by disrupting and replacing one or more of the G-tetrads, leading to unchanged or even enhanced stability. Such explanation reconciles our CD and FRET data and was first proposed by Marchand et al. on the basis of an extensive CD and mass spectrometry study [86].

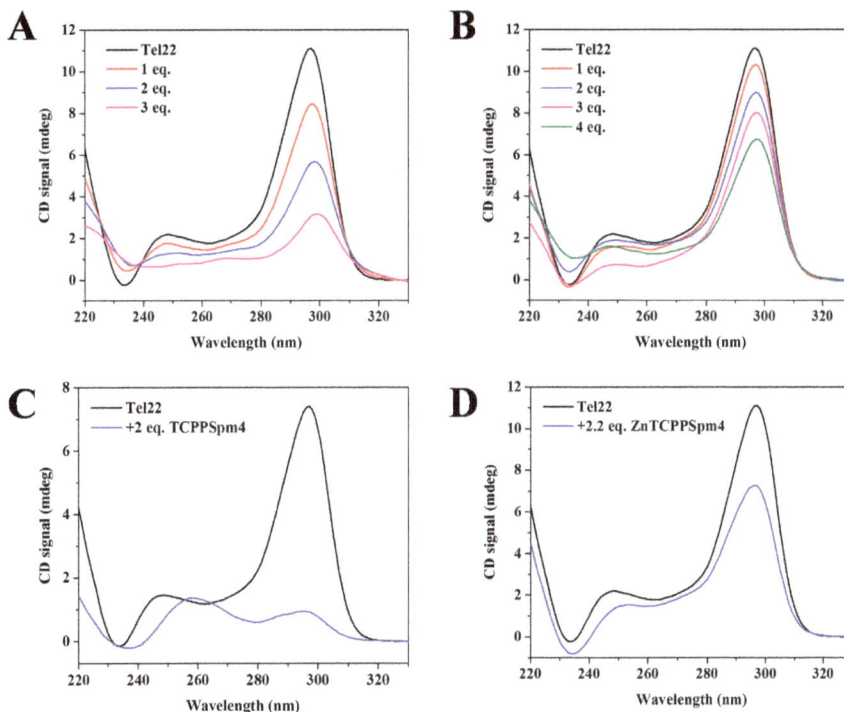

Figure 7. CD titration of 15.0 μM Tel22 with up to 4 eq. of (**A**) TCPPSpm4 and (**B**) ZnTCPPSpm4. Samples were incubated for 12 min after each addition of the porphyrin. CD annealing of (**C**) 10.0 μM Tel22 with 2.0 eq. of TCPPSpm4 and of (**D**) 15 μM Tel22 with 2.2 eq. of Zn TCPPSpm4. Data were collected at 20 °C. We have also completed CD melting on the annealed samples and saw no-to-weak stabilization (Figure S4).

2.6. The Presence of Induced CD (iCD) Confirms Close Contacts between Porphyrins and Tel22 Aromatic Systems

We further characterized porphyrin-Tel22 interactions by investigating changes in the CD Soret region. Chromophoric but achiral porphyrins produce no CD signal in this region, and the DNA CD signal is found exclusively in the UV region. However, when DNA and porphyrin interact, the complex is both chiral and chromophoric, and will produce an iCD when the π-system of a porphyrin is in close proximity to that of the DNA. For ligand binding to duplex DNA, the type of iCD has been found to correlate with the binding mode: a positive iCD corresponds to external binding and a negative one indicates intercalation [87,88]. However, a similar correlation has not yet been established for porphyrin-GQ interactions due to the scarcity of empirical data on binding modes other than end-stacking.

The addition of Tel22 to each porphyrin at stoichiometric amounts yielded a bisignate iCD with a strong positive component (Figure 8). The trough and the peak occur at 410 and 426 nm for TCPPSpm4-Tel22 and at 427 and 442 nm for ZnTCPPSpm4-Tel22, which is consistent with their Soret band positions. Once we established the presence of the iCD, we conducted CD titrations in the Soret region. Due to low iCD signal intensity, the data display high variability, but nevertheless indicate that the strongest iCD is observed for complexes with the stoichiometric quantities of porphyrins (4 eq. for TCPPSpm4 and ~12–15 eq. for ZnTCPPSpm4, Figure S5). In sum, the presence of iCD is consistent with strong binding of both porphyrins to the Tel22, and suggests close proximity of the porphyrin ring and G-tetrad(s), indicative of end-stacking. In addition, the split bisignate shape of iCD indicates that porphyrins are not disorderly distributed on Tel22 and that there is communication between the porphyrins in the assembly, in agreement with the RLS data described earlier. The iCD was likewise observed for TCPPSpm4 binding to (dTGGGAG)$_4$ GQ aptamer [73] and to poly(dG-dC) and CT DNA [89], and for ZnTCPPSpm4 binding to poly(dG-dC) in both B and Z conformations [90]. However, the shape of the iCD was different from that observed in this work, underlining differences in the binding modes.

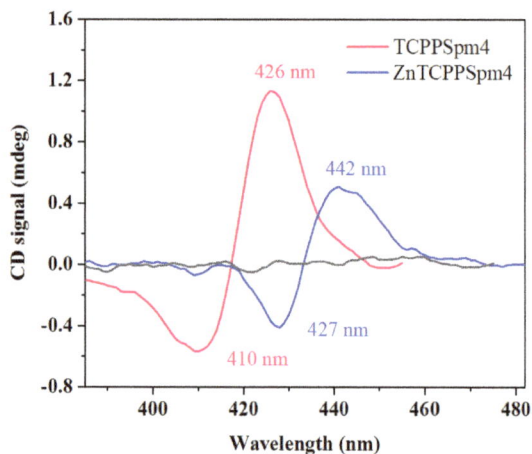

Figure 8. iCD signature of TCPPSpm4-Tel22 and ZnTCPPSpm4-Tel22 complexes prepared at stoichiometric amounts of porphyrins and DNA (4:1 for TCPPSpm4 and 12:1 for ZnTCPPSpm4). The data were scaled to 1 μM porphyrin. The CD scan of porphyrin alone is shown in grey. The data were smoothed using Savitzky–Golay smoothing filter with a 13-point quadratic function.

3. Materials and Methods

3.1. Porphyrins and Oligonucleotides

TCPPSpm4 and ZnTCPPSpm4 were synthesized as described previously [72,90] and dissolved in double-distilled water (ddH$_2$O) at 1–5 mM and stored at 4 °C in the dark. The concentration of TCPPSpm4 was determined via UV-Vis spectroscopy using the extinction coefficient of 3.0×10^5 M^{-1}cm^{-1} at 415 nm at pH 6.5 [72]. The extinction coefficient for ZnTCPPSpm4 was measured to be 1.34×10^5 M^{-1}cm^{-1} at 424 nm at pH 7 using Beer's law (Figure S1). Tel22 was purchased from Midland Certified Reagent Company (Midland, TX, USA) and dissolved in 5K buffer (10 mM lithium cacodylate, pH 7.2, 5 mM KCl and 95 mM LiCl). Calf thymus (CT) DNA was purchased from Sigma-Aldrich and dissolved in a solution of 10 mM lithium cacodylate 7.2 and 1 mM Na$_2$EDTA at a concentration of 1 mM. The solution was then equilibrated for one week, filtered, and stored at 4 °C. The fluorescently labeled oligonucleotide 5′-6-FAM-GGG(TTAGGG)$_3$-Dabcyl-3′ (F21D) was purchased from Integrated DNA Technologies (Coralville, IA, USA), dissolved at 0.1 mM in ddH$_2$O, and stored at −80 °C prior to use. The concentrations of all nucleic acids were determined through UV-Vis spectroscopy at 90 °C using the extinction coefficients $\varepsilon^{260\,nm}$ = 228.5 mM^{-1}cm^{-1} for Tel22, 247.6 mM^{-1}cm^{-1} for F21D, and 12.2 mM^{-1}cm^{-1} (per base pair) for CT DNA. Extinction coefficients were calculated with the Integrated DNA Technologies OligoAnalyzer (available at https://www.idtdna.com/calc/analyzer, accessed on November 20, 2018) which uses the nearest-neighbor approximation model [91,92].

To induce GQ structure formation, DNA samples at the desired concentrations alone or in the presence of 1–2 eq. of porphyrin were heated at 95 °C for ten minutes in 5K buffer, allowed to cool to room temperature over three hours, and equilibrated overnight at 4° C. All experiments were done in 5K buffer.

3.2. UV-Vis Titrations and Job Plot

UV-Vis experiments were performed on a Cary 300 (Varian) spectrophotometer with a Peltier-thermostated cuvette holder (error of ± 0.3 °C) using 1 cm methylmethacrylate or quartz cuvettes and dual beam detection. The sample cuvette contained 2.3–3.1 µM TCPPSpm4 or 1.0–6.4 µM ZnTCPPSpm4 and the reference cuvette contained 5K buffer. UV-Vis titrations were conducted by adding small volumes of concentrated Tel22 in a stepwise manner to a 1 mL of porphyrin solutions, mixing thoroughly, and equilibrating for at least two minutes. UV-vis scans were collected in the range of 352–500 nm. DNA was added until at least three final spectra were superimposable. All titrations were performed at least three times. All spectra were corrected mathematically for dilutions, and analyzed as described previously using a Direct Fit model [19,42] with GraphPad Prism software at 415 and 429 nm for TCPPSpm4 and 424 nm wavelengths for ZnTCPPSpm4. Job plot UV-Vis titration experiments were performed to independently determine the stoichiometry of ligand-Tel22 binding interactions. Job plot experiments were conducted for both porphyrins using the procedure and data processing described in our earlier work [19]. Both porphyrins and DNA were prepared at 3–4 µM. Job plot experiments were completed at least three times.

3.3. Fluorescence Spectroscopy

3.3.1. Resonance Light Scattering (RLS)

RLS experiments [78] were conducted using a conventional fluorimeter, Fluorolog FL-11 Jobin-Yvon Horiba. A 2.1 mL solution of 2 µM porphyrin in a 1 cm quartz cuvette was titrated with 0.5 mM annealed and equilibrated Tel22 solution at 25 °C. Final concentration of Tel22 varied between 0.05-10.0 µM, and the total volume of all additions was 42 µL (2%). After each addition of Tel22, the cuvette was equilibrated for 10 min and the data was collected with the following parameters: scan range of 380–630 nm, wavelength offset of 0 nm, increment of 1.0 nm, averaging time of 0.5 sec, number of scans 3 (averaged), and 1.5 nm slits for both excitation and emission.

3.3.2. Fluorescent Titrations

Fluorescence titrations were performed on a Photon Technology International QuantaMaster 40 spectrofluorimeter. A 2.0 mL solution of porphyrin in a 1 cm black quartz cuvette was titrated with annealed and equilibrated Tel22 solution at 20 °C. The concentration of TCPPSpm4 was 0.3 µM, and the concentration of ZnTCPPSpm4 was ~0.5 µM. Tel22 was added from three different stocks with increasing concentration: stock 1 was 3–4 µM, stock 2 was 95–150 µM, and stock 3 was 500–850 µM. Total volume of addition was ~60 µL (3%). After each addition of Tel22, the cuvette was equilibrated for at least two minutes and the scan was collected with the following parameters: excitation at 420 nm (at the isosbestic point for TCPPSpm4), emission range of 575–750 nm, increment of 1.0 nm, averaging time of 0.5 sec, one scan, and 3 nm slits both for excitation and emission.

3.4. Circular Dichroism (CD) Spectroscopy

CD scans and melting experiments were performed on an Aviv 410 spectropolarimeter equipped with a Peltier heating unit (error of ± 0.3 °C) in 1 cm quartz cuvettes. The solution of 10–15 µM Tel22 was annealed and equilibrated with 2 eq. of porphyrins and CD scans were collected with the following parameters: 220 to 330 nm spectral width, 1 nm bandwith, 1 sec averaging time, 25 °C, and 3–5 scans (averaged). CD melting was performed on the same samples with the following parameters: 294 nm wavelength, 15–90 °C temperature range, 30 sec equilibration time, and 10 sec averaging time.

CD scans were collected before and after the melt to check if the melting process is reversible. CD data were analyzed as described in our earlier work [19,42].

Two sets of CD titrations were performed. First, 7–15 μM Tel22 was titrated with up to 4 eq. of 0.44 mM TCPPSpm4 or 5.75 mM ZnTCPPSpm4 in 1 eq. increments. After each addition of the porphyrin, the sample was equilibrated for 12 min after which CD scans were collected in 220–330 nm region. Secondly, to detect induced CD signal (iCD) 2–6 μM porphyrin solution was titrated with small increments of 100–200 μM Tel22. Samples were equilibrated for 10 min and CD spectra were collected in the 375–480 nm region using 5–10 scans to obtain good signal-to-noise ratio.

3.5. Fluorescence Resonance Energy Transfer (FRET) Assays

FRET studies were conducted according to the published protocol [81]. A solution of 0.2 μM F21D was incubated in the presence of 0–8 eq. of TCPPSpm4 or 0–20 eq. of ZnTCPPSpm4 and melting curves were collected. FRET competition experiments were performed using 0.2 μM F21D in the presence of fixed amounts of TCPPSpm4 (0.75 μM, 3.7 eq.) or ZnTCPPSpm4 (2.2 μM, 11 eq.) and increasing amounts of CT DNA (up to 96 μM, 480 eq.), and analyzed as described previously [42].

4. Conclusions

There is a great need to develop ligands capable of binding to and regulating the stability of GQs strongly and selectively. In this work, we characterized interactions of novel spermine-derivatized porphyrins, TCPPSpm4 and ZnTCPPSpm4, with human telomeric DNA, Tel22. Both porphyrins bind tightly to the GQ with Ka of $(5–14) \times 10^6$ M^{-1} and provide strong stabilization, with the selectivity ratio of 200–300 over dsDNA. Interestingly, we observe a high binding stoichiometry, which may indicate multiple binding modes, the most prominent of which are end-stacking and DNA-assisted self-association of porphyrins. In addition, the spermine arms of the porphyrins likely act as four tentacles reaching into groves and stabilizing the GQ. The mild selectivity for GQ over dsDNA is likely due to strong electrostatic interactions between the polycationic ligand and negatively charged DNA backbone. Consistent with the prior work, addition of Zn(II) to the porphyrin core did not improve selectivity, in spite of the presence of fifth axial water ligand, but increased Ka three-fold.

Overall, our findings demonstrate that spermine group derivatization is a valid strategy in the design of novel GQ binders, especially given the fact that polyamines are taken up extensively by cancer cells [67,68], and thus, could be used for selective cancer targeting. Future work will focus on optimizing these porphyrins by decreasing their charge (limiting the number of spermine arms to 1–3) and adding functional groups known to improve GQ selectivity. Biological studies of the new ligands should also be a priority.

Supplementary Materials: Supplementary materials can be found at http://www.mdpi.com/1422-0067/19/11/3686/s1.

Author Contributions: Conceptualization, L.A.Y. and A.D.; experimental work, N.C.S., J.C., J.H.(J.)L., and C.M.A.G.; data analysis and interpretation, all authors; writing—original draft preparation, N.C.S., L.A.Y., and A.D.; writing—review and editing, L.A.Y. All authors have reviewed the manuscript and agreed to publish the results.

Funding: This research was funded by the National Institute of Health, grant number 1R15CA208676-01A1 (to L.A.Y.); Camille and Henry Dreyfus Teacher-Scholar Award (to L.A.Y.); an HHMI undergraduate research fellowship (to N.C.S.), a Benjamin Franklin Travel Grant (to N.C.S.); and the University of Catania, "Piano della Ricerca di Ateneo 2016–2018" (to A.D.).

Acknowledgments: We thank Roberto Purello (University of Catania) for useful discussions; we thank Jean-Louis Mergny (Institut Européen de Chimie et de Biologie) for hosting L.A.Y. and N.C.S. while some of the presented experimental work was performed; and Nick Kaplinsky (Swarthmore College) for use of his RT-PCR machine.

Conflicts of Interest: The authors declare no conflict of interest

Abbreviations

GQ	Guanine Quadruplex
FRET	Fluorescence Resonance Energy Transfer
CD	Circular Dichroism
iCD	Induced Circular Dichroism
TCPPSpm4	*meso*-tetrakis-(4-carboxysperminephenyl)porphyrin
ZnTCPPSpm4	Zn(II) *meso*-tetrakis-(4-carboxysperminephenyl)porphyrin
CT DNA	Calf Thymus DNA
F21D	5′-6-FAM-GGG(TTAGGG)$_3$-Dabcyl-3′

References

1. Yatsunyk, L.A.; Mendoza, O.; Mergny, J.L. "Nano-oddities": Unusual nucleic acid assemblies for DNA-based nanostructures and nanodevices. *Acc. Chem. Res.* **2014**, *47*, 1836–1844. [CrossRef] [PubMed]
2. Hänsel-Hertsch, R.; Di Antonio, M.; Balasubramanian, S. DNA G-quadruplexes in the human genome: Detection, functions and therapeutic potential. *Nat. Rev. Mol. Cell Biol.* **2017**, *18*, 279. [CrossRef] [PubMed]
3. Largy, E.; Mergny, J.L.; Gabelica, V. Role of Alkali Metal Ions in G-Quadruplex Nucleic Acid Structure and Stability. *Met. Ions. Life. Sci.* **2016**, *16*, 203–258. [CrossRef] [PubMed]
4. Campbell, N.H.; Neidle, S. G-quadruplexes and metal ions. *Met. Ions. Life Sci.* **2012**, *10*, 119–134. [CrossRef] [PubMed]
5. Huppert, J.L.; Balasubramanian, S. Prevalence of quadruplexes in the human genome. *Nucleic Acids Res.* **2005**, *33*, 2908–2916. [CrossRef] [PubMed]
6. Rhodes, D.; Lipps, H.J. G-quadruplexes and their regulatory roles in biology. *Nucleic Acids Res.* **2015**, *43*, 8627–8637. [CrossRef] [PubMed]
7. Todd, A.K.; Johnston, M.; Neidle, S. Highly prevalent putative quadruplex sequence motifs in human DNA. *Nucleic Acids Res.* **2005**, *33*, 2901–2907. [CrossRef] [PubMed]
8. Bedrat, A.; Lacroix, L.; Mergny, J.L. Re-evaluation of G-quadruplex propensity with G4Hunter. *Nucleic Acids Res.* **2016**, *44*, 1746–1759. [CrossRef] [PubMed]
9. Biffi, G.; Tannahill, D.; McCafferty, J.; Balasubramanian, S. Quantitative visualization of DNA G-quadruplex structures in human cells. *Nat. Chem.* **2013**, *5*, 182–186. [CrossRef] [PubMed]
10. Henderson, A.; Wu, Y.; Huang, Y.C.; Chavez, E.A.; Platt, J.; Johnson, F.B.; Brosh, R.M.; Sen, D.; Lansdorp, P.M. Detection of G-quadruplex DNA in mammalian cells. *Nucleic Acids Res.* **2014**, *42*, 860–869. [CrossRef] [PubMed]
11. Huang, W.C.; Tseng, T.Y.; Chen, Y.T.; Chang, C.C.; Wang, Z.F.; Wang, C.L.; Hsu, T.N.; Li, P.T.; Chen, C.T.; Lin, J.J.; et al. Direct evidence of mitochondrial G-quadruplex DNA by using fluorescent anti-cancer agents. *Nucleic Acids Res.* **2015**, *43*, 10102–10113. [CrossRef] [PubMed]
12. Zhang, S.; Sun, H.; Wang, L.; Liu, Y.; Chen, H.; Li, Q.; Guan, A.; Liu, M.; Tang, Y. Real-time monitoring of DNA G-quadruplexes in living cells with a small-molecule fluorescent probe. *Nucleic Acids Res.* **2018**, *46*, 7522–7532. [CrossRef] [PubMed]
13. O'Sullivan, R.J.; Karlseder, J. Telomeres: Protecting chromosomes against genome instability. *Nat. Rev. Mol. Cell Biol.* **2010**, *11*, 171–181. [CrossRef] [PubMed]
14. Dai, J.; Carver, M.; Yang, D. Polymorphism of human telomeric quadruplex structures. *Biochimie* **2008**, *90*, 1172–1183. [CrossRef] [PubMed]
15. Phan, A.T. Human telomeric G-quadruplex: Structures of DNA and RNA sequences. *FEBS J.* **2010**, *277*, 1107–1117. [CrossRef] [PubMed]
16. Li, J.; Correia, J.J.; Wang, L.; Trent, J.O.; Chaires, J.B. Not so crystal clear: The structure of the human telomere G-quadruplex in solution differs from that present in a crystal. *Nucleic Acids Res.* **2005**, *33*, 4649–4659. [CrossRef] [PubMed]
17. Heddi, B.; Phan, A.T. Structure of human telomeric DNA in crowded solution. *J. Am. Chem. Soc.* **2011**, *133*, 9824–9833. [CrossRef] [PubMed]
18. Xue, Y.; Kan, Z.Y.; Wang, Q.; Yao, Y.; Liu, J.; Hao, Y.H.; Tan, Z. Human telomeric DNA forms parallel-stranded intramolecular G-quadruplex in K+ solution under molecular crowding condition. *J. Am. Chem. Soc.* **2007**, *129*, 11185–11191. [CrossRef] [PubMed]

19. Nicoludis, J.M.; Barrett, S.P.; Mergny, J.-L.; Yatsunyk, L.A. Interaction of G-quadruplex DNA with N-methyl mesoporphyrin IX. *Nucleic Acids Res.* **2012**, *40*, 5432–5447. [CrossRef] [PubMed]

20. Nicoludis, J.M.; Miller, S.T.; Jeffrey, P.D.; Barrett, S.P.; Rablen, P.R.; Lawton, T.J.; Yatsunyk, L.A. Optimized end-stacking provides specificity of N-methyl mesoporphyrin IX for human telomeric G-quadruplex DNA. *J. Am. Chem. Soc.* **2012**, *134*, 20446–20456. [CrossRef] [PubMed]

21. Parkinson, G.N.; Lee, M.P.; Neidle, S. Crystal structure of parallel quadruplexes from human telomeric DNA. *Nature* **2002**, *417*, 876–880. [CrossRef] [PubMed]

22. Renciuk, D.; Kejnovska, I.; Skolakova, P.; Bednarova, K.; Motlova, J.; Vorlickova, M. Arrangement of human telomere DNA quadruplex in physiologically relevant K$^+$ solutions. *Nucleic Acids Res.* **2009**, *37*, 6625–6634. [CrossRef] [PubMed]

23. Wang, Y.; Patel, D.J. Solution structure of the human telomeric repeat d[AG3(T2AG3)3] G-tetraplex. *Structure* **1993**, *1*, 263–282. [CrossRef]

24. Phan, A.T.; Luu, K.N.; Patel, D.J. Different loop arrangements of intramolecular human telomeric (3+1) G-quadruplexes in K+ solution. *Nucleic Acids Res.* **2006**, *34*, 5715–5719. [CrossRef] [PubMed]

25. Phan, A.T.; Kuryavyi, V.; Luu, K.N.; Patel, D.J. Structure of two intramolecular G-quadruplexes formed by natural human telomere sequences in K+ solution. *Nucleic Acids Res.* **2007**, *35*, 6517–6525. [CrossRef] [PubMed]

26. Xu, Y.; Noguchi, Y.; Sugiyama, H. The new models of the human telomere d[AGGG(TTAGGG)3] in K+ solution. *Bioorg. Med. Chem.* **2006**, *14*, 5584–5591. [CrossRef] [PubMed]

27. Luu, K.N.; Phan, A.T.; Kuryavyi, V.; Lacroix, L.; Patel, D.J. Structure of the human telomere in K+ solution: An intramolecular (3 + 1) G-quadruplex scaffold. *J. Am. Chem. Soc.* **2006**, *128*, 9963–9970. [CrossRef] [PubMed]

28. Ambrus, A.; Chen, D.; Dai, J.; Bialis, T.; Jones, R.A.; Yang, D. Human telomeric sequence forms a hybrid-type intramolecular G-quadruplex structure with mixed parallel/antiparallel strands in potassium solution. *Nucleic Acids Res.* **2006**, *34*, 2723–2735. [CrossRef] [PubMed]

29. Lim, K.W.; Amrane, S.; Bouaziz, S.; Xu, W.; Mu, Y.; Patel, D.J.; Luu, K.N.; Phan, A.T. Structure of the human telomere in K+ solution: A stable basket-type G-quadruplex with only two G-tetrad layers. *J. Am. Chem. Soc.* **2009**, *131*, 4301–4309. [CrossRef] [PubMed]

30. Hanahan, D.; Weinberg, R.A. The hallmarks of cancer. *Cell* **2000**, *100*, 57–70. [CrossRef]

31. Neidle, S. Quadruplex Nucleic Acids as Novel Therapeutic Targets. *J. Med. Chem.* **2016**. [CrossRef] [PubMed]

32. Ohnmacht, S.A.; Neidle, S. Small-molecule quadruplex-targeted drug discovery. *Bioorg. Med. Chem. Lett.* **2014**, *24*, 2602–2612. [CrossRef] [PubMed]

33. Anantha, N.V.; Azam, M.; Sheardy, R.D. Porphyrin binding to quadrupled T4G4. *Biochemistry* **1998**, *37*, 2709–2714. [CrossRef] [PubMed]

34. Fiel, R.J.; Howard, J.C.; Mark, E.H.; Datta Gupta, N. Interaction of DNA with a porphyrin ligand: Evidence for intercalation. *Nucleic Acids Res.* **1979**, *6*, 3093–3118. [CrossRef] [PubMed]

35. D'Urso, A.; Fragalà, M.E.; Purrello, R. Non-covalent interactions of porphyrinoids with duplex DNA. In *Applications of Porphyrinoids*; Springer: Berlin/Heidelberg, Germany, 2013; pp. 139–174.

36. Georgiou, G.N.; Ahmet, M.T.; Houlton, A.; Silver, J.; Cherry, R.J. Measurement of the rate of uptake and subsellular localization of porphyrins in cells using fluorescence digital imaging microscopy. *Photochem. Photobiol.* **1994**, *59*, 419–422. [CrossRef] [PubMed]

37. Benimetskaya, L.; Takle, G.B.; Vilenchik, M.; Lebedeva, I.; Miller, P.; Stein, C.A. Cationic porphyrins: Novel delivery vehicles for antisense oligodeoxynucleotides. *Nucleic Acids Res.* **1998**, *26*, 5310–5317. [CrossRef] [PubMed]

38. Izbicka, E.; Wheelhouse, R.T.; Raymond, E.; Davidson, K.K.; Lawrence, R.A.; Sun, D.; Windle, B.E.; Hurley, L.H.; Von Hoff, D.D. Effects of cationic porphyrins as G-quadruplex interactive agents in human tumor cells. *Cancer Res.* **1999**, *59*, 639–644. [PubMed]

39. Sabharwal, N.C.; Savikhin, V.; Turek-Herman, J.R.; Nicoludis, J.M.; Szalai, V.A.; Yatsunyk, L.A. N-methylmesoporphyrin IX fluorescence as a reporter of strand orientation in guanine quadruplexes. *FEBS J.* **2014**, *281*, 1726–1737. [CrossRef] [PubMed]

40. Han, F.X.; Wheelhouse, R.T.; Hurley, L.H. Interactions of TMPyP4 and TMPyP2 with quadruplex DNA. Structural basis for the differential effects on telomerase inhibition. *J. Am. Chem. Soc.* **1999**, *121*, 3561–3570. [CrossRef]

41. Shi, D.F.; Wheelhouse, R.T.; Sun, D.; Hurley, L.H. Quadruplex-interactive agents as telomerase inhibitors: Synthesis of porphyrins and structure-activity relationship for the inhibition of telomerase. *J. Med. Chem.* **2001**, *44*, 4509–4523. [CrossRef] [PubMed]

42. Bhattacharjee, A.J.; Ahluwalia, K.; Taylor, S.; Jin, O.; Nicoludis, J.M.; Buscaglia, R.; Chaires, J.B.; Kornfilt, D.J.P.; Marquardt, D.G.S.; Yatsunyk, L.A. Induction of G-quadruplex DNA structure by Zn(II) 5,10,15,20-tetrakis(N-methyl-4-pyridyl)porphyrin. *Biochimie* **2011**, *93*, 1297–1309. [CrossRef] [PubMed]

43. Sabharwal, N.C.; Mendoza, O.; Nicoludis, J.M.; Ruan, T.; Mergny, J.-L.; Yatsunyk, L.A. Investigation of the interactions between Pt(II) and Pd(II) derivatives of 5,10,15,20-tetrakis (N-methyl-4-pyridyl) porphyrin and G-quadruplex DNA. *J. Biol. Inorg. Chem.* **2016**, *21*, 227–239. [CrossRef] [PubMed]

44. Ruan, T.L.; Davis, S.J.; Powell, B.M.; Harbeck, C.P.; Habdas, J.; Habdas, P.; Yatsunyk, L.A. Lowering the overall charge on TMPyP4 improves its selectivity for G-quadruplex DNA. *Biochimie* **2017**, *132*, 121–130. [CrossRef] [PubMed]

45. Pan, J.; Zhang, S. Interaction between cationic zinc porphyrin and lead ion induced telomeric guanine quadruplexes: Evidence for end-stacking. *J. Biol. Inorg. Chem.* **2009**, *14*, 401–407. [CrossRef] [PubMed]

46. Yao, X.; Song, D.; Qin, T.; Yang, C.; Yu, Z.; Li, X.; Liu, K.; Su, H. Interaction between G-Quadruplex and Zinc Cationic Porphyrin: The Role of the Axial Water. *Sci. Rep.* **2017**, *7*, 10951. [CrossRef] [PubMed]

47. Phan, A.T.; Kuryavyi, V.; Gaw, H.Y.; Patel, D.J. Small-molecule interaction with a five-guanine-tract G-quadruplex structure from the human MYC promoter. *Nat. Chem. Biol.* **2005**, *1*, 167. [CrossRef] [PubMed]

48. Le, V.H.; Nagesh, N.; Lewis, E.A. Bcl-2 promoter sequence G-quadruplex interactions with three planar and non-planar cationic porphyrins: TMPyP4, TMPyP3, and TMPyP2. *PLoS ONE* **2013**, *8*, e72462. [CrossRef] [PubMed]

49. Lubitz, I.; Borovok, N.; Kotlyar, A. Interaction of monomolecular G4-DNA nanowires with TMPyP: Evidence intercalation. *Biochemistry* **2007**, *46*, 12925–12929. [CrossRef] [PubMed]

50. Cavallari, M.; Garbesi, A.; Di Felice, R. Porphyrin intercalation in G4-DNA quadruplexes by molecular dynamics simulations. *J. Phys. Chem. B* **2009**, *113*, 13152–13160. [CrossRef] [PubMed]

51. Wei, C.; Wang, L.; Jia, G.; Zhou, J.; Han, G.; Li, C. The binding mode of porphyrins with cation side arms to (TG4T)4 G-quadruplex: Spectroscopic evidence. *Biophys. Chem.* **2009**, *143*, 79–84. [CrossRef] [PubMed]

52. Parkinson, G.N.; Ghosh, R.; Neidle, S. Structural basis for binding of porphyrin to human telomeres. *Biochemistry* **2007**, *46*, 2390–2397. [CrossRef] [PubMed]

53. McClure, J.E.; Baudouin, L.; Mansuy, D.; Marzilli, L.G. Interactions of DNA with a new electron-deficient tentacle porphyrin: Meso-tetrakis[2,3,5,6-tetrafluoro-4-(2-trimethylammoniumethyl-amine)phenyl]porphyrin. *Biopolymers* **1997**, *42*, 203–217. [CrossRef]

54. Mukundan, N.E.; Petho, G.; Dixon, D.W.; Kim, M.S.; Marzilli, L.G. Interactions of an electron-rich tetracationic tentacle porphyrin with calf thymus DNA. *Inorg. Chem.* **1994**, *33*, 4676–4687. [CrossRef]

55. Mukundan, N.E.; Petho, G.; Dixon, D.W.; Marzilli, L.G. DNA-tentacle porphyrin interactions: AT over GC selectivity exhibited by an outside binding self-stacking porphyrin. *Inorg. Chem.* **1995**, *34*, 3677–3687. [CrossRef]

56. Marzilli, L.G.; Petho, G.; Lin, M.; Kim, M.S.; Dixon, D.W. Tentacle porphyrins: DNA interactions. *J. Am. Chem. Soc.* **1992**, *114*, 7575–7577. [CrossRef]

57. Thomas, T.J.; Tajmir-Riahi, H.A.; Thomas, T. Polyamine-DNA interactions and development of gene delivery vehicles. *Amino Acids.* **2016**, *48*, 2423–2431. [CrossRef] [PubMed]

58. Thomas, T.J.; Thomas, T. Collapse of DNA in packaging and cellular transport. *Int. J. Biol. Macromol.* **2018**, *109*, 36–48. [CrossRef] [PubMed]

59. Ouameur, A.A.; Tajmir-Riahi, H.-A. Structural analysis of DNA interactions with biogenic polyamines and cobalt(III)hexamine studied by fourier transform infrared and capillary electrophoresis. *J. Biol. Chem.* **2004**, *279*, 42041–42054. [CrossRef] [PubMed]

60. Parkinson, A.; Hawken, M.; Hall, M.; Sanders, K.J.; Rodger, A. Amine induced Z-DNA in poly (dG-dC)· poly (dG-dC): Circular dichroism and gel electrophoresis study. *Phys. Chem. Chem. Phys.* **2000**, *2*, 5469–5478. [CrossRef]

61. Mergny, J.-L.; Lacroix, L.; Teulade-Fichou, M.-P.; Hounsou, C.; Guittat, L.; Hoarau, M.; Arimondo, P.B.; Vigneron, J.-P.; Lehn, J.-M.; Riou, J.-F.; et al. Telomerase inhibitors based on quadruplex ligands selected by a fluorescence assay. *Proc. Natl. Acad. Sci. USA* **2001**, *98*, 3062–3067. [CrossRef] [PubMed]

62. Li, G.; Huang, J.; Zhang, M.; Zhou, Y.; Zhang, D.; Wu, Z.; Wang, S.; Weng, X.; Zhou, X.; Yang, G. Bis(benzimidazole)pyridine derivative as a new class of G-quadruplex inducing and stabilizing ligand. *Chem. Commun.* **2008**, 4564–4566. [CrossRef] [PubMed]

63. Collie, G.W.; Promontorio, R.; Hampel, S.M.; Micco, M.; Neidle, S.; Parkinson, G.N. Structural basis for telomeric G-quadruplex targeting by naphthalene diimide ligands. *J. Am. Chem. Soc.* **2012**, *134*, 2723–2731. [CrossRef] [PubMed]

64. Guyen, B.; Schultes, C.M.; Hazel, P.; Mann, J.; Neidle, S. Synthesis and evaluation of analogues of 10H-indolo[3,2-*b*]quinoline as G-quadruplex stabilizing ligands and potential inhibitors of the enzyme telomerase. *Org. Biomol. Chem.* **2004**, *2*, 981–988. [CrossRef] [PubMed]

65. Schultes, C.M.; Guyen, B.; Cuesta, J.; Neidle, S. Synthesis, biophysical and biological evaluation of 3,6-bis-amidoacridines with extended 9-anilino substituents as potent G-quadruplex-binding telomerase inhibitors. *Bioorg. Med. Chem. Letters* **2004**, *14*, 4347–4351. [CrossRef] [PubMed]

66. Mandal, S.; Mandal, A.; Johansson, H.E.; Orjalo, A.V.; Park, M.H. Depletion of cellular polyamines, spermidine and spermine, causes a total arrest in translation and growth in mammalian cells. *Proc. Natl. Acad. Sci. USA* **2013**, *110*, 2169–2174. [CrossRef] [PubMed]

67. Gerner, E.W.; Meyskens, F.L. Polyamines and cancer: Old molecules, new understanding. *Nat. Rev. Cancer* **2004**, *4*, 781–792. [CrossRef] [PubMed]

68. Pegg, A.E.; Casero, R.A. Current status of the polyamine research field. *Methods Mol. Biol.* **2011**, *720*, 3–35. [CrossRef] [PubMed]

69. Carlisle, D.L.; Devereux, W.L.; Hacker, A.; Woster, P.M.; Casero, R.A. Growth status significantly affects the response of human lung cancer cells to antitumor polyamine-analogue exposure. *J. Clin. Cancer Res.* **2002**, *8*, 2684–2689.

70. Cullis, P.M.; Green, R.E.; Merson-Davies, L.; Travis, N. Probing the mechanism of transport and compartmentalisation of polyamines in mammalian cells. *Nat. Chem. Biol.* **1999**, 717–729. [CrossRef]

71. Wang, C.; Delcros, J.-G.; Biggerstaff, J.; Phanstiel, O.I. Synthesis and biological evaluation of N1-(anthracen-9-ylmethyl)triamines as molecular recognition elements for the polyamine transporter. *J. Med. Chem.* **2003**, 2663–2671. [CrossRef] [PubMed]

72. Gangemi, C.M.A.; Randazzo, R.; Fragala, M.E.; Tomaselli, G.A.; Ballistreri, F.P.; Pappalardo, A.; Toscano, R.M.; Sfrazzetto, G.T.; Purrello, R.; D'Urso, A. Hierarchically controlled protonation/aggregation of a porphyrin–spermine derivative. *New J. Chem.* **2015**, *39*, 6722–6725. [CrossRef]

73. D'Urso, A.; Randazzo, R.; Rizzo, V.; Gangemi, C.; Romanucci, V.; Zarrelli, A.; Tomaselli, G.; Milardi, D.; Borbone, N.; Purrello, R. Stabilization vs. destabilization of G-quadruplex superstructures: The role of the porphyrin derivative having spermine arms. *Phys. Chem. Chem. Phys.* **2017**, *19*, 17404–17410. [CrossRef] [PubMed]

74. Pasternack, R.F.; Briganid, R.A.; Abrams, M.J.; Williams, A.P.; Gibbs, E.J. Interactions of porphyrins and metalloporphyrins with single-stranded poly(dA). *Inorg. Chem.* **1990**, *29*, 4483–4486. [CrossRef]

75. Job, P. Formation and Stability of Inorganic Complexes in Solution. *Annali di Chimica Applicata* **1928**, *9*, 113–203.

76. Huang, C.Y. Determination of binding stoichiometry by the continuous variation method: The Job plot. *Methods Enzymol.* **1982**, *87*, 509–525. [PubMed]

77. Boschi, E.; Davis, S.; Taylor, S.; Butterworth, A.; Chirayath, L.A.; Purohit, V.; Siegel, L.K.; Buenaventura, J.; Sheriff, A.H.; Jin, R.; et al. Interaction of a Cationic Porphyrin and Its Metal Derivatives with G-Quadruplex DNA. *J. Phys. Chem. B* **2016**, *120*, 12807–12819. [CrossRef] [PubMed]

78. Pasternack, R.; Collings, P. Resonance light scattering: A new technique for studying chromophore aggregation. *Science* **1995**, *269*, 935–939. [CrossRef] [PubMed]

79. Kelly, J.M.; Tossi, A.B.; McConnell, D.J.; OhUigin, C. A study of the interactions of some polypyridylruthenium (II) complexes with DNA using fluorescence spectroscopy, topoisomerisation and thermal denaturation. *Nucleic Acids Res.* **1985**, *13*, 6017–6034. [CrossRef] [PubMed]

80. Kelly, J.M.; Murphy, M.J.; McConnell, D.J.; OhUigin, C. A comparative study of the interaction of 5,10,15,20-tetrakis (N-methylpyridinium-4-yl)porphyrin and its zinc complex with DNA using fluorescence spectroscopy and topoisomerisation. *Nucleic Acids Res.* **1985**, *13*, 167–184. [CrossRef] [PubMed]

81. De Cian, A.; Guittat, L.; Kaiser, M.; Saccà, B.; Amrane, S.; Bourdoncle, A.; Alberti, P.; Teulade-Fichou, M.-P.; Lacroix, L.; Mergny, J.-L. Fluorescence-based melting assays for studying quadruplex ligands. *Methods* **2007**, *42*, 183–195. [CrossRef] [PubMed]

82. DuPont, J.I.; Henderson, K.L.; Metz, A.; Le, V.H.; Emerson, J.P.; Lewis, E.A. Calorimetric and spectroscopic investigations of the binding of metallated porphyrins to G-quadruplex DNA. *Biochim. Biophys. Acta.* **2016**, *1860*, 902–909. [CrossRef] [PubMed]

83. Morris, M.J.; Wingate, K.L.; Silwal, J.; Leeper, T.C.; Basu, S. The porphyrin TMPyP4 unfolds the extremely stable G-quadruplex in MT3-MMP mRNA and alleviates its repressive effect to enhance translation in eukaryotic cells. *Nucleic Acids Res.* **2012**, *40*, 4137–4145. [CrossRef] [PubMed]

84. Waller, Z.A.E.; Sewitz, S.A.; Hsu, S.-T.D.; Balasubramanian, S. A small molecule that disrupts G-quadruplex DNA structure and enhances gene expression. *J. Am. Chem. Soc.* **2009**, *131*, 12628–12633. [CrossRef] [PubMed]

85. Kaluzhny, D.; Ilyinsky, N.; Shchekotikhin, A.; Sinkevich, Y.; Tsvetkov, P.O.; Tsvetkov, V.; Veselovsky, A.; Livshits, M.; Borisova, O.; Shtil, A.; et al. Disordering of human telomeric G-quadruplex with novel antiproliferative anthrathiophenedione. *PLoS ONE* **2011**, *6*, e27151. [CrossRef] [PubMed]

86. Marchand, A.; Granzhan, A.; Iida, K.; Tsushima, Y.; Ma, Y.; Nagasawa, K.; Teulade-Fichou, M.-P.; Gabelica, V. Ligand-induced conformational changes with cation ejection upon binding to human telomeric DNA G-quadruplexes. *J. Am. Chem. Soc.* **2015**, *137*, 750–756. [CrossRef] [PubMed]

87. Pasternack, R.F. Circular dichroism and the interactions of water soluble porphyrins with DNA—A minireview. *Chirality* **2003**, *15*, 329–332. [CrossRef] [PubMed]

88. Pasternack, R.F.; Gibbs, E.J.; Villafranca, J.J. Interactions of porphyrins with nucleic acids. *Biochemistry* **1983**, *22*, 2406–2414. [CrossRef] [PubMed]

89. Gangemi, C.M.; D'Agostino, B.; Randazzo, R.; Gaeta, M.; Fragalà, M.E.; Purrello, R.; D'Urso, A. Interaction of spermine derivative porphyrin with DNA. *J. Porphyr. Phthalocyanines* **2018**, *2*, 1–7. [CrossRef]

90. Gangemi, C.M.A.; D'Urso, A.; Tomaselli, G.A.; Berova, N.; Purrello, R. A novel porphyrin-based molecular probe ZnTCPPSpm4 with catalytic, stabilizing and chiroptical diagnostic power towards DNA B-Z transition. *J. Inorg. Biochem.* **2017**, *173*, 141–143. [CrossRef] [PubMed]

91. Tataurov, A.V.; You, Y.; Owczarzy, R. Predicting ultraviolet spectrum of single stranded and double stranded deoxyribonucleic acids. *Biophys. Chem.* **2008**, *133*, 66–70. [CrossRef] [PubMed]

92. Cantor, C.R.; Warshaw, M.M.; Shapiro, H. Oligonucleotide interactions. 3. Circular dichroism studies of the conformation of deoxyoligonucleotides. *Biopolymers* **1970**, *9*, 1059–1077. [CrossRef] [PubMed]

International Journal of
Molecular Sciences

MDPI

Article

Co^II(Chromomycin)2 Complex Induces a Conformational Change of CCG Repeats from i-Motif to Base-Extruded DNA Duplex

Yu-Wen Chen [1], Roshan Satange [2,3], Pei-Ching Wu [2], Cyong-Ru Jhan [4], Chung-ke Chang [5], Kuang-Ren Chung [6], Michael J. Waring [7], Sheng-Wei Lin [8], Li-Ching Hsieh [2,9,*] and Ming-Hon Hou [1,2,3,4,*]

[1] Institute of Biotechnology, National Chung-Hsing University, Taichung 402, Taiwan; dodochinchin@gmail.com
[2] Institute of Genomics and Bioinformatics, National Chung-Hsing University, Taichung 402, Taiwan; roshan.satange@gmail.com (R.S.); jane871057@gmail.com (P.-C.W.)
[3] Ph.D. Program in Medical Biotechnology, National Chung Hsing University, Taichung 402, Taiwan
[4] Department of Life Sciences, National Chung-Hsing University, Taichung 402, Taiwan; f810409@gmail.com
[5] Institute of Biomedical Sciences, Academia Sinica, Taipei 115, Taiwan; chungke@ibms.sinica.edu.tw
[6] Department of Plant Pathology, National Chung-Hsing University, Taichung 402, Taiwan; krchung@nchu.edu.tw
[7] Department of Biochemistry, University of Cambridge, Cambridge CB2 1GA, UK; mjw11@cam.ac.uk
[8] Institute of Biological Chemistry, Academia Sinica, Taipei 115, Taiwan; sanway@gate.sinica.edu.tw
[9] Advanced Plant Biotechnology Center, National Chung-Hsing University, Taichung 402, Taiwan
* Correspondence: liching@dragon.nchu.edu.tw (L.-C.H.); mhho@nchu.edu.tw (M.-H.H.)

Received: 20 August 2018; Accepted: 7 September 2018; Published: 17 September 2018

Abstract: We have reported the propensity of a DNA sequence containing CCG repeats to form a stable i-motif tetraplex structure in the absence of ligands. Here we show that an i-motif DNA sequence may transition to a base-extruded duplex structure with a GGCC tetranucleotide tract when bound to the (Co^II)-mediated dimer of chromomycin A3, Co^II(Chro)2. Biophysical experiments reveal that CCG trinucleotide repeats provide favorable binding sites for Co^II(Chro)2. In addition, water hydration and divalent metal ion (Co^II) interactions also play a crucial role in the stabilization of CCG trinucleotide repeats (TNRs). Our data furnish useful structural information for the design of novel therapeutic strategies to treat neurological diseases caused by repeat expansions.

Keywords: i-motif; CCG repeats; trinucleotide repeat DNA; chromomycin A3; neurological disease; X-ray crystallography; nucleotide flip-out; DNA deformation

1. Introduction

The formation of expanded repeat sequences has long been known to correlate with the etiology of many human diseases [1–3]. Tandem repeats can form unusual DNA structures, resulting in consecutive GpC sites that are flanked by mismatched G:G or C:C base pairs in the X chromosome [2,4,5]. Fragile X syndrome (FXS) is a genetic disorder caused by an expansion of CGG/CCG tandem repeats in the Fragile X Mental Retardation 1 gene (*FMR1*) on the X chromosome [6,7]. The repeats in *FMR1* result in a defective protein that has been associated with symptoms of FXS. Recently, our studies have suggested that the expansion of $(CCG)_n$ trinucleotide repeats (TNRs) may be attributed to the slippage of DNA strands along the hairpin structures, forming a four-stranded helical structure that is stabilized by intertwining i-motifs during DNA replication [8].

Small molecules that specifically bind to TNR DNA conformations could have applications as diagnostic tools as well as therapeutic agents against these genetic diseases. For example, naphthyridine derivatives can inhibit DNA polymerases during replication because they can selectively recognize and stabilize the CNG repeat hairpin structures formed by a single-strand DNA expansion [9]. Moreover, several well-known DNA-binding drugs including actinomycin D, doxorubicin and mitomycin C have been demonstrated to prevent the amplification of abnormal CNG trinucleotide repeats [10–12]. Chromomycin A3 (Chro), produced by some strains of *Streptomyces griseus*, is an anthraquinone glycoside antibiotic belonging to the aureolic acid family [13]. Chro contains di- and trisaccharide components linked to a β-ketophenol chromophore (anthracene ring) via *O*-glycosidic bonds at position 2 and 6, respectively. Chro can bind to divalent metal ions and form a dimer, (Chro)$_2$, that has a unique fluorescent emission under different environmental conditions [14]. Previously, Ni^{II}(Chro)$_2$ has been shown to bind specifically to CCG TNRs via a "forced" induced-fit mechanism [15]. Upon binding to TNRs, Ni^{II}(Chro)$_2$ exhibits a unique fluorescence signature which can potentially be used to identify fragile X syndrome in clinical specimens.

However, previous studies utilized different DNA sequences, prompting the question of whether chromomycin compounds are capable of "transforming" an i-motif sequence to a base-extruded sequence. The identity of the metal ion might also be significant; e.g., Ni may or may not be important for binding. In the presence of divalent cobalt ions, Chro can also form a metal-coordinated dimer Co^{II}(Chro)$_2$, which binds selectively to GpC sequences in the minor groove of DNA. In the current study, experiments were conducted to gain a better understanding of the effects of Co^{II}(Chro)$_2$ on the i-motif structure of CCG TNRs. The crystal structure of the dT(CCG)$_3$A sequence has been solved in the presence and absence of the cobalt-containing Chro dimer. These studies revealed that the CCG repeats can fold into a hairpin structure with tetraplex i-motif formation. Co^{II}(Chro)$_2$ can alter hairpin formation of CCG repeat DNA and is responsible for the formation of a double-helical conformation of CCG repeat DNA with dual cytosine flipping. The results also revealed that water-mediated interactions and divalent cobalt ions are essential to maintain the conformational integrity and stability of the Chro-DNA complex.

2. Results

2.1. A Non-Canonical DNA Structure of the dT(CCG)$_3$A Sequence Contains an i-Motif Tetraplex Core

The dT(CCG)$_3$A sequence in the absence of Co^{II}(Chro)$_2$ was crystallized in a slightly acidic environment (pH 6.0) to yield a high resolution structure of 1.71 Å. The initial phase for the dT(CCG)$_3$A was solved using the previously reported coordinates of PDB ID: 4PZQ. All atoms present in the DNA molecule were included in the refined structure and exhibited clear electron density, as shown in the Supplementary Materials Figure S1A. The crystal structure revealed that each asymmetric unit contained a single-strand dT(CCG)$_3$A molecule, which could form a CCG loop by folding back within the central CCG unit to generate a hairpin-like structure (Figure 1A,B). Two symmetrical dT(CCG)$_3$A hairpins joined together by hydrogen bonds to form a tetraplex structure with an i-motif core, which includes four intercalated C:C$^+$ base pairs flanked by two G:G homopurine base pairs. Moreover, several stacking interactions were observed in the two symmetrical dT(CCG)$_3$A hairpins, which were important for maintenance of the i-motif conformation. These stacking interactions included a 5′ cytosine residue (C5), which protruded into the centre of the i-motif core to form a stacking interaction with G4 of the other strand. Two flipped-out nucleotides (C6 and G7) in each of the central CCG loops stacked together with the 5′-end T1 base, which was tilted out into the wide groove. The 3′ ends of two CGA oligonucleotides were aligned in parallel, resulting in C:C$^+$, G:G$^+$ and A:A$^+$ base pairs forming a right-handed duplex stem. Along with overwound twist angles at the G4–C3 and C9–G10 steps, CGA oligonucleotides formed two symmetrical hairpins that were tightly twisted in a clockwise direction to produce a right-handed tetrahelix.

2.2. Stabilization of the i-Motif Tetraplex by Water Hydration

In total, fifteen bridging water molecules were identified as mediating the interactions between the two hairpin structures of dT(CCG)$_3$A (Figure 1C,D). Six water molecules (W102, W106, W109, W127, W128, and W133) mediated the DNA-DNA inter-strand interactions, while ten water molecules (W101, W110, W112, W113, W117, W118, W123, W130, W132 and W133) mediated the DNA-DNA intra-strand interactions. Interestingly, W133 was found to mediate both inter- and intra-strand interactions. W110, W113, and W117 stabilized the central CCG loop structure. Two flipped-out C5 bases in opposite dT(CCG)$_3$A strands were linked by W128 located at the top of the structure. W102, W109 and W127 water molecules mediated the interactions between the cytosine residues (C2 and C8) at the i-motif core. W106 linked the pyrimidine base C3 and the purine base G4 by bridging the N2 of G4 and the O2 of C3 in the opposite chain. Moreover, W133 played a key role in stabilizing the structure by mediating the inter-strand G10-A11 interaction as well as inter-strand C9-G10 interaction of the dT(CCG)$_3$A hairpin structure. Analysis of the high-resolution crystal structure of the i-motif tetraplex revealed that water was positioned so as to hold and stabilize the dimeric hairpins via hydrogen bonds. A list of intra- and inter-strand water-mediated interactions with their respective distances between the atoms in the two symmetrical dT(CCG)$_3$A hairpins is provided in Table S2. Interestingly, the structure reported here did not involve metal ions as reported in the previous structure [8], instead relying exclusively on water-mediated interactions to stabilize the i-motif.

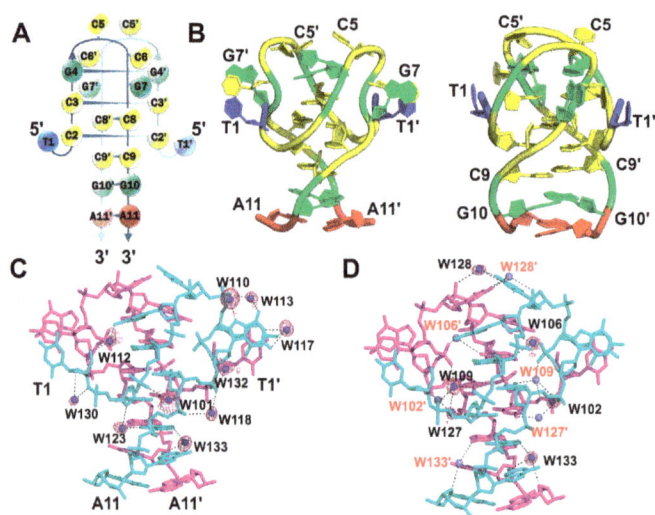

Figure 1. The structural features of dT(CCG)$_3$A i-motifs: (**A**) Schematic diagram of the crystal structure of two symmetrical dT(CCG)$_3$A strands that fold into a tetraplex i-motif as shown on the left. Guanine bases are colored in green, adenine in red, thymine in blue, and cytosine in yellow. (**B**) Representation of the dT(CCG)$_3$A final refined structure viewed from the narrow-groove (middle) and wide-groove (right) directions. (**C**) Inter-strand water cluster stabilizing the single-stranded hairpin structure. (**D**) View of the water cluster formed between two hairpins of dT(CCG)$_3$A from the narrow groove. Each DNA strand is colored magenta or cyan. 2Fo-Fc electron density map of the coordinated waters (blue spheres) in the refined structure is contoured at 1.0 σ, while the waters coordinating from the other asymmetric unit are shown as slate-coloured spheres. The hydrogen bonds are represented by dashed lines within the distance of 3.5 Å.

2.3. Co^{II}(Chro)$_2$ Complex Induces Conformational Changes in the d[T(CCG)$_3$A]$_2$ DNA Duplex

Chro bound to cobalt divalent cations to form a [Co^{II}(Chro)$_2$] dimer (Figure 2A). To understand the structure of the dT(CCG)$_3$A sequence in the presence of dimer, Co^{II}(Chro)$_2$ bound to the DNA sequence was crystallized in a similar manner to that described for the formation of the dT(CCG)$_3$A i-motif crystals. The electron density map with a resolution of 1.87 Å revealed that all atoms in the refinement structure had a clear electron density, except for two cytosines (C2 and C13), which were extruded from the structure due to poor mapping and had to be modelled with an energy minimization module using Accelrys Discovery Studio Client (v2.5.0.9164) (Figure S1C) [16]. The extruded cytosines thus form an e-motif structure which might stabilize the packing of the complex within the crystal lattice. The presence of such structures has been reported previously [17–19]. Analysis of the crystal structures revealed that Co^{II}(Chro)$_2$ altered the formation of the i-motif tetraplex, which was composed of two hairpin-like structures. Co^{II}(Chro)$_2$ bound to the pseudo-palindromic duplex DNA sequence in the minor groove, resulting in the formation of a Co^{II}(Chro)$_2$-d[T(CCG)$_3$A]$_2$ complex (Figure 2B). The binding resulted in deformation of the DNA resembling the Ni^{II}(Chro)$_2$ dimer compounds. A central d(GGCC) motif was formed due to the extrusion of four cytosines (C5, C6, C16, and C17) upon binding to a Chro dimer. Unlike the i-motif adopted by dT(CCG)$_3$A dimeric hairpins, the guanine base of the second CCG unit of each DNA strand was not flipped out, resulting in the re-formation of a GGCC tetranucleotide tract that provided the flexibility in the DNA to better accommodate Co^{II}(Chro)$_2$. The four projected cytosines (C5, C6, C16, and C17) interacted with the disaccharide B ring of (Chro)$_2$ so as to enhance the stability of the extruded residues via hydrogen bonds and van der Waals forces (Figure 2C). We suppose that the observed conformation for the DNA within the complex is effectively induced by Co^{II}(Chro)$_2$ binding. Although we cannot exclude the possibility that both the i-motif and duplex structure co-exist in solution, and Co^{II}(Chro)$_2$ may "choose" the duplex structure through conformational selection, we consider the possibility to be low for the following reasons: (1) the i-motif has been shown to be more stable than its duplex counterpart [20] and should be the favoured conformation, and (2) circular dichroism studies have also shown that the CCG repeats adopt spectra indicative of the i-motif conformation in solution [21,22].

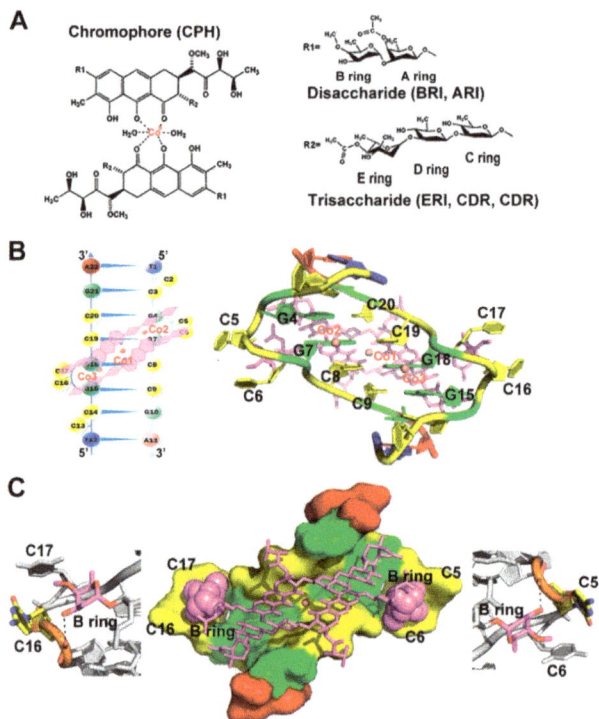

Figure 2. Crystal structure of the Co^{II}(Chro)$_2$-d[T(CCG)$_3$A]$_2$ complex. (**A**) Chemical structure of the Co^{II}(Chro)$_2$ dimer. (**B**) Schematic diagram of the Co^{II}(Chro)$_2$-d[T(CCG)$_3$A]$_2$ complex shown on the left. The cobalt(II) ions and the Co^{II}(Chro)$_2$ complex are drawn in salmon and pink. The refined structure of the Co^{II}(Chro)$_2$-d[T(CCG)$_3$A]$_2$ complex viewed from the major groove is shown on the right. The Co^{II}(Chro)$_2$ complex binds to the central G4-C20, G7-C19, C8-G18 and C9-G15 base pairs of the d[T(CCG)$_3$A]$_2$ DNA structure accompanied by extrusion of four cytosine bases. (**C**) The extruded cytosine residues stabilized by the disaccharide B ring of (Chro)$_2$ via hydrogen bonds are shown at sides with hydrogen bonds indicated by dashed lines. In the middle of the complex, stabilized by van der Waals forces, is represented with the disaccharide B ring and d[T(CCG)$_3$A]$_2$ duplex viewed from the minor-groove direction (sphere and solvent-accessible surface respectively).

2.4. Stabilization of the Co^{II}(Chro)$_2$-d[T(CCG)$_3$A]$_2$ Complex by Interacting with Cobalt(II) Ions and Water Hydration

There were three cobalt ions (Co1, Co2, and Co3, Figure 2B) present in the Co^{II}(Chro)$_2$-d[T(CCG)$_3$A]$_2$ complex. The Co1 ion formed an octahedral coordination with two oxygen atoms, O1 and O9, of each chromophore moiety and two oxygen atoms of water in the centre of the (Chro)$_2$ complex (Figure 3A). The two water oxygen atoms were also involved in the formation of a hydrogen bond network between the cytosine base of the GGCC tetranucleotide tract and the chromophore of the (Chro)$_2$ complex. In the major groove, the Co2 and Co3 ions interacted with N7 of guanine G7 and G18, also forming an octahedral coordination with guanine and five water molecules (Figure 3B). Water molecules were further involved in the interactions between the G7 and G4 bases and the G18 and G15 bases. Water bound to the GpG steps of the tetranucleotide tract, which was formed by extruding two cytosine bases of each DNA strand (Figure 3C). Previous studies have shown that water molecules commonly play a profound role in groove binding [23,24]. In the Co^{II}(Chro)$_2$-d[T(CCG)$_3$A]$_2$ complex structure, a total of four water molecules were identified as being directly involved in the interaction between $Co^{II}(Chro)_2$ and the CCG TNRs (Figure 3C). These four water molecules (W15, W19, W29, and W36) stabilized the extruded cytosine residues and the last guanine in the complex via specific water-mediated interactions with $(Chro)_2$. W15 and W19 bridged the interactions between the N4 amine on the extruded cytosines and the O3 oxygen atoms of the disaccharide B ring. W29 and W36 formed water-mediated hydrogen bonds with the phosphate oxygen atom of the last guanine (G10 and G21) and the O1 oxygen atoms of the trisaccharide C ring. These interactions mediated by water molecules could also be found in the $Ni^{II}(Chro)_2$-d[TT(CCG)_3AA]_2 complex structure, indicating that they were indispensable for recognition of the $(Chro)_2$ ligand by the CCG DNA repeat as well as the stability of the complex.

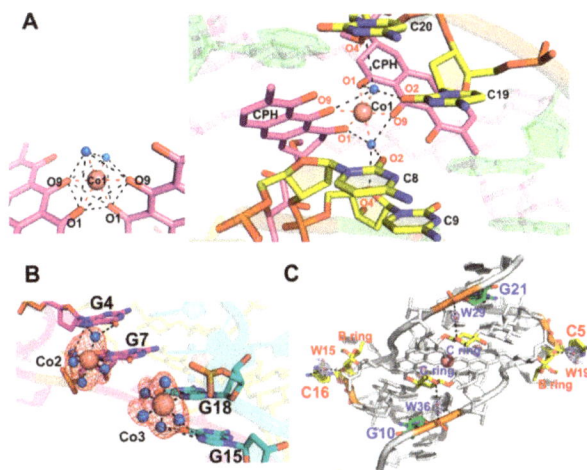

Figure 3. Coordination of cobalt ions and water mediate interactions in the $Co^{II}(Chro)_2$-d[T(CCG)3A]_2 complex structure. (**A**) Close-up view from the major groove of the $Co^{II}(Chro)_2$-d[T(CCG)_3A]_2 complex representing the specific interactions between $Co^{II}(Chro)_2$ and central GGCC steps. The octahedral coordination of Co1 is also shown on the left. The cobalt ion and two water molecules that mediate Chro and DNA interactions are represented as salmon and blue spheres, respectively. Coordination and hydrogen bonds are shown by dashed lines. (**B**) Close-up view of the $Co^{II}(Chro)_2$-d[T(CCG)_3A]_2 structure showing the octahedral coordination of Co2 and Co3 ions interacting with the unpaired N7 [G7] and N7 [G18] bases, respectively, and the coordinated water molecules also involved in mediating the cobalt(II) ions and G4 (G15) base interactions. (**C**) The bridging water molecules mediate the interaction between the $Co^{II}(Chro)$ and DNA viewed from the minor groove. The *2Fo-Fc* electron density map is contoured at a 1.0 σ level.

2.5. The Cobalt–Chro Complex Specifically Recognizes the Hairpin Structure of CCG TNRs

Surface plasmon resonance (SPR) was employed to analyse the binding affinity of $Co^{II}(Chro)_2$ to $d(CCG)_n$. The results revealed that the interactions of $Co^{II}(Chro)_2$ with $d(CCG)_3$ or $d(CCG)_4$ resulted in a high resonance unit (RU) (Figure 4A), which is indicative of strong binding. However, $Co^{II}(Chro)_2$ failed to bind to $d(CCG)_2$, resulting in a low RU (Figure S2). The binding rate constant (k_a) and the dissociation rate constant (k_d) were calculated according to the kinetic 1:1 Langmuir binding model, and the binding constant (K_a) was calculated from the values of k_a and k_d respectively (Table 1). We found that the binding affinity of $Co^{II}(Chro)_2$ for DNA could be enhanced by increasing the number of CCG repeat units. To determine the selectivity and stabilization effects of $Co^{II}(Chro)_2$ on various TNR DNA sequences, the melting temperature differences (ΔT_m) of a duplex DNA were measured. Various $d(CXG)_n$ repeats, where X could be any base and n was the repeat number, were synthesized and used for binding assays with $Co^{II}(Chro)_2$ at a ratio of 1:4. The results indicated that the binding of $Co^{II}(Chro)_2$ to $d(CXG)_n$ increases the overall stability of $d(CXG)_n$ compared to controls. The type of trinucleotide repeat also considerably impacted ΔT_m (Figure 4B). The binding of $Co^{II}(Chro)_2$ to $d(CAG)_{3or4}$ or $d(CTG)_{3or4}$ resulted in a low ΔT_m, indicative of a poor stabilization effect. By contrast, binding $Co^{II}(Chro)_2$ to $d(CCG)_{3or4}$ or $d(CGG)_{3or4}$ resulted in high ΔT_m values. It appears that $Co^{II}(Chro)_2$ binds preferentially to $d(CCG)_{3or4}$ duplexes compared to other $d(CXG)_n$ repeats, clearly indicating the selective ligand binding.

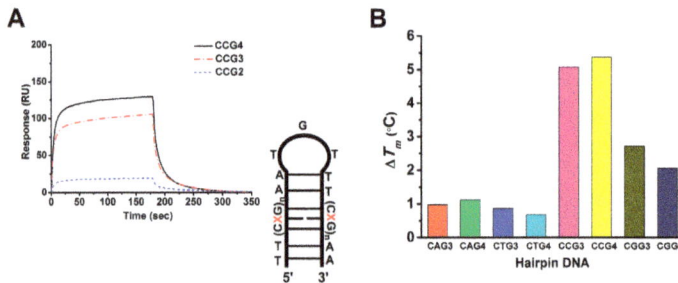

Figure 4. Binding affinity and stabilizing effects of the $Co^{II}(Chro)_2$ complex on various $d(CXG)_n$ trinucleotide repeats, where X could be any base and n is the repeat number, as shown in the schematic diagram. (**A**) Surface plasmon resonance (SPR) sensorgrams representing the binding of the $Co^{II}(Chro)_2$ complex to the immobilized 5′ biotin-labelled hairpin DNAs (CCG2, CCG3, and CCG4). The reactions were carried out in 50 mM sodium cacodylate buffer (pH 7.3) containing 50 mM NaCl. The resonance unit (RU) is defined as 1 RU = 1 pg/mm^2. (**B**) Effects of $Co^{II}(Chro)_2$ on the T_m values of various hairpin DNA fragments measured in 50 mM sodium cacodylate buffer (pH 7.3) containing 50 mM NaCl, with DNA and $Co^{II}(Chro)_2$ at a 1:4 molar ratio. ΔT_m values were obtained by subtracting the T_m value in the presence of the ligand.

Table 1. Binding parameters for $Co^{II}(Chro)_2$ complexes and various $(CCG)_n$ trinucleotide repeats.

Drugs	DNA Forms	k_a (M^{-1}s^{-1})	k_d (s^{-1})	K_a (M^{-1})
	CCG2	null [a]	null [a]	null [a]
$Co^{II}(Chro)_2$	CCG3	4.19×10^3	4.54×10^{-2}	9.12×10^4
	CCG4	8.15×10^3	5.68×10^{-2}	1.43×10^5

[a] Undetermined.

3. Discussion

Fragile X syndrome (FXS) is a genetic disorder caused by the expansion of CGG/CCG tandem repeats in the Fragile X Mental Retardation 1 gene (*FMR1*) on the X chromosome [6,25]. The CGG/CCG tandem repeats are scattered along the X chromosome, with a high percentage around *FMR1* (Figure 5A). Examination of the coding and noncoding sequences of *FMR1* revealed the presence of two CGG/CCG tandem repeats with 9 or 10 copies in the 5′ untranslated region (UTR) (Figure 5B). No tandem repeats were found in the coding region. The expansion of repeats in *FMR1* results in a defective protein that is known to be associated with the symptoms of FXS [26]. Previous studies have shown that under physiological conditions, both the G-rich and C-rich single-stranded d(CGG)·d(CCG) repeats are able to form secondary structures and cause unusual expansions [8,27,28]. The propensity to form the secondary structure is more pronounced for $d(CCG)_n$ than it is for $d(CGG)_n$ repeats [29]. This observation suggests that the $d(CCG)_n$ strand is more likely to form a hairpin or slippage structure and exhibit an asymmetric strand expansion during DNA replication. In addition to hairpin structures, CCG repeats have been reported to adopt a tetraplex structure based on two parallel-oriented hairpins that are held together by hemiprotonated intermolecular C:C+ pairs [30].

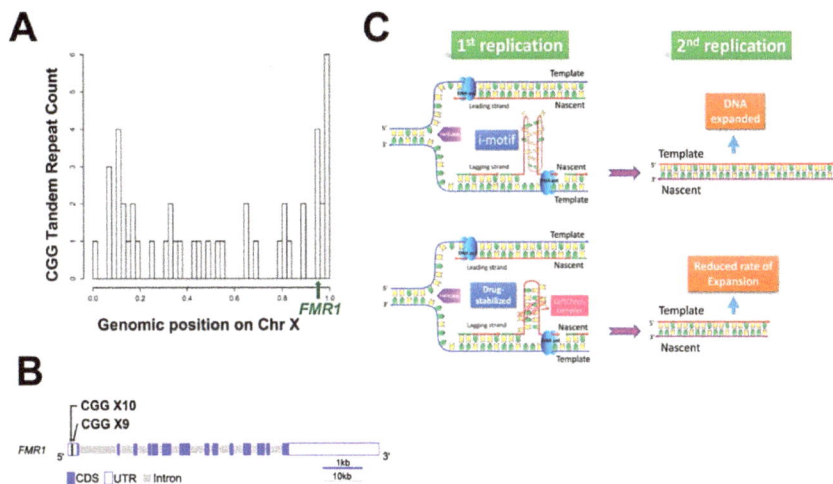

Figure 5. CGG/CCG tandem repeats on the *FMR1* gene and DNA expansion. (**A**) The distribution of the CGG/CCG tandem repeats with more than five on chromosome X and the location of the *FMR1* gene (the length of chromosome X scaled to 1). There are several CGG/CCG tandem repeats around the locations of the *FMR1* gene in the normal human genome. (**B**) The gene structure for human *FMR1*. There are 2 CGG/CCG tandem repeats with 9 and 10 copies, respectively, found on the 5′ UTR of *FMR1* in the normal human genome. (**C**) A proposed model for the biological consequences that occur following binding of the $Co^{II}(Chro)_2$ complex to the slipped CCG repeats at the nascent lagging strand during DNA replication. In the first cycle of DNA replication, if slipped CCG repeats fold into an unstable tetraplex i-motif structure on the nascent strand, it would lead to a subsequent expansion when using the nascent lagging strand as a template in the next cycle of replication. In contrast, when the $Co^{II}(Chro)_2$ complex recognizes the slipped CCG repeats, the resulting DNA conformational change may help to stabilize the slipped DNA and lead to the newly generated DNA duplex, with no length changes in the next cycle of DNA replication.

Recently, researchers have focused on designing novel compounds that can bind to expanded CNG repeat DNA in a sequence-specific manner. The propensity of small molecules to bind to the substrate with expanded CNG repeats can inhibit abnormal DNA replication as well as transcription [31–33]. In the present study, we have found that two $d(T(CCG)_3A)_2$ sequences can form hairpin structures that stack together in parallel to form a tetraplex, with a core i-motif surrounded by two G:G homo-base pairs. A similar structure has been reported previously in which it was shown that divalent cobalt ions are crucial for maintaining the i-motif structure [8]. However, the i-motifs reported in this work are exclusively stabilized by water molecules. Many of the water molecules occupy similar positions to the cobalt ions observed in the previous work. Furthermore, we observed here that several interactions, including stacking forces, stabilize the extruded residues in the i-motif tetraplex. On the other hand, the structure of two $d(T(CCG)_3A)_2$ sequences in the presence of the $Co^{II}(Chro)_2$ complex clearly shows a double-stranded helical conformation, which implies a crucial role of $Co^{II}(Chro)_2$ in binding to CCG. Evidently, the binding of $Co^{II}(Chro)_2$ to the $d(T(CCG)_3A)_2$ causes the extrusion of four cytosines in the repeat motif. It is conspicuous that the chromomycin A3 restores the double-stranded helical conformation by preventing the repeat DNA motif from forming i-motif structures (Figure S3).

Previously, the $Ni^{II}(Chro)_2$ complex has been shown to bind specifically to CCG TNRs via a "forced" induced-fit mechanism [15]. The $Co^{II}(Chro)_2$ complex structure with DNA shows significant similarity in overall structure with that of the $Ni^{II}(Chro)_2$ complex. However, the stability of these structures in terms of melting temperature is quite different, which clearly implies the occurrence of local differences. Analysis of the central GpGpCpC segment surrounding the Chro dimer binding site revealed considerable differences between the helical parameters of the Ni^{II} and Co^{II} complexes. Although the unwinding parameters are similar for both structures, prominent variations in the DNA roll and rise parameters were observed for the central GpC step (Figure S4) [34]. Based on detailed analysis of the structures, we propose a model to explain how tandem repeats could lead to sequence expansion during DNA replication (Figure 5C). The binding of the Co^{II}(Chro)$_2$ complex to the slipped CCG repeats between the nascent and lagging strands during DNA replication could result in two different situations. If the CCG repeats fold into a tetraplex i-motif structure in the absence of the Co^{II}(Chro)$_2$ complex during the first DNA replication cycle, DNA polymerase would use the nascent lagging strand as a template for DNA synthesis, leading to sequence expansion during the second replication cycle. On the other hand, recognition and binding of Co^{II}(Chro)$_2$ complex to the slipped CCG repeats would change the DNA conformation, which might slow down the rate of DNA expansion in the newly generated DNA duplex. After DNA unwinding during replication or transcription, the two strands become separated. Each single strand possesses CGG and CCG repeats, and the C-rich ends may again form i-motif structures consisting of double hairpins. This provides Co^{II}(Chro)$_2$ with binding sites on both single-strand DNA, thus further reducing the rate of DNA slippage and consequent DNA expansion. The CCG-Co^{II}(Chro)$_2$ complex, therefore, could prevent the formation of i-motifs and force the DNA to return to a double helical form.

Metal ions often play an important role in either stabilizing DNA structures or stabilizing crystals [35,36]. Many studies have provided detailed information about the interactions of divalent metal ions stabilizing DNA duplexes [37–40]. Previous work has also highlighted the role of Co(II) metal ions in stabilizing drug-DNA complexes [41]. The divalent cations form a tetrahedral or octahedral coordination complexes with the drug chromophore. It has also been shown that the N7 atom of the purine or N3 of pyrimidine residues as well as exocyclic oxygen atoms and the phosphate O atoms are the preferential sites of metal binding to stabilize the DNA structures in the complex. Furthermore, other divalent metal ions, such as Cu(II) and Hg(II), are also known to alter DNA conformation [42,43]. In the current study, we found that two Co(II) ions can interact with the DNA backbone to stabilize the overall structure. The two metal ions formed coordination bonds with the G7-N7/G18-N7 and five water molecules in a six-coordinate octahedral geometry. The coordinated water molecules were also involved in mediating the interactions between cobalt(II) ions and the G4 (G15) base. Hydrogen bonding between the water molecules and the oxygen atoms of the phosphate backbone can also stabilize the DNA. Based on these interactions, it can be concluded that the inclusion of Co(II) is crucial for maintaining the Co^{II}(Chro)$_2$-d(T(CCG)$_3$A)$_2$ complex structure. Since Co^{II}(Chro)$_2$ has been shown to be less toxic to cells compared to the more potent Ni^{II}(Chro)$_2$, it may represent a viable alternative for chromomycin-based drug appropriation or development. In addition, the DNA-Co^{II}(Chro)$_2$ complex has been shown to display extreme resistance to polyamine-mediated extraction of the divalent cation, making it an attractive ligand for exploring its therapeutic potential against CCG-repeat diseases. The potential of other types of metal complexes, such as ruthenium-based ligands, opens up the possibility of studying the interaction of the complexes with unusual DNA structures including i-motifs, triplexes and quadruplexes in future [44].

With this work, we complete our overview of chromomycin-based ligands containing a transition metal ion and their binding effects on repetitive DNA [15,33,45,46]. Our results provide additional clues to piece together the complete flow of how CCG DNA repeat amplification may arise, and provide a structural basis to speed the development and screening of specific drugs to treat diseases caused by the abnormal expansion of repeat DNA motifs.

4. Materials and Methods

4.1. Chemicals and Oligonucleotides

All chemicals used were of reagent grade, obtained from Sigma Chemical Co. (St. Louis, MO, USA). Deionised water from a Milli-Q system was used for all experimental procedures. Absorbance measurements to determine oligonucleotide concentrations were performed in quartz cuvettes using a Hitachi U-2000 spectrophotometer. The oligonucleotide concentrations were determined by Beer's law (A = ε.b.c; A: optical density at 260 nm; ε: extinction coefficient; b: cell path length, 1 cm; and c: DNA concentration in M). Synthetic oligodeoxynucleotides were purified by gel electrophoresis. The oligomer extinction coefficients were calculated on the basis of tabulated values for monomer and dimer extinction coefficients with reasonable assumptions as specified in ref. [47].

4.2. Melting Temperature Measurements

T_m values for the i-motif tetraplex sequence and the hairpin DNAs complexed with $Co^{II}(Chro)_2$ were determined as previously described using a JASCO UV-VIS spectrophotometer to monitor the sample absorbance (O.D.) at 260 nm and 295 nm [14,48]. The hairpin DNA, d(TT(CXG)$_{3-4}$AA<u>TGT</u>TT(CXG)$_{3-4}$AA, (X = A, T, C, or G), purified from polyacrylamide gel, was the substrate for the T_m experiments (the hairpin loop is underlined). The experiments were performed by increasing the temperature from 5 to 95 °C at a rate of 0.5 °C/min and recording the temperature every 30 s. T_m values (temperature corresponding to the dissociation of half of the DNA structures) were determined from polynomial fitting of the observed curves. The first derivative of the absorbance with respect to the temperature (dA/dT) of the melting curve was computed and used to determine the T_m value.

4.3. SPR Analysis

The affinity, association and dissociation between the drug and the DNA duplexes were measured using a BIAcore 3000 A surface plasmon resonance (SPR) instrument (Pharmacia, Uppsala, Sweden) equipped with a sensor chip SA5 from Pharmacia that monitored changes in the refractive index at the surface of the sensor chip. These changes were generally assumed to be proportional to the mass of the molecules bound to the chip and are recorded in resonance units (RU) [49]. The 5′-biotin-labelled hairpin DNA, biotin-d(TT(CCG)$_{2-4}$AA<u>TGT</u>TT(CCG)$_{2-4}$AA), purified from polyacrylamide gel electrophoresis, was used in the SPR experiments (the hairpin loop is underlined). To control the amount of DNA bound to the chip surface, the biotinylated oligomer was manually immobilized onto the surface of a streptavidin chip. Solutions of the metal-derived Chro complexes buffered with 50 mM sodium cacodylate at pH 7.3 in 50 mM NaCl were used. Different concentrations of the complexes were passed over the surface of the chip for 180 s at a flow rate of 10 µL min^{-1} to reach equilibrium; one of the flow cells remained blank as a control. Blank buffer solution was then passed over the chip to initiate the dissociation reaction, and this procedure was continued for 300 s to complete the reaction. The surface was then recovered by washing it with 10 µL of a 10 mM HCl solution. The sensorgrams for the interactions between the hairpin DNA duplex and the drug were analysed using version 3 of the BIAcore evaluation software.

4.4. Crystallization of d(T(CCG)₃A) and Co^{II}(Chro)₂-d[T(CCG)₃A]₂ Complex

Crystals yielding both the structures of the d(T(CCG)$_3$A) i-motif and Co^{II}(Chro)$_2$-d[T(CCG)$_3$A]$_2$ complex were obtained in a similar fashion. Both were crystallized using the sitting drop vapour diffusion method. The crystals of d(T(CCG)$_3$A) were obtained from a solution of 1.0 mM single-stranded DNA, 50 mM sodium cacodylate (pH 6.0), 1 mM magnesium chloride, 3% 2-methylpentane-2,4-diol, and 0.5 mM cobalt(II) chloride. The solution for crystallization was equilibrated against 500 µL of 30% MPD at 4 °C. Cylinder-shaped crystals of d(T(CCG)$_3$A) appeared after 4 weeks. Crystals of the Co^{II}(Chro)$_2$-d[T(CCG)$_3$A]$_2$ complex were grown by co-crystallizing 0.75 mM single-stranded DNA, 1.5 mM Chro, and 3 mM CoCl$_2$, in 50 mM sodium cacodylate buffer (pH 6.0), 1 mM MgCl$_2$, 1 mM spermine and 1% MPD, equilibrated against 500 µL of 30% MPD. Because Chro is yellow, the yellowish colour and the rod-shaped morphology of the crystals implied formation of the d[(TT(CCG)$_3$AA)]$_2$-Co^{II}(Chro)$_2$ complex; these were harvested after 4 weeks.

4.5. Data Collection, Processing, and Refinement of d(T(CCG)₃A) and Co^{II}(Chro)₂-d[T(CCG)₃A]₂ Complex Structures

The diffraction data for the d(T(CCG)$_3$A) crystal in space group $P4_32_12$ with unit-cell parameters a = b = 38.23, c = 54.23 Å, were collected at 110 K on an ADSC Q315r detector at beamline 13B1 of the National Synchrotron Radiation Research Center (Taiwan). The software package HKL2000 was used to index, integrate, and scale the X-ray diffraction data [50]. The reported resolution of 1.71 Å, at which the structure was refined, is based on the correlation coefficient (CC*) between the data and the model (PDB ID: 4PZQ) using the *PHENIX* suite (v1.8.4-1496) [51]. The nucleotides in d(T(CCG)$_3$A) are numbered from T1 to A11 in each strand. The structure was refined using the Refmac5 program in the CCP4 suite [52]. The DNA force field parameters reported by Parkinson et al. were used [53]. The diffraction data in space group $P3_212$ with unit-cell parameters a = b = 46.4, c = 73.8 Å, for the d[(T(CCG)$_3$A)]$_2$-Co^{II}(Chro)$_2$ complex crystal were collected using the same equipment at 100 K. Fluorescence scanning revealed a strong peak at the Co^{II} wavelength, consistent with the presence of Co^{II} ions in the structure. Multiple-wavelength anomalous diffraction (MAD) data were collected from three wavelengths using cobalt as the anomalous scattering atom. The diffraction spots were indexed, integrated and scaled using the HKL-2000 software package, followed by Co^{II} substructure localization using SHELX C/D/E. The resulting well-defined MAD electron density maps were used to build initial models using the program Coot. These structures were refined using the *PHENIX* program (v1.8.4-1496) using the high remote wavelength data for subsequent refinements. Most of the atoms in the structure were well-resolved and readily assigned in the density map, revealing a clear conformation of the DNA duplex in complex with Co^{II}(Chro)$_2$, except the uninterpretable region of the bases at the 5′ end of the DNA. B-factor analysis also suggested that the bases at the 5′ end of DNA were thermally less well ordered. Moreover, a well-defined MAD electron density map at a resolution of 1.87 Å was used to build the initial models for d(T(CCG)$_3$A) DNA alone. The force field of Chro was generated using the atomic coordinates of a 0.89 Å resolution crystal structure of Co^{II}(Chro)$_2$. The DNA nucleotide geometry parameters reported by Parkinson et al. were used. The full data collection and refinement statistics are given in Table S1. Coordinates and experimental data can be downloaded from www.wwpdb.org using the PDB IDs in Table S1.

4.6. Bioinformatics Analysis

The human genome sequence and the accompanying information pertaining to the gene structure of human *FMR1* were obtained from Ensembl version 91 (with genome assembly GRCh38.p10). We developed in-house software to search the sequence pattern including not less than six CGG tandem repeats against the chromosome X sequence. Statistical analysis was conducted using R Statistical Software (version 3.5.1) (R Foundation for Statistical Computing, Vienna, Austria) [54]. The *FMR1* gene structure was displayed using GSDS 2.0 [55].

5. Conclusions

In this report we have demonstrated the propensity of CCG repeats to undergo base pairing between the hemiprotonated cytosine residues of one C-rich hairpin duplex and the cytosine residues of a second hairpin duplex to form a stable i-motif tetraplex structure. The i-motif tetraplex was found to be stabilized by water molecules. The formation of i-motif tetramers may lead to DNA expansion during replication due to the presence of both matched and mismatched base pairs in the CCG repeat region in hairpin or slipped structures. In addition, we found that specific binding of Co^{II}(Chro)$_2$ to d(CCG)$_n$ sequences induced conformational changes of the CCG repeat DNA from i-motif to DNA duplex with cytosine-cytosine flip out. The specificity of Co^{II}(Chro)$_2$ towards (CCG)$_n$ may be partly due to the intrinsic instability and flexibility of C–C mismatches, sufficient to allow adoption of the geometrically optimal conformation that causes cytosines to extrude out of the helix and form the GGCC tetranucleotide patch. Extending the concept further, we hypothesize that TNR-binding compounds may induce a variety of sequences to form specific cognate structural motif(s) representing a substantial step towards the development of new therapeutic or diagnostic agents to treat these neurological diseases.

Supplementary Materials: Supplementary materials can be found at http://www.mdpi.com/1422-0067/19/9/2796/s1.

Author Contributions: Conceptualization, M.-H.H.; Methodology, M.-H.H. and L.-C.H.; Software, M.-H.H. and L.-C.H.; Validation, R.S. and P.-C.W.; Investigation, Y.-W.C. and C.-R.J.; Resources, M.-H.H., L.-C.H. and S.-W.L.; Writing-Original Draft Preparation, R.S., P.-C.W., L.-C.H. and M.-H.H.; Writing-Review & Editing, M.-H.H., C.-K.C., K.-R.C., and M.J.W.; Visualization, R.S. and P.-C.W.; Funding Acquisition, M.-H.H. and L.-C.H.

Funding: This work was supported by the grants from the Ministry of Science and Technology, Taiwan [106-2628-M-005-001-MY3 to M.-H.H.] and from the Taichung Veterans General Hospital/National Chung Hsing University Joint Research Program [TCVGH-NCHU10776010 to L.-C.H.]. Funding for open access charge: Ministry of Science and Technology, Taiwan.

Acknowledgments: We thank NSRRC staff for X-ray data collection.

Conflicts of Interest: The authors declare no conflict of interest.

Abbreviations

TNR	Trinucleotide repeats
FMR1	Fragile X Mental Retardation 1 gene
FXS	Fragile X syndrome
SPR	Surface plasmon resonance
RU	Resonance Units
PHENIX	Python-based Hierarchical ENvironment for Integrated Xtallography

References

1. Budworth, H.; McMurray, C.T. A brief history of triplet repeat diseases. *Methods Mol. Biol.* **2013**, *1010*, 3–17. [PubMed]

2. Mirkin, S.M. DNA structures, repeat expansions and human hereditary disorders. *Curr. Opin. Struct. Biol.* **2006**, *16*, 351–358. [CrossRef] [PubMed]

3. Huang, T.Y.; Chang, C.K.; Kao, Y.F.; Chin, C.H.; Ni, C.W.; Hsu, H.Y.; Hu, N.J.; Hsieh, L.C.; Chou, S.H.; Lee, I.R.; et al. Parity-dependent hairpin configurations of repetitive DNA sequence promote slippage associated with DNA expansion. *Proc. Natl. Acad. Sci. USA* **2017**, *114*, 9535–9540. [CrossRef] [PubMed]

4. Iyer, R.R.; Pluciennik, A.; Napierala, M.; Wells, R.D. DNA triplet repeat expansion and mismatch repair. *Annu. Rev. Biochem.* **2015**, *84*, 199–226. [CrossRef] [PubMed]

5. Satange, R.; Chang, C.K.; Hou, M.H. A survey of recent unusual high-resolution DNA structures provoked by mismatches, repeats and ligand binding. *Nucleic Acids Res.* **2018**, *46*, 6416–6434. [CrossRef] [PubMed]

6. Verkerk, A.J.; Pieretti, M.; Sutcliffe, J.S.; Fu, Y.H.; Kuhl, D.P.; Pizzuti, A.; Reiner, O.; Richards, S.; Victoria, M.F.; Zhang, F.P.; et al. Identification of a gene (FMR-1) containing a CGG repeat coincident with a breakpoint cluster region exhibiting length variation in fragile X syndrome. *Cell* **1991**, *65*, 905–914. [CrossRef]

7. Fu, Y.H.; Kuhl, D.P.; Pizzuti, A.; Pieretti, M.; Sutcliffe, J.S.; Richards, S.; Verkerk, A.J.; Holden, J.J.; Fenwick, R.G., Jr.; Warren, S.T.; et al. Variation of the CGG repeat at the fragile X site results in genetic instability: Resolution of the Sherman paradox. *Cell* **1991**, *67*, 1047–1058. [CrossRef]

8. Chen, Y.W.; Jhan, C.R.; Neidle, S.; Hou, M.H. Structural basis for the identification of an i-motif tetraplex core with a parallel-duplex junction as a structural motif in CCG triplet repeats. *Angew. Chem. Int. Ed.* **2014**, *53*, 10682–10686. [CrossRef] [PubMed]

9. Nakatani, K.; Hagihara, S.; Goto, Y.; Kobori, A.; Hagihara, M.; Hayashi, G.; Kyo, M.; Nomura, M.; Mishima, M.; Kojima, C. Small-molecule ligand induces nucleotide flipping in (CAG)n trinucleotide repeats. *Nat. Chem. Biol.* **2005**, *1*, 39–43. [CrossRef] [PubMed]

10. Hashem, V.I.; Pytlos, M.J.; Klysik, E.A.; Tsuji, K.; Khajavi, M.; Ashizawa, T.; Sinden, R.R. Chemotherapeutic deletion of CTG repeats in lymphoblast cells from DM1 patients. *Nucleic Acids Res.* **2004**, *32*, 6334–6346. [CrossRef] [PubMed]

11. Lo, Y.S.; Tseng, W.H.; Chuang, C.Y.; Hou, M.H. The structural basis of actinomycin D-binding induces nucleotide flipping out, a sharp bend and a left-handed twist in CGG triplet repeats. *Nucleic Acids Res.* **2013**, *41*, 4284–4294. [CrossRef] [PubMed]

12. Hou, M.H.; Robinson, H.; Gao, Y.G.; Wang, A.H. Crystal structure of actinomycin D bound to the CTG triplet repeat sequences linked to neurological diseases. *Nucleic Acids Res.* **2002**, *30*, 4910–4917. [CrossRef] [PubMed]

13. Slavik, M.; Carter, S.K. Chromomycin A3, mithramycin, and olivomycin: Antitumor antibiotics of related structure. *Adv. Pharmacol. Chemother.* **1975**, *12*, 1–30. [PubMed]

14. Hsu, C.W.; Chuang, S.M.; Wu, W.L.; Hou, M.H. The crucial role of divalent metal ions in the DNA-acting efficacy and inhibition of the transcription of dimeric chromomycin A3. *PLoS ONE* **2012**, *7*, e43792. [CrossRef] [PubMed]

15. Tseng, W.H.; Chang, C.K.; Wu, P.C.; Hu, N.J.; Lee, G.H.; Tzeng, C.C.; Neidle, S.; Hou, M.H. Induced-fit recognition of CCG trinucleotide repeats by a nickel-chromomycin complex resulting in large-scale DNA deformation. *Angew. Chem. Int. Ed.* **2017**, *56*, 8761–8765. [CrossRef] [PubMed]

16. Dassault Systèmes BIOVIA. *Discovery Studio Client*; 2.5.0.9164; Dassault Systèmes: San Diego, CA, USA, 2005.

17. Gao, X.; Huang, X.; Smith, G.K.; Zheng, M.; Liu, H. New antiparallel duplex motif of DNA CCG repeats that is stabilized by extrahelical bases symmetrically located in the minor groove. *J. Am. Chem. Soc.* **1995**, *117*, 8883–8884. [CrossRef]

18. Zheng, M.; Huang, X.; Smith, G.K.; Yang, X.; Gao, X. Genetically unstable CXG repeats are structurally dynamic and have a high propensity for folding. An NMR and UV spectroscopic study. *J. Mol. Biol.* **1996**, *264*, 323–336. [CrossRef] [PubMed]

19. Pan, F.; Zhang, Y.; Man, V.H.; Roland, C.; Sagui, C. E-motif formed by extrahelical cytosine bases in DNA homoduplexes of trinucleotide and hexanucleotide repeats. *Nucleic Acids Res.* **2018**, *46*, 942–955. [CrossRef] [PubMed]

20. Konig, S.L.; Huppert, J.L.; Sigel, R.K.; Evans, A.C. Distance-dependent duplex DNA destabilization proximal to G-quadruplex/i-motif sequences. *Nucleic Acids Res.* **2013**, *41*, 7453–7461. [CrossRef] [PubMed]

21. Fojtik, P.; Vorlickova, M. The fragile X chromosome (GCC) repeat folds into a DNA tetraplex at neutral pH. *Nucleic Acids Res.* **2001**, *29*, 4684–4690. [CrossRef] [PubMed]

22. Vorlickova, M.; Zimulova, M.; Kovanda, J.; Fojtik, P.; Kypr, J. Conformational properties of DNA dodecamers containing four tandem repeats of the CNG triplets. *Nucleic Acids Res.* **1998**, *26*, 2679–2685. [CrossRef] [PubMed]

23. Wei, D.; Wilson, W.D.; Neidle, S. Small-molecule binding to the DNA minor groove is mediated by a conserved water cluster. *J. Am. Chem. Soc.* **2013**, *135*, 1369–1377. [CrossRef] [PubMed]

24. Erlitzki, N.; Huang, K.; Xhani, S.; Farahat, A.A.; Kumar, A.; Boykin, D.W.; Poon, G.M.K. Investigation of the electrostatic and hydration properties of DNA minor groove-binding by a heterocyclic diamidine by osmotic pressure. *Biophys. Chem.* **2017**, *231*, 95–104. [CrossRef] [PubMed]

25. Jin, P.; Warren, S.T. Understanding the molecular basis of fragile X syndrome. *Hum. Mol. Genet.* **2000**, *9*, 901–908. [CrossRef] [PubMed]

26. Grigsby, J. The fragile X mental retardation 1 gene (FMR1): Historical perspective, phenotypes, mechanism, pathology, and epidemiology. *Clin. Neuropsychol.* **2016**, *30*, 815–833. [CrossRef] [PubMed]

27. Zamiri, B.; Mirceta, M.; Bomsztyk, K.; Macgregor, R.B., Jr.; Pearson, C.E. Quadruplex formation by both G-rich and C-rich DNA strands of the C9orf72 (GGGGCC)8*(GGCCCC)8 repeat: Effect of CpG methylation. *Nucleic Acids Res.* **2015**, *43*, 10055–10064. [PubMed]

28. Moore, H.; Greenwell, P.W.; Liu, C.P.; Arnheim, N.; Petes, T.D. Triplet repeats form secondary structures that escape DNA repair in yeast. *Proc. Natl. Acad. Sci. USA* **1999**, *96*, 1504–1509. [CrossRef] [PubMed]

29. Fojtik, P.; Kejnovska, I.; Vorlickova, M. The guanine-rich fragile X chromosome repeats are reluctant to form tetraplexes. *Nucleic Acids Res.* **2004**, *32*, 298–306. [CrossRef] [PubMed]

30. Mirkin, S.M. Expandable DNA repeats and human disease. *Nature* **2007**, *447*, 932–940. [CrossRef] [PubMed]

31. Sheng, J.; Gan, J.; Huang, Z. Structure-based DNA-targeting strategies with small molecule ligands for drug discovery. *Med. Res. Rev.* **2013**, *33*, 1119–1173. [CrossRef] [PubMed]

32. Hou, M.-H.; Satange, R.; Chang, C.-K. Chapter 6 Binding of small molecules to trinucleotide DNA repeats associated with neurodegenerative diseases. In *DNA-targeting Molecules as Therapeutic Agents*; Waring, M.J., Ed.; Royal Society of Chemistry: London, UK, 2018; pp. 144–174.

33. Chang, C.K.; Jhan, C.R.; Hou, M.H. The interaction of DNA-binding ligands with trinucleotide-repeat DNA: Implications for therapy and diagnosis of neurological disorders. *Curr. Top. Med. Chem.* **2015**, *15*, 1398–1408. [CrossRef] [PubMed]

34. Zheng, G.; Lu, X.J.; Olson, W.K. Web 3DNA—A web server for the analysis, reconstruction, and visualization of three-dimensional nucleic-acid structures. *Nucleic Acids Res.* **2009**, *37*, W240–W246. [CrossRef] [PubMed]

35. Turel, I.; Kljun, J. Interactions of metal ions with DNA, its constituents and derivatives, which may be relevant for anticancer research. *Curr. Top. Med. Chem.* **2011**, *11*, 2661–2687. [CrossRef] [PubMed]

36. Gao, Y.G.; Sriram, M.; Wang, A.H. Crystallographic studies of metal ion-DNA interactions: Different binding modes of cobalt(II), copper(II) and barium(II) to N7 of guanines in Z-DNA and a drug-DNA complex. *Nucleic Acids Res.* **1993**, *21*, 4093–4101. [CrossRef] [PubMed]

37. Morris, D.L., Jr. DNA-bound metal ions: Recent developments. *Biomol. Concepts* **2014**, *5*, 397–407. [CrossRef] [PubMed]

38. Theophanides, T.; Anastassopoulou, J. The effects of metal ion contaminants on the double stranded DNA helix and diseases. *J. Environ. Sci. Health Part A Toxic/Hazard. Subst. Environ. Eng.* **2017**, *52*, 1030–1040. [CrossRef] [PubMed]

39. Egli, M. DNA-cation interactions: Quo vadis? *Chem. Biol.* **2002**, *9*, 277–286. [CrossRef]

40. Eichhorn, G.L.; Shin, Y.A. Interaction of metal ions with polynucleotides and related compounds. XII. The relative effect of various metal ions on DNA helicity. *J. Am. Chem. Soc.* **1968**, *90*, 7323–7328. [CrossRef] [PubMed]

41. Gochin, M. A high-resolution structure of a DNA-chromomycin-Co(II) complex determined from pseudocontact shifts in nuclear magnetic resonance. *Structure* **2000**, *8*, 441–452. [CrossRef]

42. Day, H.A.; Wright, E.P.; MacDonald, C.J.; Gates, A.J.; Waller, Z.A. Reversible DNA i-motif to hairpin switching induced by copper(II) cations. *Chem. Commun.* **2015**, *51*, 14099–14102. [CrossRef] [PubMed]

43. Kondo, J.; Yamada, T.; Hirose, C.; Okamoto, I.; Tanaka, Y.; Ono, A. Crystal structure of metallo DNA duplex containing consecutive Watson-Crick-like T-Hg(II)-T base pairs. *Angew. Chem. Int. Ed.* **2014**, *53*, 2385–2388. [CrossRef] [PubMed]

44. Cardin, C.J.; Kelly, J.M.; Quinn, S.J. Photochemically active DNA-intercalating ruthenium and related complexes–insights by combining crystallography and transient spectroscopy. *Chem. Sci.* **2017**, *8*, 4705–4723. [CrossRef] [PubMed]

45. Chen, Y.W.; Hou, M.H. The binding of the Co(II) complex of dimeric chromomycin A3 to GC sites with flanking G:G mismatches. *J. Inorg. Biochem.* **2013**, *121*, 28–36. [CrossRef] [PubMed]

46. Hou, M.H.; Robinson, H.; Gao, Y.G.; Wang, A.H. Crystal structure of the [Mg^{2+}-(chromomycin A3)2]-d(TTGGCCAA)2 complex reveals GGCC binding specificity of the drug dimer chelated by a metal ion. *Nucleic Acids Res.* **2004**, *32*, 2214–2222. [CrossRef] [PubMed]

47. Cantor, C.R.; Tinoco, I., Jr. Absorption and optical rotatory dispersion of seven trinucleoside diphosphates. *J. Mol. Biol.* **1965**, *13*, 65–77. [CrossRef]

48. Hou, M.H.; Lu, W.J.; Huang, C.Y.; Fan, R.J.; Yuann, J.M. Effects of polyamines on the DNA-reactive properties of dimeric mithramycin complexed with cobalt(II): Implications for anticancer therapy. *Biochemistry* **2009**, *48*, 4691–4698. [CrossRef] [PubMed]

49. Yuann, J.M.; Tseng, W.H.; Lin, H.Y.; Hou, M.H. The effects of loop size on Sac7d-hairpin DNA interactions. *Biochim. Biophys. Acta Proteins Proteom.* **2012**, *1824*, 1009–1015. [CrossRef] [PubMed]

50. Otwinowski, Z.; Minor, W. Processing of X-ray diffraction data collected in oscillation mode. *Methods Enzymol.* **1997**, *276*, 307–326. [PubMed]

51. Adams, P.D.; Afonine, P.V.; Bunkoczi, G.; Chen, V.B.; Davis, I.W.; Echols, N.; Headd, J.J.; Hung, L.W.; Kapral, G.J.; Grosse-Kunstleve, R.W.; et al. PHENIX: A comprehensive Python-based system for macromolecular structure solution. *Acta Crystallogr. Sect. D Biol. Crystallogr.* **2010**, *66*, 213–221. [CrossRef] [PubMed]

52. Winn, M.D.; Ballard, C.C.; Cowtan, K.D.; Dodson, E.J.; Emsley, P.; Evans, P.R.; Keegan, R.M.; Krissinel, E.B.; Leslie, A.G.; McCoy, A.; et al. Overview of the CCP4 suite and current developments. *Acta Crystallogr. Sect. D Biol. Crystallogr.* **2011**, *67*, 235–242. [CrossRef] [PubMed]

53. Parkinson, G.; Vojtechovsky, J.; Clowney, L.; Brunger, A.T.; Berman, H.M. New parameters for the refinement of nucleic acid-containing structures. *Acta Crystallogr. Sect. D Biol. Crystallogr.* **1996**, *52*, 57–64. [CrossRef] [PubMed]

54. R Core Team. R: A Language and Environment for Statistical Computing. R Foundation for Statistical Computing. Available online: http://www.R-project.org/ (accessed on 2 July 2018).

55. Hu, B.; Jin, J.; Guo, A.Y.; Zhang, H.; Luo, J.; Gao, G. GSDS 2.0: An upgraded gene feature visualization server. *Bioinformatics* **2015**, *31*, 1296–1297. [CrossRef] [PubMed]

International Journal of
Molecular Sciences

MDPI

Article

Cellular Pharmacology of Palladinum(III) Hematoporphyrin IX Complexes: Solution Stability, Antineoplastic and Apoptogenic Activity, DNA Binding, and Processing of DNA-Adducts

Georgi Momekov [1,*], Iva Ugrinova [2], Evdokia Pasheva [2], Daniela Tsekova [3] and Galina Gencheva [3,*]

[1] Department of Pharmacology, Pharmacotherapy and Toxicology, Faculty of Pharmacy, Medical University-Sofia, 2 Dunav Str., BG1000 Sofia, Bulgaria
[2] Laboratory of Chromatin Structure and Functions, Institute of Molecular Biology, Bulgarian Academy of Sciences, G. Bonchev Str. Bl.21, BG1113 Sofia, Bulgaria; ugryiva@gmail.com (I.U.); eva@bio21.bas.bg (E.P.)
[3] Department of Analytical Chemistry, Faculty of Chemistry and Pharmacy, Sofia University, 1 J. Bourchier Str., BG1164 Sofia, Bulgaria; dtsekova@chem.uni-sofia.bg
* Correspondence: gmomekov@gmail.com (G.M.); ggencheva@chem.uni-sofia.bg (G.G.); Tel.: +359-895776792 (G.M.); +359-877879088 (G.G.)

Received: 21 July 2018; Accepted: 12 August 2018; Published: 19 August 2018

Abstract: Two paramagnetic Pd^{III} complexes of hematoporphyrin IX ((7,12-bis(1-hydroxyethyl)-3,8,13,17-tetramethyl-21H-23H-porphyn-2,18-dipropionic acid), Hp), namely a dinuclear one $[Pd^{III}_2(Hp_{-3H})Cl_3(H_2O)_5] \cdot 2PdCl_2$, **Pd1** and a mononuclear metalloporphyrin type $[Pd^{III}(Hp_{-2H})Cl(H_2O)] \cdot H_2O$, **Pd2** have been synthesized reproducibly and isolated as neutral compounds at different reaction conditions. Their structure and solution stability have been assayed by UV/Vis and EPR spectroscopy. The compounds researched have shown in vitro cell growth inhibitory effects at micromolar concentration against a panel of human tumor cell lines. A DNA fragmentation test in the HL-60 cell line has indicated that **Pd1** causes comparable proapoptotic effects with regard to cisplatin but at substantially higher concentrations. **Pd1** and cisplatin form intra-strand guanine bis-adducts as the palladium complex is less capable of forming DNA adducts. This demonstrates its cisplatin-dissimilar pharmacological profile. The test for efficient removal of DNA-adducts by the NER synthesis after modification of pBS plasmids with either cisplatin or **Pd1** has manifested that the lesions induced by cisplatin are far better recognized and repaired compared those of **Pd1**. The study on the recognition and binding of the HMGB-1 protein to cisplatin or **Pd1** modified DNA probes have shown that HMG proteins are less involved in the palladium agent cytotoxicity.

Keywords: palladium(III) complexes; hematoporphyrin IX; antiproliferative activity; DNA binding and repair; HMGB-1 protein; apoptosis

---

1. Introduction

Cisplatin and the clinically accepted platinum drugs have a great importance for the cancer treatment. They have been applied in most anticancer chemotherapeutic regimens [1–4]. Intensive studies in this area date back to the mid-sixties of the 20th century, with Rosenberg's remarkable discovery of the medicinal power of the inorganic coordination compound "nicknamed" cisplatin [1,5–7]. Since then, cisplatin has revolutionized cancer treatment converting the formerly fatal disease, largely curable [8]. Nowadays, cisplatin is still the most successful anticancer drug in the world, and is widely used in the treatment of a multitude of different cancers. Still, regardless of the achievements of cisplatin and the related platinum-based drugs—carboplatin—a second generation, and oxaliplatin—third generation, and those of regional use in specific countries, such as nedaplatin, lobaplatin, and heptaplatin, their application is limited [9]. The restrictions are due to their major drawbacks: effectiveness against a limited range of cancers and its intrinsic resistance, development of acquired resistance [10,11], severe side-effects [12,13], and low solubility. Hence, the efforts in the field of antitumor drugs design target introduction of new formulas possessing both a widened spectrum of chemotherapy and an improved clinical profile [3,4,9,14]. Intensive research on drug development processes during the past years shows that the invention of universal compounds active against many cancer types is a task difficult to accomplish. The right way is to develop drugs effective against a small group or a subgroup of cancers. Thus, the successful route leading to new efficient drugs inducing a better tumor response in individual patients is to design a pharmacological agent targeting specific abnormalities in particular cancer cells [4].

There are several approaches to develop new metal-based antitumor agents. Historically, the first one follows the correspondence between anticancer activity and the molecular structure of cisplatin [3,4,15]. Thus, a variety of cisplatin-similar platinum(II) complexes with different ligands have been synthesized and tested. The numerous results achieved have led to a set of rules for constructing a molecular structure that appeared to be required in order to manifest antitumor activity [7]. According to the rules, the neutral platinum complex with square-planar geometry, containing two cis-am(m)ine carrier ligands (capable of participating in a hydrogen bonding formation) and two cis-coordinated leaving groups (Cl^- ligands) possesses the necessary structural features for intravenous administration into the blood stream. It is considered that the complex should remain largely unchanged during circulation and after entering the cells a substitution of one or two more labile ligands occurs thereby activating the complex. This requirement underlines the necessity to render the complex inactive during the transport and determines that the ligand exchange rates should be compatible to the rates of cell division processes [1,16]. The platinum(II) complexes whose structure follows these rules represent the so-called classical platinum chemotherapeutics. Numerous investigations have been conducted in order to determine the mechanism by which these drugs carry out their anticancer action [17] and most results concerns cisplatin. In general, the main steps cover: cellular uptake by passive diffusion [18] or active transport via the copper transporters CTR1 and CTR2 [19,20]; aquation followed by activation; DNA platination and cellular processing of Pt-DNA lesion leading to apoptosis or to the cell survival. Nuclear DNA is considered the ultimate target of cisplatin and related platinum therapeutics and their capability to form bifunctional DNA crosslinks causing the DNA distortion. Thus, the platinum induced kink in the DNA molecules leads to a chain of events including protein recognition and eventual apoptosis. Unfortunately, some of the platinum compounds can be activated in biological milieu before reaching the tumor cell and they can also interact with nontarget biomolecules. Furthermore, it is clear that DNA is the ultimate pharmacological target of the platinum drugs, but the important issue here is to discern tumor cells from healthy cells. Now it is accepted that the classical platinum compounds can enter in each cell, however the healthy cells and some of the tumor cells can reverse the damage and remove the platinum compounds. Therefore, it is clear that the compounds belonging to this group cannot offer any advantages over cisplatin. Indeed, more than ten other cisplatin related derivatives are currently in clinical trials and the experimental data show [21,22] that they have not overcome considerably many of the disadvantages arising from this common structure. However, the achievement analysis in this field demonstrates that metal coordination compounds can play an important part in anticancer treatment regimes.

Another route to create new metal-based anticancer agents is focused on the mechanism of their antitumor action [23]. In this respect one could distinguish compounds capable of interacting in two different mechanisms: compounds that cause DNA distortion, here including cisplatin- similar and cisplatin dissimilar interactions [3,24–26] and compounds interacting with the key protein targets (including enzymes) that are selective for the specific malignancy and/or that regulate apoptosis, and/or that are responsible for cell invasion and metastasis [27,28]. The classical approach based on cell viability assay and characterization of the compounds that bind to DNA and adducts formed has been applied for more than forty years and now this is the first stage for evaluation. Nowadays, the drug design concepts are based on one side on the chemical nature of the compounds and on the other side—on the cancer cells biochemistry. Thus, the new strategies for design and synthesis result in the development of new classes of antitumor agents, the so called "classical nonplatinum metal compounds" and "nonclassical metal compounds" [3] and thus several compounds (e.g., NKP-1339 [29], NAMI-A [30], Satraplatin [31], etc.) are proposed that are on the "verge" of clinical application. These strategies are consistent with requirements for: a comparatively soluble prodrug's form; existence of an inactivated form in the blood stream; selective cellular uptake; activation of the drug form in the tumor cells; appropriate tumor cell damage.

Metal coordination compounds with their inherent properties can meet these requirements to a great extent. Their specific three-dimensional structures combined with suitable electronic and ligand exchange properties are attractive for the creation of a successful formula [3,4]. The transition series metal complexes having variable oxidation states and coordination numbers, and the ability to bind, of different strength, to a wide variety of donor functional groups are some of the most tested compounds. A special attention is paid to the metals from second and third transition series. They usually exchange their ligands slowly, on the same time scale as the cell division processes. Their comparatively inert behavior keeps them as prodrugs, which have to be activated by ligand substitution reaction in the cells. In general, from a chemical point of view, the successful drug should be a well soluble compound with lipophilicity and acid-base properties determining a selective uptake. The drug should be activated inside the tumor cells and attack the target molecules by ligand substitution reactions. It is necessary to interact predominantly with target molecules and the formed adducts should be stable enough in order to provoke proper cell death. It means that a key element in the design processes is the control on the kinetics of ligand substitution reactions in vivo and the thermodynamic stability of the initial, intermediate, and final complexes produced during the antitumor action. In addition, the redox properties determined in terms of both metal and ligand are of special interest. As a part of antitumor mechanism, the change of the metal oxidation state can trigger ligand release and the participation of the ligands in in vivo redox reactions could produce reactive oxygen species. Also, the possibility to achieve light-triggered activation of an excited-state of the metal complexes in tumor cells instead of their ground-state gives additional advantages in the drugs design. All these considerations give the reason to conclude that a properly selected metal ion with regard to its nature, oxidation state, and coordination polyhedron with right ligands in the inner coordination sphere can create a drug able to overcome some of cisplatin's disadvantages.

Recently, a series of hematoporphyrin IX (7,12-bis(1-hydroxyethyl)-3,8,13,17-tetramethyl-21H-23H-porphyn-2,18-dipropionic acid, Hp) complexes [32–39] of platinum, palladium, and gold (Scheme 1) have been tested as tumor growth inhibitors. Hematoporphyrin IX and its derivatives are well known for their widespread application in the photodynamic therapy and diagnosis [40]. It is considered that, due to their acid-base and hydrophobic properties, the porphyrins could preferably accumulate in the neoplastic tissues [41–43]. In fact the selective uptake of porphyrins in malignant tissue cells is due to complex mechanisms, the most important of them being the LDL-receptor mediated endocytosis of porphyrin—lipoprotein complexes formed in the systemic circulation [40].

A new strategy for cytotoxic agents design with improved properties has been proposed in our group, based on the capabilities of this ligand to stabilize unusual oxidation states of the metals. Thus, three Pt^{III} (d^7) [32,33], two Pd^{III} (d^7) [34], and one Au^{II} (d^9) [35,36] complexes of Hp were obtained and characterized. The metal ions in the complexes have distorted octahedral coordination. Because of the metals' electronic configuration, the complexes manifest paramagnetic behavior. Further they show comparatively high cytotoxicity in in vitro tests against a panel of human cell lines [32–34,36]. Great efforts have been made in order to understand the effect of the ligand coordination to the antitumor behavior of the complexes. The ligand Hp has a polydentate nature. Three different modes of coordination have been established [37] studying its interaction with metal ions representatives of first [38,39], second [34], and third transition series [32,33,35,36]. Coordination via the four pyrrole N-atoms in the porphyrin framework and forming of metalloporphyrin-type complexes is the most widespread mode of binding, typical of all metal ions under investigation. Coordination via two N-atoms of adjacent pyrrole rings to the metal ions in cis-position leads to the so called "sitting atop" type complexes (SAT). Coordination via the side chains deprotonated propionic COO⁻ groups, outside the porphyrin macrocycle is a less common mode of coordination. The nitrogen donor atoms of the imino (>N) and aza (=N-) groups of the pyrrole rings, as well as the outside COO⁻ groups determine the nature and the size of three different coordination modes. The metal ions choose a different mode of coordination as a function of their nature and properties.

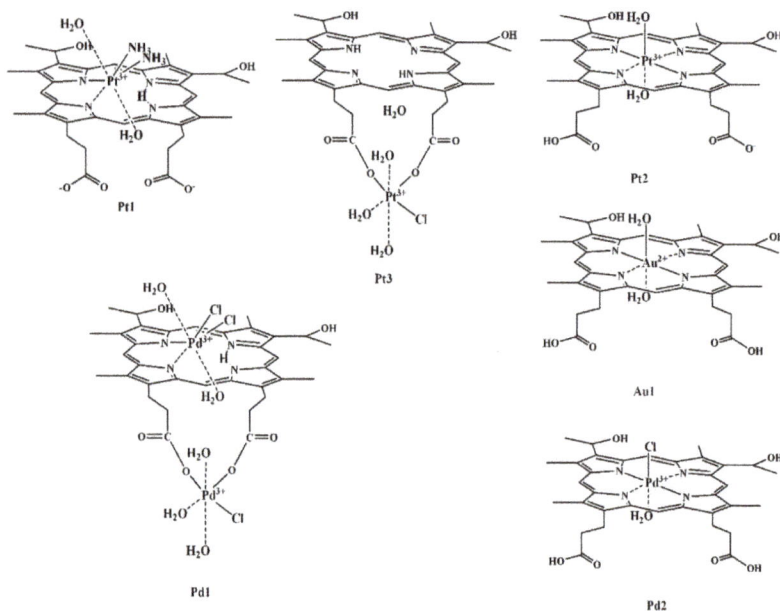

Scheme 1. Structures of the hematoporphyrin IX complexes of platinum(III), gold(II) and palladium(III).

Due to its intrinsic inertness, Pt^{III} forms stable complexes with the three different modes of Hp coordination (Scheme 1) in the proper reaction conditions [32]. All these complexes with distinct coordination patterns exert concentration dependent antiproliferative activity against a spectrum of cell lines representing some important types of neoplastic diseases in humans. They have also proven to be far less cytotoxic against the human embryonal kidney cell line HEK-293T compared to cisplatin. It has been discovered that the "sitting atop" complex **Pt1**, with PtN$_4$ coordination plane formed by two adjacent porphyrin pyrrole nitrogens and two NH$_3$-molecules in cis-position as well as the metalloporphyrin-type complex **Pt2** possess higher potency compared to **Pt3**. In the latter, Pt^{III} is coordinated to the deprotonated propionic carboxylic groups from the side chains of hematoporphyrin IX. The better solubility of **Pt2** compared to **Pt1** makes it preferable in research. Thus, its characteristics, such as superior proapoptotic activity that strongly correlates with its cytotoxicity and significantly higher degree of accumulation in tumor cells compared to the reference drug cisplatin, make it a reliable representative of the group of the "metal-based drugs that break the rules" [4].

Another promising representative of this group of compounds is the octahedral gold(II) (d^9) complex of hematoporphyrin IX, **Au1** [35] that is structurally similar to **Pt2**. The comparison of its cytotoxicity to that of **Pt2** manifests that the Au^{II} complex exerts superior activity exactly against the T-cell leukaemia SKW3. These data correlate well with the estimated specific inhibiting effect of gold species upon immune cells and T-cells [36]. Nevertheless, the compound exerts well-pronounced proapoptotic properties against malignant cells and it is less cytotoxic than cisplatin for the human kidney. Furthermore, this compound demonstrated significant intracellular accumulation presumably mediated by formation of FCS-lipoprotein complexes and subsequent endocytosis.

Two relatively new members of this group compounds are the Pd^{III} (d^7) coordination compounds of hematoporphyrin IX [34]. The dinuclear $[Pd^{III}_2(Hp_{-3H})Cl_3(H_2O)_5] \cdot 2PdCl_2$, **Pd1** and mononuclear $[Pd^{III}(Hp_{-2H})Cl(H_2O)] \cdot H_2O$, **Pd2** have been obtained during the interaction of the ligand with $Pd^{II}Cl_4^{2-}$ in alkaline-aqueous medium. In the dinuclear complex, **Pd1**, one of the Pd^{III} ions is coordinated to the deprotanated COO^- groups from the side chains of the porphyrin ligand and the second Pd^{III} ion—to two adjacent pyrrole N-atoms on the top of the porphyrin ring. The compound is spontaneously obtained at a large metal excess from the reaction in alkaline-aqueous medium. It is accepted that because of the greater kinetic lability of Pd^{III} compared to Pt^{III} both places for coordination are occupied simultaneously and thus a dinuclear Pd^{III}-Hp-Pd^{III} system is formed. The Pd^{III} ion in the mononuclear complex, **Pd2** is incorporated in the porphyrin core. The Pd^{III} centers in both complexes have a distorted octahedral coordination filled with additional donor species such as Cl^- and H_2O.

As a member of the platinum group metals, palladium, with its coordination compounds, is also extensively tested for antitumor activity [44–47]. Because of lanthanoid contraction, both metals palladium and platinum in oxidation state +2 have close ionic radii. They adopt a square-planar geometry and behave like soft acids, forming strong bonds with nitrogen and sulfur-containing ligands [48]. While the equilibrium constants, characterizing the stability of their isostructural complexes differ slightly (only about ten times higher for Pt^{II} complexes), their kinetic behavior with respect of the ligand substitution is completely dissimilar as a consequence of the lower electron density of Pd^{II}. The complexes of Pd^{II} simple analogues of Pt^{II}-antitumor drugs undergo aquation and ligand exchange reaction 10^4 to 10^5 times more rapidly. The fast hydrolysis of the leaving groups leads to the formation of very reactive Pd^{II}-aquated-species, unable to reach their pharmacological targets as active compounds. It is accepted that the toxic side effects are a result of inactivation of certain enzymes due to binding to the thiol groups of cysteine residues and obviously the much higher reaction rates typical of Pd^{II} compounds are favorable for general toxicity.

In order to tune the kinetic behavior and thermodynamic stability of the new compounds proposed as antitumor agents different approaches can be applied. Here, beside the construction of special coordination polyhedra, the intermediate oxidation state +3 of palladium centers has been applied as a factor of great importance. The less common oxidation state of palladium +3 provides many advantages of its complexes, such as a controlled delay of the ligand substitution reactions and reactivity, an octahedral geometry with axial ligands that could alter the redox potential and lipophilicity. The promising antiproliferative activity in micromolar concentration range that has been shown by the Pd^{III}-Hp complexes [34] together with the remarkable cytotoxicity against the K-562 cells with more than 4 fold lower IC_{50} value compared to cisplatin, characteristic for the dinuclear compound **Pd1**, is the basis for further detailed biological investigation on the mechanism of action of these new proposed cisplatin-dissimilar agents.

2. Results

2.1. Synthesis and Characterization in Solution

The Pd^{III}-complexes that undergo extensive biological screening have been obtained from the interaction of the initial palladium(II) salt ($[Pd^{II}Cl_4^{2-}]$) and hematoporphyrin IX. The coordination reaction proceeds in alkaline-aqueous medium on air. The ligand used for the syntheses is dissolved in 5×10^{-2} M KOH solution. The initial acidity of the reaction system has been adjusted within the range of pH values from 11.2 to 11.5 (by adding KOH). The reactions always start by spontaneous increase of the reaction acidity ($\Delta pH = 3$–4) within 3 to 4 h. The creation of paramagnetic complex species during the interaction has been followed by the EPR method. The EPR spectra were registered of samples taken from the reaction systems and frozen. A very wide signal (ΔHpp ~250 G) with g ~2.12 has been observed in the EPR spectrum of the 1:1 molar ratio reaction system during the first few hours. Further, the interaction continues with appearance of two different signals of variable intensities (Figure 1a): a narrow signal (ΔHpp = 7.2 G) with g = 2.000, typical for a radical formation and a wide low-intensive signal (ΔHpp ~45 G) with g ~2.06 without a hyperfine structure. Two days later, together with a radical's singlet, an additional third signal was observed with increasing intensity displaying a hyperfine structure. The final spectrum registered (Figure 1b) consists mainly of two signals: a singlet with parameters proving the presence of a stable radical overlaying a signal of two-component axial anisotropy and principal values of the g-tensor determined by the experiment as $g_{\perp}$ = 2.037 and g_{II} = 1.979. The parameters measured for the superhyperfine coupling tensor are $a_{\perp}$ = 17.8 $\times$ 10^{-4} cm^{-1} and a_{II} = 14.8 $\times$ 10^{-4} cm^{-1}.

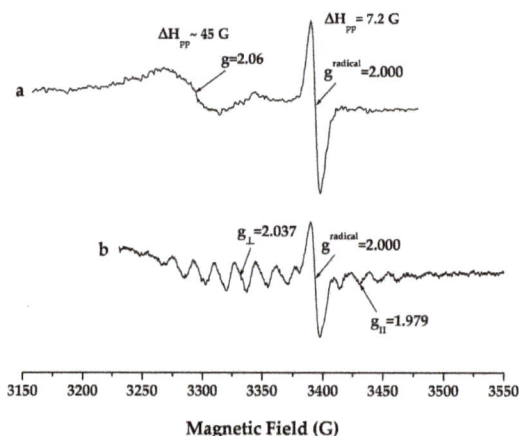

Figure 1. X-band EPR spectra of frozen solutions of the reaction system Pd:Hp = 1 (130 K): (**a**) a day from the beginning of the reaction; (**b**) a week later (operating frequency 9.500 GHz).

The spectral changes in the electronic-absorption spectra during the interaction in alkaline-aqueous medium refer to the bands of the free ligand spectrum as follows [49]: an intensive band belonging to a transition $S_0 \Rightarrow S_2$ at 376 nm (B or Soret) and four Q-bands at 505, 540, 567, and 621 nm from the transitions to nondegenerated orbitals of the first exited state S_1, with intensities IV > III > II > I, typical of etio-type porphyrins. Two separate stages during the interaction can be distinguished. A slight hypsochromic shift of the Soret band has been observed in the beginning followed by a decrease of its intensity, widening, and a batochromic shift by ~15 nm. During this period of the reaction, the four Q-bands have been shifted batochromically. They have changed their intensities as IV > III ~II > I and have reduced to three (~516, 558, and 620sh nm) several hours later with intensities: III ~II > I. The UV/Vis spectra measured in the second stage of the interaction have been characterized by the Soret band at 390 nm and two Q-bands at 520 and 559 nm; the long-wave one being the more intensive. Regardless of the metal-to-ligand ratios, in the final spectrum of the interaction, the Q-bands have always reduced to two.

The reaction has been directed to obtaining the particular Pd^{III}-complexes as neutral compounds at different metal-to-ligand ratios. The electronic absorption spectra of the reaction systems before the isolation of the complexes are presented on Figure 2.

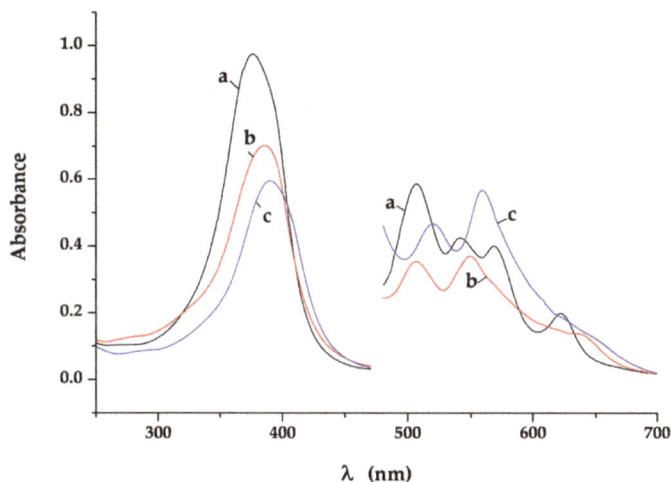

Figure 2. Electronic absorption spectra of: (**a**) Hp; (**b**) reaction system Pd:Hp = 4 before isolating complex **Pd1**; (**c**) reaction system 1:1 before isolating complex **Pd2**.

A dinuclear Pd^{III} complex with a composition $[Pd^{III}_2(Hp_{-3H})Cl_3(H_2O)_5]\cdot 2PdCl_2$, **Pd1** is the main product of the interaction at a metal excess (Pd:Hp $\geq$ 4). The compound was separated spontaneously from the reaction system with coprecipitated 2 molecules of the initial $PdCl_2$ at pH ~8. A mononuclear complex with a composition $[Pd^{III}(Hp_{-2H})Cl(H_2O)]\cdot H_2O$, **Pd2**, was the main product from the interaction at equimolar ratio of the reagents or a slight excess of one of the reagents. The compound was isolated as a neutral one by addition of hydrochloric acid (5×10^{-2} M HCl). The solid-state structure of the complexes and the mode of ligand coordination were studied in detail using magnetic measurements, EPR spectra, thermal and elemental analyses, and IR and UV/Vis spectroscopy [34].

The EPR and UV/Vis spectra were also used to characterize the solution-structure of the complexes and to assay their stability. Both complexes are slightly soluble in water and their solubility is pH dependent. Moderate solubility was achieved by increasing the pH value with KOH or using solvents such as DMSO (dimethyl sulfoxide) or DMF (*N,N*-dimethyl formamide). In the UV/Vis spectrum of hematoporphyrin IX recorded in DMSO, the Soret band (400 nm, lgε = 5.02) with a shoulder (376sh, lgε = 4.89) and the four Q-bands (502 nm, lgε = 3.77; 536 nm, lgε = 3.61; 572 nm, lgε = 3.84; 623 nm, lgε = 3.17) can be readily distinguished (Figure 3a). The UV/Vis spectra of DMSO solutions of the complexes display the characteristic three component (for Pd1: 507 nm (lgε = 3.68), 546 nm (lgε = 3.68), 567 nm (lgε = 3.63)) and two component (for **Pd2**: 512 nm (lgε = 3.58), 548 (lgε = 3.74)) pattern in the region of Q-bands (Figure 3b,c).

The EPR spectrum of the DMSO solution of the complex **Pd1** (Figure 4a) contains more than two signals. Two of them possess axial symmetry with principal values of g-tensors, respectively for the first: $g_\perp$ = 2.057 and g_{II} = 2.496 with readily observed superhyperfine structure in perpendicular region (a = 15.31×10^{-4} cm^{-1}) and for the second—$g_\perp$ = 2.029 and g_{II} = 2.354. The wide signal at ~3050 G was assigned to an exchange interaction of two paramagnetic centers in the molecule. In addition, a low intensive signal with g = 1.990 and H_{pp} = 8 G was assigned to a stable radical. The EPR spectrum of the complex **Pd2** dissolved in DMSO (Figure 4b) shows only a signal with two-component axial anisotropy and principal values of the g-tensor measured from the experiment as $g_\perp$ = 2.038 and g_{II} = 1.979. The superhyperfine structure measured from the experimental spectrum is well resolved both in perpendicular and parallel regions. The determined principal values of the axial superhyperfine coupling tensor are $a_\perp$ = 16.65×10^{-4} cm^{-1} and a_{II} = 13.85×10^{-4} cm^{-1}.

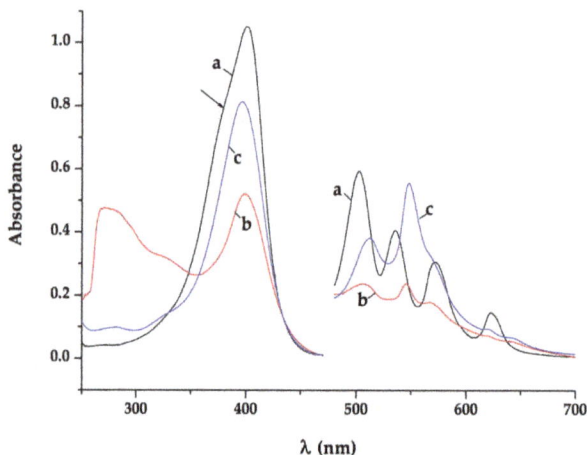

Figure 3. Electronic absorption spectra of Hp and its complexes dissolved in DMSO: (a) Hp; (b) complex **Pd1**; (c) complex **Pd2**.

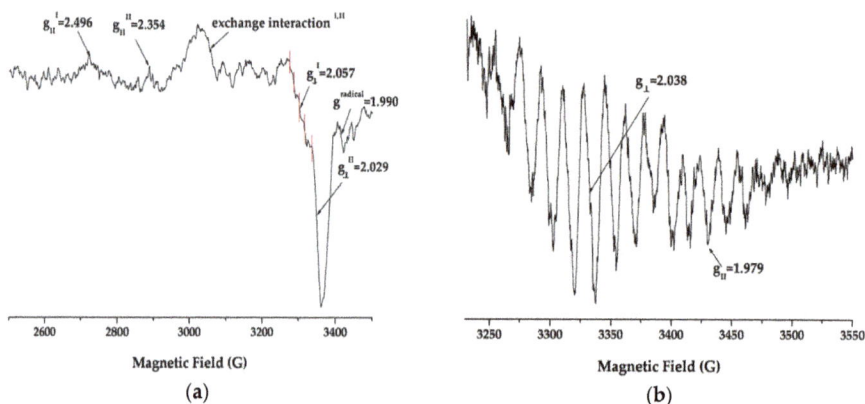

Figure 4. X-band EPR spectra of Hp complexes dissolved in DMSO, frozen solutions at 130 K (a) **Pd1**, operating frequency 9.513 GHz; (b) **Pd2**, operating frequency 9.502 GHz.

2.2. In Vitro Tumor Cell Growth Inhibition

The cell growth inhibitory effects of the complexes subjected to detailed biological screening were evaluated in a panel of human tumor cell lines with distinct cell type and origin in order to determine the most sensitive class of cell lines. The cytotoxicity was assessed using standard MTT-dye reduction assay [50,51]. The pilot studies covered the cell lines: SKW-3 (T-cell leukemia), LAMA-84 and K-562 cells (chronic myeloid leukemia), and 5637 (urinary bladder cancer) and the data were published in a previous research [34]. In addition, several leukemic and solid cell lines have been explored, namely: HL-60 (Acute myeloid leukemia), HD-MY-Z (Hodgkin-lymphoma), EJ (Urinary bladder carcinoma), and MCF-7 (Mammary gland carcinoma). The tested compounds inhibited the growth of tumor cells in a concentration-dependent manner. The IC_{50} values were calculated using nonlinear regression analysis and are summarized in Table 1. Normally, the tested palladium compounds exert lower activity compared to that of the referent drug cisplatin. The results showed also that as a rule the dinuclear complex **Pd1** demonstrated superior activity in contrast to the metalloporphyrin type complex **Pd2** (Table 1). Furthermore, it should be underlined, that **Pd1** exerts remarkable activity against K-562 cells, as the compound tested caused 50% cell growth inhibition at more than 4-fold lower concentration with regard to the reference agent [34]. Noteworthy, its activity against the HD-MY-Z cell line is comparable to that of cisplatin.

Table 1. Comparative antiproliferative activity of the investigated palladium complexes (**Pd1** and **Pd2**) and cisplatin in a panel of tumor cell lines after 72 h continuous exposure (MTT-dye reduction assay).

Cell Line	Cell Type	IC_{50} (μM) [1]		
		Pd1	Pd2	Cisplatin
SKW-3 [2]	T-cell leukemia	34.4 ± 2.2	77.9 ± 5.1	10.1 ± 1.4
K-562 [2]	Chronic myeloid leukemia	2.1 ± 0.9	31.9 ± 4.3	9.4 ± 2.1
LAMA-84 [2]	Chronic myeloid leukemia	40.0 ± 3.4	94.0 ± 6.2	18.3 ± 2.6
HL-60	Acute myeloid leukemia	75.2 ± 2.9	>200	8.2 ± 1.9
HD-MY-Z	Hodgkin-lymphoma	14.1 ± 4.5	>200	10.2 ± 3.4
5637 [2]	Urinary bladder carcinoma	96.0 ± 1.1	177.2 ± 10.2	4.4 ± 1.7
EJ	Urinary bladder carcinoma	117.2 ± 6.9	>200	9.3 ± 1.8
MCF-7	Mammary gland carcinoma	79.4 ± 4.2	>200	6.5 ± 1.1

[1] Arithmetic mean $\pm$ standard deviation of at least eight independent experiments. [2] Data have been published in [34].

2.3. Induction of Apoptosis

The ability of the palladium complexes to induce programmed cell death was evaluated using the more active complex **Pd1** compared to cisplatin in HL-60 cells. The assessment has been made with a commercially available DNA-fragmentation kit allowing semi-quantitative determination of the degree of oligonucleosomal genomic DNA fragmentation. The results concerning the HL-60 cell line (Figure 5) proved that both, **Pd1** and cisplatin cause significant increase in the apoptotic histone-associated DNA fragments, but the comparison of their behavior shows that the novel dinuclear compound causes comparable proapoptotic effects at substantially higher concentrations, which is in line with the tumor cell line chemosensitivity bioassay (Figure 5).

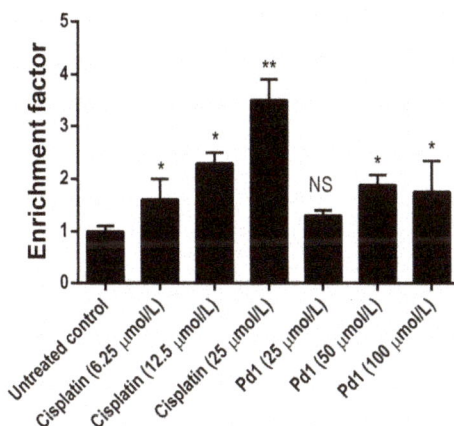

Figure 5. Apoptotic DNA-fragmentation in HL-60 cells after 24 h exposure to complex **Pd1** and cisplatin. The cytosolic mono- and oligo-nucleosomzal enrichment was determined using a commercially available ELISA kit. Each bar is representative for three independent experiments. Asterisks indicate statistical significance vs. the untreated control with $p \leq 0.05$ (*) or $p \leq 0.01$ (**) taken as significance levels; NS = non-significant (paired Student's *t*-test).

2.4. DNA-Modification

In order to evaluate in parallel the ability of **Pd1** and cisplatin to form intra-strand guanine bis-adducts, we investigated their binding to a synthetic 40-base DNA fragment bearing a single GG sequence [52]. Optimal binding to the target DNA molecule was observed at molar ratio 1:50 for cisplatin and molar ratio 1:100 for **Pd1**, respectively, after overnight incubation (Figure 6). Cisplatin treatment led to a total inhibition of the BamH1-mediated fragmentation of the DNA-probe indicating a high metallation ability. The dinuclear palladium complex **Pd1** also inhibited the nuclease activity as is evident by the low quantity of digested fragment but failed in the totally hamper the fragmentation of the target DNA-molecule.

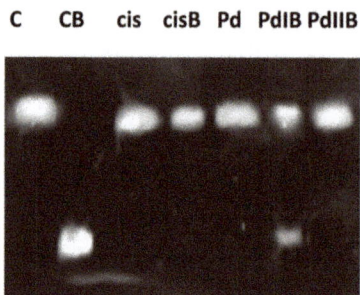

Figure 6. Metallation of a 40-base DNA fragment following treatment with cisplatin (molar ratio 1:50) and **Pd1** (molar ratios 1:100 and 1:200). The level of metal binding following interactions between the tested complexes and the DNA probe was analyzed after BamH1 treatment, electrophoretic analysis in 5% native polyacrilamide gel, and ethidium bromide staining. On the picture: **C**, unmodified DNA; **CB**, BamH1-digested DNA; **cis**, cisplatin-modified DNA; **cisB**, cisplatin-modified DNA (1:50) BamH1-digested; **Pd**, **Pd1**-modified DNA; **PdIB**, **Pd1**-modified DNA (1:100) BamH1-digested, **PdIIB**, **Pd1**-modified DNA (1:200) BamH1-digested.

2.5. DNA-Repair Synthesis

The elucidation of the cellular processing of metallodrug-induced DNA-adducts is of crucial importance for delineation of the pharmacological behavior of the antitumor agents [53,54]. The results of the DNA-repair experiments run under optimal conditions: plasmid:drug ratios 1:750, overnight are shown on Figure 7. The adducts induced by cisplatin or **Pd1** treatment were recognized and efficiently repaired by the nucleotide excision repair (NER) enzymes. It was found that the level of DNA-repair of pBS plasmid after treatment with **Pd1** was far less efficient than that encountered by the cisplatin-modified DNA probe. These findings indicate that the **Pd1** would have advantageous behavior against tumor cells, characterized by an overexpression of the NER-enzymatic machinery, mediating one of the most important mechanisms of post-target resistance to platinum drugs.

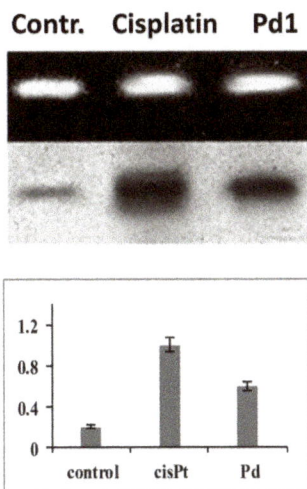

Figure 7. Relative efficiency of the nucleotide excision repair process in pBS plasmids modified by treatment with cisplatin or **Pd1** (plasmid:drug ratio 1:750, overnight). Repair of cisplatin-modified pBS is set as 1.

2.6. HMGB-1 Binding to Pd1 or Cisplatin-Modified DNA

The ability of high mobility group box (HMGB)-1 protein to bind metallated 40-base DNA fragments was investigated in a cell-free system by electrophoretic mobility shift assay (EMSA). The modification of the DNA probe with the reference cytotoxic drug cisplatin resulted in the emergence of a more slowly migrating band indicating the specific binding of HMGB-1 to the platinated 40-base DNA fragments [55,56] (Figure 8). As seen on the electrophoregram depicted in Figure 8 of the DNA probe modification with the reference cytotoxic drug cisplatin resulted in the emergence of a more slowly migrating band indicating the specific binding of HMGB-1 to the platinated 40-base DNA fragments. In a dissimilar manner, the **Pd1**-modified DNA was not recognized as evidenced by the similar mobility of both **Pd1**-lesioned and untreated DNA probes. Thus it would be expected that the **Pd1**-modified DNA would not be shielded by the HMGB-1, indicating a further discrepancy in the molecular pharmacology of the presented class of agents vs. the classical cisplatin analogs.

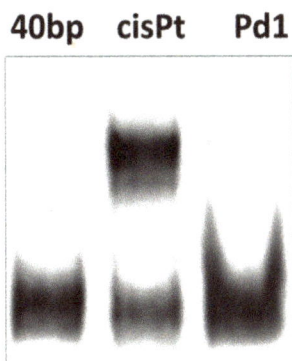

Figure 8. Binding of HMGB-1 protein to DNA-modified by either cisplatin or **Pd1** as determined by electrophoretic mobility shift assay (EMSA). DNA:protein molar ratio 1:9.

3. Discussion

In the framework of the design of new antitumor agents, two paramagnetic Pd^{III}—complexes of hematoporphyrin IX were obtained and biologically tested as representatives of the group of cisplatin-dissimilar metal-based coordination compounds. The reaction conditions for their synthesis were chosen to provide obtaining of Pd^{3+} species [57] and their stabilization in solution and solid state through formation of Hp-complexes [32–35]. The complexes were synthesized during the interaction of an initial Pd^{II} compound as a chloride complex ([PdCl$_4$$^{2-}$]) with the twofold deprotonated at peripheral propionic acid groups hematoporphyrin IX ligand ([Hp$_{-2H}$]$^{2-}$) in alkaline-aqueous medium achieved by adding KOH. All measurements performed during the interaction manifested that a redoxy process takes place together with the coordination reaction. The large spontaneous increase of the acidity of the reaction system together with the appearance of the EPR signals (Figure 1) proved that the studied process is a very complicated coordination reaction with production of a mixture of paramagnetic metal complexes and stable radicals. The appearance of two-component anisotropic EPR signals with different parameters during the interaction indicates also formation of complex species with different inner coordination sphere of the paramagnetic metal centers. The mode of Hp coordination can be distinguished by using UV/Vis characterization of the reaction system. Coordination via the side deprotonated propionic COO$^-$ groups could be supposed because of the hypsochromic shift of the Soret band owing to the porphyrin plane distortion. The spectral changes that follow are most probably connected to a distortion of the porphyrin ring symmetry owing to the coordination through two adjacent pyrrole N-atoms and formation of lower symmetry complex species. The reduction of the Q-bands number to two at the end of the reaction is owing to degeneration of the exited state S$_1$ orbitals [58] and proves the formation of a metalloporphyrin type complex (D$_{4h}$ symmetry).

Two of the complex species formed during the interaction PdCl$_4$$^{2-}$-Hp$_{-2H}$ have been isolated at proper reaction conditions. A dinuclear compound **Pd1** was the main product from the interaction at a metal excess (Pd:Hp $\geq$ 4) and was spontaneously precipitated at pH ~8 and a mononuclear metalloporphyrin type complex **Pd2** was the main product of the interaction at an equimolar ratio of the reagents and was precipitated by adding hydrochloric acid (5 × 10^{-2} M HCl). The composition of the complexes was derived from the elemental analyses and the content of H$_2$O and Cl$^-$ in the inner or the outer coordination sphere was proven by studying their thermal behavior [34]. The molecular structure of the complexes in solid state was deduced based on detailed investigations of their magnetic properties and spectroscopic characterization, published in [34].

A crucial issue in developing a new drug formulation is to study the structure of the compound in solution and to establish the relationship with the solid state structure. The data presented here relate to the behavior of the complexes in DMSO solutions. The low temperature EPR spectra (Figure 4a,b) of the complexes dissolved in DMSO show characteristic EPR spectral patterns of paramagnetic Pd-compounds. The EPR spectra recorded several hours (5–8 h) after the dissolving correspond well to the solid state EPR spectra [34]. The two anisotropic EPR signals observed in the spectrum of **Pd1** possess axial symmetry. The principal values of the g-tensors of the two anisotropic signals g$_{II}$ > g$_\perp$ > 2.0023 are consistent with formation of an elongated octahedral coordination with (dz^2)^1-ground state. A five-component superhyperfine structure is observed only in the perpendicular region of the low-field signal. The principal value of the superhyperfine coupling tensor a$_\perp$(N) = 15.31 × 10^{-4} cm^{-1} (I(^{14}N) = 1) and the number of superhyperfine lines typical for interaction of the uncoupled electron with two ^{14}N nuclei proved coordination of one Pd^{3+} via two of the pyrrole nitrogen donors of the Hp ligand. The absence of a superhyperfine structure from ^{14}N nuclei on the upfield EPR signal is due to coordination out of the porphyrin ring through the propionic acid groups. The wide signal observed at ~3050 G could be assigned to exchange singlet–triplet interaction between the two unpaired electrons of the differently coordinated Pd^{3+}-centers in the molecule of **Pd1**. Hence the EPR spectra proved the formation of a Pd^{3+}-L-Pd^{3+} system where each of the Pd-ions possesses different coordination. The presence of a signal for a radical could be explained with significant delocalization of the unpaired electron density because of electron-acceptor properties and significant flexibility of the porphyrin moiety. The complex **Pd2** features a completely different EPR spectrum pattern in solution containing exactly one two-component anisotropic signal. The signal displays axial anisotropy with principal values of the g-tensor $g_\perp > 2.0023 > g_{II}$ proving formation of a compressed octahedral structure with $(dx^2 - y^2)^1$ ground state. Superhyperfine lines were observed both in perpendicular and parallel regions. The principal values of the axially symmetrical superhyperfine coupling tensor are typical for the interaction of an uncoupled electron with ^{14}N (I = 1) nuclei. The nine superhyperfine lines are readily distinguished in the perpendicular region and prove coordination of Pd^{3+} into the four pyrrole N-donors in the porphyrin ring. Thus the spectrum confirms the existence of a stable metalloporphyrin type complex in solution.

It is well-known that changes in the conjugation and the symmetry of the Hp-ligand can affect the UV/Vis absorption spectra and can be used to obtain data about the mode of Hp coordination at complex species in solution [32–39]. While the metal binding through inner nitrogen atoms induces strong changes in the visible region of the spectra, the variations in the peripheral substituents causes minor changes owing to geometrical deformation of the porphyrin ring. The UV/Vis spectra were recorded of the studied compounds dissolved in DMSO. It was found that the position of the Soret band is sensitive to the processes such as acid-base equilibria and aggregation, and also to the modes of metal coordination. It is accepted that the Soret band appears at 376 nm if the propionic acid groups are deprotonated and shifts in accordance with the electronic density distribution caused by protonation and metal-coordination. The major component of the Soret band in the spectrum of the free ligand dissolved in DMSO (400 nm, $\lg\varepsilon = 5.02$) is batochromically shifted compared to the spectrum registered in alkaline-aqueous medium. The shoulder observed at 376 nm is due to the protolitic equilibrium $Hp + 2Solv \Leftrightarrow [Hp_{-2H}]^{2-} + 2[HSov]^+$, which is clearly shifted to the neutral Hp molecules in the weakly proton-acceptor solvent DMSO. In the spectra of the complexes this band is observed at 399 nm (**Pd1**) and 396 nm (**Pd2**), respectively. Thus it indicates that the propionic acid groups are engaged in the processes of coordination or protonation.

The spectrum pattern in the area of Q-bands is determined by the symmetry group of the porphyrin macrocycle. The two diagonally located hydrogen atoms at pyrrolic nitrogens in the molecule of the free ligand define a D_{2h} symmetry and hence a four number Q-bands spectrum. The loose of a proton gives a monoprotic ligand form. Many metal ions at proper pH value (6 < pH < 10) interact with the monoprotic ligand anion and bind with its two cis-disposed pyrrole N-donors on the top of the macrocycle forming sitting-atop (SAT) complex. The metal in these complexes is located out of the porphyrin plane and distorts it. The symmetry of the structure is lower (approximately between $C_{4v} \rightarrow C_1$) than that of both metalloporphyrins (D_{4h}) where the two ammine hydrogens are substituted by the metal located coplanar on the porphyrin ring and the free Hp ligand (D_{2h}). This reflects on the Q-bands number reduction and changes their intensities. Hence the observed spectrum of the complex **Pd1** in the area of Q-bands containing mainly three Q-bands (507, 546, and 567 nm) with intensity IV ~III > II is consistent with coordination through the two cis-disposed pyrrole N-donors on the top of the porphyrin macrocycle substituting one of the pyrrolic nitrogen. Further on, the slight red shift of the Q-bands: IV and III and a blue shift of the Q band II with degeneration of the Q_x and Q_y orbitals connected with vibronic structure supports simultaneous coordination on the top of the porphyrin ring and out-side the porphyrin macrocycle through the peripheral deprotonated carboxylic groups. The typical two Q-band spectrum of the complex **Pd2** indicates unambiguously a D_{4h} symmetry achieved at coordination through the four *N*-heteronuclei in the porphyrin macrocycle. The intensive bands at 270 ($\lg\varepsilon = 4.98$) and 323 nm ($\lg\varepsilon = 4.81$) in the spectrum of **Pd1** as well as the one at 280 ($\lg\varepsilon = 3.99$) in the spectrum of **Pd2** were assigned to the ligand-to-metal charge transfer bands of the Pd-Cl bonds. This proves the presence of Cl^- ions in the palladium ions inner coordination sphere.

The spectroscopic characteristics observed were unchangeable within more than five days. Hence in this period the compounds, dissolved in DMSO solution are paramagnetic palladium-hematoporphyrin IX complexes as follows: (1) for **Pd1**—a dinuclear Pd-Hp-Pd system, with two Pd^{III} ions that occupied two coordination places—at the porphyrin ring binding out-of-plan through two cis-disposed N-donors and at the two peripheral propionic acid groups; (2) for **Pd2**—normal metalloporphyrin of Pd^{III}. Furthermote, this is the reason to accept that during the biological investigation the active species in solution are these palladium complexes of hematoporphyrin IX.

The cytotoxic screening of the two palladium(III) complexes was conducted on a wide spectrum of cell lines—representatives of the main human cancer types. The results of the MTT-dye reduction assay [50,51] unambiguously indicate that the two compounds exert concentration-dependent antiproliferative effects against the chosen spectrum of cell lines. Data analysis shows that the tested palladium compounds are generally less active than cisplatin, causing half-maximal inhibition of cell viability at generally higher concentrations (Table 1). It has also been proven that the dinuclear palladium complex **Pd1** demonstrates superior activity as compared to the metalloporphyrin type complex **Pd2**. The latter **Pd2** compound exerts only marginal activity and fails to cause 50% inhibition of malignant cell growth against a half of the cell lines under evaluation. Throughout the panel investigated the anticancer drug cisplatin proved to outclass the novel palladium complexes, with the only exception of K-562 leukemia whereby **Pd1** showed remarkable cytotoxicity [34].

These findings gave us a reason to conduct more detailed pharmacodynamic evaluation of **Pd1** especially regarding its ability to induce apoptosis and to modify DNA. Significant apoptotic fragmentation of genomic DNA was established after treatment of HL-60 cells with both cisplatin and **Pd1**, whereby the proapoptotic effect of **Pd1** required the cell exposure to concentrations significantly higher than its IC$_{50}$. These data indicate that the cytotoxicity of both compounds is mediated by induction of apoptosis, although its threshold level was significantly higher for the palladium compound.

Further experiments were carried out in order to characterize the adduct-forming ability of **Pd1** and the cellular processing of the DNA-lesions. The N7 position of guanine is considered as the ultimate pharmacological target of platinum drugs, leading to formation of intrastrand adducts whose recognition and processing leads to activation of the cell death signaling pathways [59,60]. On this ground we studied the metallation of a single strand 40-base DNA fragment in a cell free system. The palladium compound **Pd1** was less capable of forming DNA adducts under the experimental conditions, thus further demonstrating its cisplatin-dissimilar pharmacological properties. These findings indicate that although the DNA-modification plays an important role for the mode of antiproliferative action of **Pd1**; its capacity to modify DNA is lower as compared to that of cisplatin.

The structural perturbations induced by the DNA-metallation are recognized by diverse proteins including the DNA-repair enzymatic machinery [59,61,62]. Removal of platinum-DNA adducts by the nucleotide excision repair (NER) is one of the crucial mechanisms of cellular resistance to cisplatin. This prompted us to evaluate the efficiency of NER repair synthesis after modification of pBS plasmids with either cisplatin or **Pd1**. The lesions induced by cisplatin were far better recognized and repaired as compared to those of **Pd1**, implying that, by virtue of the significant structural differences between the tested complexes, they induce highly dissimilar alterations of DNA conformation. The lower level of NER-mediated removal and repair of **Pd1** modified DNA are an advantageous future of the novel compound as this would condition retained activity against malignant cells overexpressing the NER-enzymatic system.

Apart from the DNA-repair enzymes other proteins are also capable of recognizing and binding cisplatin-modified DNA [59,61,63]. Among these special attention has been paid to the high mobility group domain (HMG) proteins [55,59,61,64]. They are considered to play crucial role for the cytotoxicity of platinum drugs, whereby the proposed mechanisms include: (i) shielding of platinated DNA and steric hindrance against NER-mediated repair of metal-adducts; (ii) "hijacking" i.e., binding of transcription factors or other regulatory proteins to HMG-associated platinum adducts, thus deviating them from their normal targets and compromising their role in signal transduction [59,61,63]. On this ground we evaluated the recognition and binding of HMGB-1 protein to cisplatin or **Pd1** modified DNA probes. As made evident by the results obtained **Pd1**-induced modification conditions lower level of HMGB-1 binding, as compared to cisplatin-lesioned DNA. This implies that the high mobility group proteins are most probably less involved in the cytotoxicity of the palladium agent.

4. Materials and Methods

4.1. Chemicals and Tested Compounds

All chemicals used were of analytical grade and were obtained by commercial sources and used without further purification. Agarose, ethanol, DMSO, formic acid, 2-propanol, methanol, EDTA, ethidium bromide, sodium chloride, Tris hydrochloride, Triton® X-100, L-glutamine were purchased from AppliChem GmbH, Darmstadt, Germany. The tetrazolium salt 3-(4,5-dimethylthiazol-2-yl)-2,5-diphenyltetrazolium bromide (MTT) and TE buffer were supplied by Merck, Darmstadt, Germany. Fetal calf serum (FCS), powdered RPMI 1640 medium, the amino acid, sodium pyruvate, human insulin, and polyacrylamide gel were purchased from Sigma-Aldrich GmbH, Steinheim, Germany. The referent cytotoxic drug cis-DDP was used as a commercially available sterile dosage form for clinical application (Platidiam®, Lachema, Czech Republic). The two complexes tested, namely: cis-$[Pd^{III}_2(Hp_{-3H})Cl_3(H_2O)_5] \cdot 2PdCl_2$ (**Pd1**) and, $[Pd^{III}(Hp_{-2H})Cl(H_2O)] \cdot H_2O$ (**Pd2**) were synthesized, characterized and purified as previously described [34]. UV-Visible (UV-Vis) (solv. DMSO) λ_{max}/nm, ($\log \varepsilon$): complex **Pd1**: 270 (4.98), 323 (4.81), 399 (5.02), 507 (3.68), 546 (3.68), 567 (3.63); complex **Pd2**: 280 (3.99), 396 (4.91), 512 (3.58), 548 (3.74).

4.2. Solution Stability Assays

The absorption electronic spectra were recorded on a "Carry" 100 UV-Vis spectrometer. The EPR spectra were obtained on an X-band "Bruker B-EPR 420" spectrometer at 130 K. The UV-Vis and EPR spectra were measured after dissolving the Pd-complexes samples in DMSO within the concentration intervals of $(5 \times 10^{-6}) \div (1 \times 10^{-4})$ mol/L for UV-Vis spectra and 5×10^{-4}–5×10^{-3} mol/L for EPR spectra, respectively.

4.3. Cell Lines and Culture Conditions

In this study the following cell lines were used: SKW3 (T-cell leukemia), K-562 and LAMA-84 (chronic myeloid leukemia), HL-60 (acute myelocyte leukemia), HD-MY-Z (Hodgkin lymphoma), 5637, EJ (urinary bladder carcinomas), and MCF-7 (mammary gland adenocarcinoma). The cells were maintained in controlled environment—cell culture flasks at 37 °C in an incubator 'BB 16-Function Line' Heraeus (Kendro, Hanau, Germany) with humidified atmosphere and 5% CO$_2$. They were kept in log phase by supplementing with fresh medium, two or three times a week. SKW3, K-562, LAMA-84, HL-60, HD-MY-Z, 5637, and EJ cells were grown in RPMI-1640 medium supplemented with 10% fetal bovine serum (FBS) and 2 mM L-glutamine. MCF-7 cells were grown as monolayer adherent cultures in 90% RPMI-1640 supplemented with 10% FBS, non-essential amino acids, 1 mM sodium pyruvate, and 10 mg/mL human insulin.

4.4. MTT-Dye Reduction Assay

The tumor cell growth inhibitory effects were assessed using the standard 3-[4-dimethylthiazol-2-yl]-2,5-diphenyl-2H-tetrazolium bromide (MTT)-dye reduction assay as described by Mosmann [50] with minor modifications [51]. Exponentially growing cells were seeded in 96-well flat-bottomed microplates (100 µL/well) at a density of 1×10^5 cells per mL and after 24 h incubation at 37 °C they were exposed to various concentrations of the tested compounds for 72 h. At least 8 wells were used for each concentration. After incubation with the tested compounds 10 µL MTT solution (10 mg/mL in PBS) aliquots per well were added. The microplates were further incubated for 4 h at 37 °C and the MTT-formazan crystals formed were dissolved by adding 100 µL/well 5% formic acid (in 2-propanol). The absorption was measured using a microprocessor controlled microplate reader (Labexim LMR1) at 580 nm. The cell survival data were normalized to percentage of the untreated control and were fitted to sigmoidal dose/response curves. The corresponding IC_{50} values were calculated using non-linear regression analysis.

4.5. Apoptosis Assay

The typical apoptosis oligonucleosomal DNA fragmentation was examined using a commercially available "Cell-death detection" ELISA kit (Roche Applied Science, Mannheim, Germany). Cytosolic fractions of 1×10^4 cells per group (treated or untreated) served as antigen source in a sandwich ELISA, utilizing primary anti-histone antibody-coated microplate and a secondary peroxidase-conjugated anti-DNA antibody. The photometric immunoassay for histone-associated DNA fragments was executed according to the manufacturer's instructions. The results are expressed as the oligonucleosome enrichment factor (representing a ratio between the absorption in the treated vs. the untreated control samples).

4.6. DNA-Binding

The DNA binding of the dinuclear palladium agent and cisplatin was assessed as previously described [52]. A 40 n.b. fragment (5′ CGCTATCGCTACCTATTGGATCCTTATGCGTTAGTGTA TG 3′), whereby the GG-motif is the recognition sequence of the restriction nuclease BamH1 was used as a target DNA molecule. The level of DNA modification following interactions between the tested complex and the 40 n.b. fragment was analyzed after BamH1 treatment, electrophoretic analysis in 5% native polyacrylamide gel, and ethidium bromide staining, as previously described [52].

4.7. In Vitro DNA Repair

Synthetic pBS plasmids (2.69 kb) served as a DNA probe in this study. They were propagated in *Escherichia coli* and extensively purified as closed circular DNA by means of a FlexiPrep Kit (Pharmacia Biotech, Uppsala, Sweden). Modification of plasmid DNA (200 µg/mL) with the tested palladium complex or cisplatin was carried out in TE buffer (10 mM Tris, 1 mM EDTA), pH = 7.4, in the dark at 37 °C for 16 h. Following ethanol precipitation, DNA was washed twice in 70% ethanol and redissolved in TE buffer, and the superhelical form of the plasmid DNA was purified by ethidium bromide/cesium chloride gradient centrifugation. Cell-free extract (CFE) from exponentially growing Guerin ascites tumor cells was prepared using a previously described protocol [53], adapted for in vitro DNA repair studies [54] and stored at −80 °C until use. Repair of DNA lesions induced by the metal complexes was assayed as described elsewhere [52]. Briefly, the standard 50 µL reaction mixture contained 400 ng of metal complex-treated repair substrate pBS, 400 ng, 45 mM HEPES-KOH, pH = 7.8, 70 mM KCl, 5 mM MgCl$_2$, 1 mM dithiothreitol, 0.4 mM EDTA, 2 mM ATP, 20 µM each of dTTP, dGTP, dATP, 2µCi [^{32}P]dCTP (Amersham, 3000Ci/mmol, Amersham Biscinces, Freiburg, Germany), 40 mM phosphocreatine, 2.5 µg of creatine phosphokinase, 3% glycerol, 20 µg of bovine serum albumin, and 80–120 µg of cell-free extract at 30 °C for 1 h. Reactions were stopped by adding EDTA to 20 mM, and mixtures were incubated for 20 min with RNase A (80 µg/mL) followed by another 20 min with proteinase K (200 µg/mL) in the presence of 0.5% SDS. Plasmid DNA was purified with phenol/chloroform (1:1) and precipitated by 2 vol of ethanol in the presence of glycogen (Stratagene, La Jolla, CA, USA, 1 mg/mL) at −70 °C. Thereafter the plasmids were linearized by digestion with EcoRI and resolved by electrophoresis in 0.8% agarose gel containing ethidium bromide (0.5 µg/mL). After electrophoresis, the gel was photographed under UV illumination, dried under vacuum, and exposed to Kodak XAR-5 film for 12 h at −70 °C. The autoradiograph was scanned with Gel-Pro Analyser (Media Cybernetics, Bethesda, MD, USA).

4.8. Electrophoretic Mobility Shift Assay (EMSA)

DNA binding assay of HMGB-1 and its truncated form with ^{32}P-labeled cisplatinated DNA was performed as described previously [55]. In brief, nonlabeled sonicated salmon sperm DNA was added as a competitor in all experiments except those designed to determine the dissociation constants. On completion of electrophoresis the gel was dried and exposed to Amersham hyperfilm. Quantification of band densities was performed by scanning the autoradiographs with Gel-Pro Analyzer. In some assays, the reaction mixture prepared for EMSA was supplemented with CFE preincubated with nonplatinated DNA (500-fold molar excess over the platinated probe) and loaded on the gel. Isolation and purification of recombinant HMGB-1 were carried out as described elsewhere [56].

5. Conclusions

Two newer members of the group of paramagnetic transition metal complexes of hematoporhyrin IX [32–39] have been proposed for extended biological screening: a dinuclear $[Pd^{III}{}_2(Hp_{-3H})Cl_3(H_2O)_5]\cdot 2PdCl_2$ and a mononuclear $[Pd^{III}(Hp_{-2H})Cl(H_2O)]\cdot H_2O$. The complexes were obtained reproducibly in alkaline-aqueous medium and were isolated as neutral compounds changing the acidity of the reaction system and the M:L molar ratios. Their structure and stability were studied in DMSO solutions in details. It was found that the active species in solution are a dinuclear complex (**Pd1**) and a mononuclear (**Pd2**) complex. In the dinuclear complex **Pd1**, one Pd^{III} ion is coordinated to the deprotonated COO^- groups from the side chains of the porphyrin ligand and the second Pd^{III} ion—to two adjacent pyrrole N-atoms on the top of the porphyrin ring and thus a dinuclear Pd^{III}-Hp-Pd^{III} system is created. Pd^{III} in the mononuclear complex, **Pd2**, is located in the plane of the porphyrin ring and thus a metalloporphyrin type complex is formed. Pd^{III} centers in both complexes have a distorted octahedral coordination filled with additional donor species such as Cl^- and solvent molecules.

The compounds tested manifested cell growth inhibitory effects at micromolar concentration against tumor cell lines with distinct cell type and origin. The calculated IC_{50} values proved that in general, palladium complexes exhibit lower activity compared to that of the referent drug cisplatin and as a rule the metalloporphyrin type complex **Pd2** is less active than the dinuclear compound **Pd1**. Contrarily to the general trend the **Pd1** complex exerts remarkable activity against K-562 cells, with 50% cell growth inhibition at more than 4-fold lower concentration compared to cisplatin. It also displays relatively close activity against the HD-MY-Z cell line with regard to the referent drug. The palladium complexes' ability to induce programmed cell death was evaluated in a comparative experiment of **Pd1** and cisplatin in HL-60 cells. The two compounds cause significant increase in the apoptotic histone-associated DNA fragments. However, the novel dinuclear compound causes comparable proapoptotic effects at substantially higher concentrations and that corresponds to the tumor cell line chemosensitivity bioassay. The parallel evaluation of **Pd1** and cisplatin's ability to form intra-strand guanine bis-adducts shows that **Pd1** also inhibits the nuclease activity, but failed to totally hamper the fragmentation of the target DNA-molecule. Hence, although the DNA-modification plays an important role for the mode of antiproliferative action of **Pd1**, its capacity to modify DNA is lower compared to that of cisplatin. The elucidation of DNA-adducts cellular processing by the NER enzymes demonstrated that the lesions induced by cisplatin were far better recognized and repaired as compared to those of **Pd1**. The lower level of NER-mediated removal and repair of **Pd1** modified DNA are an advantageous characteristic of the novel compound and that means that **Pd1** would retain the activity against malignant cells, overexpressing the NER-enzymatic system. The ability of HMGB-1 protein to bind metalled 40-base DNA fragment with the reference cytotoxic drug cisplatin resulted in a specific binding. In a dissimilar manner, the **Pd1**-modified DNA was not recognized and it could be expected that the **Pd1**-modified DNA would not be shielded by the HMGB-1.

The data analysis of the in-depth biological study unambiguously highlights the differences in molecular pharmacology of the presented "applicants" for antitumor agents in respect to cisplatin. Moreover, the advantages of the new compounds provide grounds for joining them to the "nonclassical metal compounds" group. Their unique structure, based on the octahedral coordination of palladium(III) stabilized with a ligand with favorable properties, is a prerequisite for constructing a new formula with a potential of controlling its kinetic behavior as well as the strength of the M-L bonds of the adducts formed in the biological milieu.

Author Contributions: Conceptualization, G.G. and G.M.; Methodology, E.P., G.M., D.T., and G.G.; Software, G.M., I.U., and D.T.; Validation, E.P., G.M., and G.G.; Formal Analysis, D.T., I.U., and G.M.; Investigation, G.M., I.U., E.P., D.T., and G.G.; Writing-Original Draft Preparation, G.G. and G.M.; Writing-Review & Editing, G.G. and D.T.; Supervision, G.G.; Project Administration, G.G. and G.M.

Funding: This research was funded by the National Science Fund (Bulgaria), grant number DN09/16, Bulgarian Ministry of Education and Science, available online: https://www.fni.bg.

Acknowledgments: We gratefully acknowledge the financial support from the National Science Fund (Project DN09/16) of Bulgarian Ministry of Education and Science.

Conflicts of Interest: The authors declare no conflict of interest.

Abbreviations

DMSO	Dimethyl sulfoxide
DMF	*N,N*-dimethyl formamide
SAT	Sitting atop
DNA	Deoxyribonucleic acid
EMSA	Electrophoretic mobility shift assay
NER	Nucleotide excision repair
HMGB1	High mobility group protein B1
FCS	Fetal calf serum
MTT	Tetrazolium salt 3-(4,5-dimethylthiazol-2-yl)-2,5-diphenyltetrazolium bromide

References

1. Reedijk, J. Increased understanding of platinum anticancer chemistry. *Pure Appl. Chem.* **2011**, *83*, 1709–1717. [CrossRef]
2. Sava, G.; Bergamo, A.; Dyson, P.J. Metal-based antitumour drugs in the post-genomic era: What comes next? *Dalton Trans.* **2011**, *40*, 9069–9075. [CrossRef] [PubMed]
3. van Rijt, S.H.; Sadler, P.J. Current applications and future potential for bioinorganic chemistry in the development of anticancer drugs. *Drug Discov. Today* **2009**, *14*, 1089–1097. [CrossRef] [PubMed]
4. Allardyce, C.S.; Dyson, P.J. Metal-based drugs that break the rules. *Dalton Trans.* **2016**, *45*, 3201–3209. [CrossRef] [PubMed]
5. Rosenberg, B.; VanCamp, L.; Trosko, J.E.; Mansori, V.H. Platinum compounds: A new class of potent antittumour agents. *Nature* **1969**, *222*, 385–386. [CrossRef] [PubMed]
6. Rosenberg, B. Platinum complexes for the treatment of cancer: Why the research goes on? In *Cisplatin, Chemistry and Biochemistry of a Leading Anticancer Drug*, 1st ed.; Lippert, B., Ed.; Wiley-VCN: Weinheim, Germany, 1999; pp. 3–28, ISBN 3-906390-20-9.
7. Cleare, M.J.; Hoeschele, J.D. Studies on the antitumor activity of group VIII transition metal complexes. Part I. Platinum(II) complexes. *Bioinorg. Chem.* **1972**, *2*, 187–210. [CrossRef]
8. Desoize, B.; Madoulet, C. Particular aspects of platinum compounds used at present in cancer tretment. *Crit. Rev. Oncol. Hematol.* **2002**, *42*, 317–325. [CrossRef]
9. Johnston, T.C.; Suntharalingam, K.; Lippard, S.J. The next generation of platinum drugs: Target Pt(II) agents, nanoparticle delivery and Pt(IV) prodrugs. *Chem. Rev.* **2016**, *116*, 3436–3486. [CrossRef] [PubMed]
10. Martin, L.P.; Hamilton, T.C.; Schilder, R.J. Platinum resistance: The role of DNA repair pathways. *Clin. Cancer Res.* **2008**, *14*, 1291–1295. [CrossRef] [PubMed]
11. Heffeter, P.; Jungwirth, U.; Jakupec, M.; Hartinger, C.; Galanski, M.; Elbling, L.; Micksche, M.; Keppler, B.; Berger, W. Resestance against novel anticancer metal compounds: Differences and similarities. *Drug Resist. Update* **2008**, *11*, 1–16. [CrossRef] [PubMed]
12. Pabla, N.; Dong, Z. Curtailing side effects in chemotherapy: A tale of PKCδ in cisplatin treatment. *Oncotarget* **2012**, *3*, 107–111. [CrossRef] [PubMed]
13. Wang, X.; Guo, Z. Targeting and delivery of platinum-based anticancer drugs. *Chem. Soc. Rev.* **2013**, *42*, 202–224. [CrossRef] [PubMed]
14. Johnstone, T.C.; Park, G.Y.; Lippard, S.J. Understanding and improving platinum anticancer drugs—Phenanthriplatin. *Anticancer Res.* **2014**, *34*, 471–476. [CrossRef] [PubMed]

15. Wilson, J.J.; Lippard, S.J. Synthetic Methods for the preparation of platinum anticancer complexes. *Chem. Rev.* **2014**, *114*, 4470–4495. [CrossRef] [PubMed]

16. Reedijk, J. Metal-ligand exchange kinetics in platinum and ruthenium complexes. *Platin. Met. Rev.* **2008**, *52*, 2–11. [CrossRef]

17. Johnstone, T.C.; Wilson, J.J.; Lippard, S.J. Monofunctional and Higher-Valent Platinum Anticancer Agents. *Inorg. Chem.* **2013**, *52*, 12234–12249. [CrossRef] [PubMed]

18. Hall, M.D.; Okabe, M.; Shen, D.-W.; Liang, X.-J.; Gottesman, M.M. The role of cellular accumulation in determining sensitivity of platinum-based chemotherapy. *Annu. Rev. Pharmacol. Toxicol.* **2008**, *48*, 495–535. [CrossRef] [PubMed]

19. Howell, S.B.; Safaei, R.; Larson, C.A.; Sailor, M.J. Copper transporters and the cellular pharmacology of the platinum-containing cancer drugs. *Mol. Pharmacol.* **2010**, *77*, 887–894. [CrossRef] [PubMed]

20. Harrach, S.; Ciarimboli, G. Role of transporters in distribution of platinum-based drugs. *Front. Pharmacol.* **2015**, *6*, 1–8. [CrossRef] [PubMed]

21. Dilruba, S.; Kalayda, G.V. Platinum-based drugs: Past, present and future. *Cancer Chemother. Pharmacol.* **2016**, *77*, 1103–1124. [CrossRef] [PubMed]

22. Galanski, M.; Jakupec, M.A.; Keppler, B.K. Update of the preclinical situation of anticancer platinum complexes: Novel design strategies and innovative analytical approaches. *Curr. Med. Chem.* **2005**, *12*, 2075–2094. [CrossRef] [PubMed]

23. Wang, D.; Lippard, S.J. Cellural processing of platinum anticancer drugs. *Nat. Rev. Drug Discov.* **2005**, *4*, 307–320. [CrossRef] [PubMed]

24. Bruijnincx, P.C.A.; Sadler, P.J. New trends for metal complexes with anticancer activity. *Curr. Opin. Chem. Biol.* **2008**, *12*, 197–206. [CrossRef] [PubMed]

25. Pages, B.J.; Garbutcheon-Singh, K.B.; Aldrich-Wright, J.R. Platinum Intercalators of DNA as Anticancer Agents. *Eur. J. Inorg. Chem.* **2017**, 1613–1624. [CrossRef]

26. Rademaker-Lakhai, J.M.; van den Bongard, D.; Pluim, D.; Beijnen, J.H.; Schellens, J.H.M. A phase I and pharmacological study with imidazolium-*trans*-DMSO-imidazole-tetrachlororuthenate, a novel ruthenium anticancer agent. *Clin. Cancer Res.* **2004**, *10*, 3717–3727. [CrossRef] [PubMed]

27. Ang, W.H.; Khalaila, I.; Allardyce, C.S.; Juillerat-Jeanneret, L.; Dyson, P.J. Rational design of platinum(IV) compounds to overcome glutathione-S-transferase mediated drug resistance. *J. Am. Chem. Soc.* **2005**, *127*, 1382–1383. [CrossRef] [PubMed]

28. Kenny, R.G.; Chuah, S.W.; Crawford, A.; Marmion, C.J. Platinum(IV) prodrugs—A step closer to Ehrlich's vision? *Eur. J. Inorg. Chem.* **2017**, 1596–1612. [CrossRef]

29. Trondl, R.; Heffeter, P.; Kowol, C.R.; Jakupec, M.A.; Berger, W.; Keppler, B.K. NKP-1339, the first ruthenium-based anticancer drug on the edge to clinical application. *Chem. Sci.* **2014**, *5*, 2925–2932. [CrossRef]

30. Leijen, S.; Burgers, S.A.; Baas, P.; Pluim, D.; Tibben, M.; van Werkhoven, E.; Alessio, E.; Sava, G.; Beijnen, J.H.; Schellens, J.H. Phase I/II study with ruthenium compound NAMI-A and gemcitabine in patients with non-small cell lung cancer after first line therapy. *Investig. New Drugs* **2015**, *33*, 201–214. [CrossRef] [PubMed]

31. Bhargava, A.; Vaishampayan, U.N. Satraplatin: Leading the new generation of oral platinum agents. *Expert Opin. Investig. Drugs* **2009**, *18*, 1787–1797. [CrossRef] [PubMed]

32. Gencheva, G.; Tsekova, D.; Gochev, G.; Momekov, G.; Tyueliev, G.; Skumryev, V.; Karaivanova, M.; Bontchev, P.R. Synthesis, structural characterization and cytotoxic activity of novel paramagnetic platinum hematoporphyrin IX complexes: Potent antitumor agents. *Met. Based Drugs* **2007**, *2007*. [CrossRef] [PubMed]

33. Momekov, G.; Karaivanova, M.; Ugrinova, I.; Pasheva, E.; Gencheva, G.; Tsekova, D.; Arpadjan, S.; Bontchev, P.R. In vitro pharmacological study of monomeric platinum(III) hematoporphyrin IX complexes. *Investig. New Drugs* **2011**, *29*, 742–751. [CrossRef] [PubMed]

34. Tsekova, D.; Gorolomova, P.; Gochev, G.; Skumryev, V.; Momekov, G.; Momekova, D.; Gencheva, G. Synthesis, structure and in vitro cytotoxic studies of novel paramagnetic palladium(III) complexes with hematoporphyrin IX. *J. Inorg. Biochem.* **2013**, *124*, 54–62. [CrossRef] [PubMed]

35. Gencheva, G.; Tsekova, D.; Gochev, G.; Mehandjiev, D.; Bontchev, P.R. Monomeric Au(II) complex with hematoporphyrin IX. *Inorg. Chem. Commun.* **2003**, *6*, 325–328. [CrossRef]

36. Momekov, G.; Ferdinandov, D.; Konstantinov, S.; Arpadjan, S.; Tsekova, D.; Gencheva, G.; Bontchev, P.R.; Karaivanova, M. In vitro evaluation of a stable monomeric gold(II) complex with hematoporphyrin IX: Cytotoxicity against tumor kidney cells, cellular accumulation and induction of apoptosis. *Bioinorg. Chem. Appl.* **2008**, *2008*. [CrossRef] [PubMed]

37. Tsekova, D.T.; Gencheva, G.G.; Bontchev, P.R. Mode of coordination of the polydentate ligand hematoporphyrin IX with Pt(III), Pd(III), Au(II) and Cu(II). An overview. *C. R. Acad. Bulg. Sci.* **2008**, *61*, 731–738. [CrossRef]

38. Tsekova, D.T.; Gochev, G.P.; Gencheva, G.G.; Bontchev, P.R. Magnetic and spectroscopic methods for structural characterization of paramagnetic hematoporphyrin IX complex with Cu(II). *Eurasian J. Anal. Chem.* **2008**, *3*, 79–90.

39. Tsekova, D.; Ilieva, V.; Gencheva, G. An NMR study on the solution behavior of series of hematoporphyrin IX complexes. In *Topics in Chemistry and Material Science. Current Issues in Organic Chemistry 2*; Nikolava, R.D., Simova, S., Denkova, P., Vayssilov, G.N., Eds.; Heron Press Ltd.: Sofia, Bulgaria, 2011; Volume 5, pp. 52–65, ISBN 1314-0795.

40. Chang, J.-E.; Yoon, I.-S.; Sun, P.-L.; Yi, E.; Jheon, S.; Shim, C.-K. Anticancer efficacy of photodynamic therapy with hematoporphyrin-modified, doxorubicin-loaded nanoparticles in liver cancer. *J. Photochem. Photobiol. B Biol.* **2014**, *140*, 49–56. [CrossRef] [PubMed]

41. Lottner, C.; Knuechel, R.; Bernhardt, G.; Brunner, H. Combined chemotherapeutic and photodynamic treatment on human bladder cells by hematoporphyrin–platinum(II) conjugates. *Cancer Lett.* **2004**, *203*, 171–180. [CrossRef] [PubMed]

42. Lottner, C.; Bart, K.-C.; Bernhardt, G.; Brunner, H. Soluble tetraarylporphyrin-platinum conjugates as cytotoxic and phototoxic antitumor agents. *J. Med. Chem.* **2002**, *45*, 2079–2089. [CrossRef] [PubMed]

43. Lottner, C.; Bart, K.-C.; Bernhardt, G.; Brunner, H. Hematoporphyrin-derived soluble porphyrin-platinum conjugates with combined cytotoxic and phototoxic antitumor activity. *J. Med. Chem.* **2002**, *45*, 2064–2078. [CrossRef] [PubMed]

44. Alam, M.N.; Huq, F. Comprehensive review on tumour active palladium compounds and structure-activity relationships. *Coord. Chem. Rev.* **2016**, *316*, 36–67. [CrossRef]

45. Fanelli, M.; Formica, M.; Fusi, V.; Giorgi, L.; Micheloni, M.; Paoli, P. New trends in platinum and palladium complexes as antineoplastic agents. *Coord. Chem. Rev.* **2016**, *310*, 41–79. [CrossRef]

46. Garoufis, A.; Hadjikakou, S.K.; Hadjiliadis, N. Palladium coordination compounds as anti-viral, anti-fungal, anti-microbial and anti-tumor agents. *Coord. Chem. Rev.* **2009**, *253*, 1384–1397. [CrossRef]

47. Garoufis, A.; Hadjikakou, S.K.; Hadjiliadis, N. The use of palladium complexes in medicine. In *Metallotherapeutic Drugs and Metal-Based Diagnostic Agents. The Use of Metals in Medicine*; Gielen, M., Tiekink, E.R.T., Eds.; Wiley VC: Chichester, UK, 2005; pp. 399–419, ISBN 0-470-86403-6.

48. Lippert, B.; Miguel, P.J.S. Comparing Pt^{II}- and Pd^{II}-nucleobase coordination chemistry: Why Pd^{II} not always is a good substitute for Pt^{II}. *Inorg. Chim. Acta* **2017**, *472*, 207–213. [CrossRef]

49. Valicsek, Z.; Horváth, O. Application of the electronic spectra of porphyrins for analytical purposes: The effects of metal ions and structural distortions. *Microchem. J.* **2013**, *107*, 47–62. [CrossRef]

50. Mosmann, T. Rapid colorimetric assay for cellular growth and survival: Application to proliferation and cytotoxicity assays. *J. Immunol. Methods* **1983**, *65*, 55–63. [CrossRef]

51. Konstantinov, S.M.; Eibl, H.; Berger, M.R. BCR-ABL influences the antileukaemic efficacy of alkylphosphocholines. *Br. J. Haematol.* **1999**, *107*, 365–374. [CrossRef] [PubMed]

52. Mitkova, E.; Ugrinova, I.; Pashev, I.; Pasheva, E.A. The inhibitory effect of HMGB-1 protein on the repair of cisplatin-damaged DNA is accomplished through the acidic domain. *Biochemistry* **2005**, *44*, 5893–5898. [CrossRef] [PubMed]

53. Li, L.; Liu, X.; Glassmann, A.B.; Keating, M.J.; Stros, M.; Plunkett, W.; Yang, L.-Y. Fludarabine triphosphate inhibits nucleotide excision repair of cisplatin-induced DNA adducts in Vitro. *Cancer Res.* **1997**, *57*, 1487–1494. [PubMed]

54. Biade, S.; Sobol, R.W.; Wilson, S.H.; Matsumoto, Y. Impairment of proliferating cell nuclear antigen-dependent apurinic/apyrimidinic site repair on linear DNA. *J. Biol. Chem.* **1998**, *273*, 898–902. [CrossRef] [PubMed]

55. Pasheva, E.A.; Pashev, I.G.; Favre, A. Preferential binding of high mobility group 1 protein to UV-damaged DNA. Role of the COOH-terminal domain. *J. Biol. Chem.* **1998**, *273*, 24730–24736. [CrossRef] [PubMed]

56. Pasheva, E.; Sarov, M.; Bidjekov, K.; Ugrinova, I.; Sang, B.; Linder, H.; Pashev, I.G. In vitro acetylation of HMGB-1 and -2 proteins by CBP: The role of the acidic tail. *Biochemistry* **2004**, *43*, 2935–2940. [CrossRef] [PubMed]

57. Mincheva, N.; Gencheva, G.; Mitewa, M.; Gochev, G.; Mehandjiev, D. Synthesis of new monomeric Pd(III) and dimeric (Pd(III),Pd(II)) complexes with biuret formed in basic medium. *Synth. React. Inorg. Met.-Org. Chem.* **1997**, *27*, 1191–1203. [CrossRef]

58. Čunderlíková, B.; Bjørklund, E.G.; Pettersen, E.O.; Moan, J. pH-Dependent spectral properties of HpIX, TPPS$_{2a}$, *m*THPP and *m*THPC. *Photochem. Photobiol.* **2001**, *74*, 246–252. [CrossRef]

59. Brabec, V.; Kasparkova, J. Platinum-based drugs. In *Metallotherapeutic Drugs and Metal-Based Diagnostic Agents*; Gielen, M., Tiekink, E.R.T., Eds.; Wiley VC: Chichester, UK, 2005; pp. 489–506, ISBN 0-470-86403-6.

60. Boulikas, T.; Vougiouka, M. Cisplatin and platinum drugs at the molecular level. *Oncol. Rep.* **2003**, *10*, 1663–1682. [CrossRef] [PubMed]

61. Brabec, V.; Kasparkova, J. Modification of DNA by platinum complexes. Relation to resistance of tumors to platinum antitumor drugs. *Drug Resist. Update* **2005**, *8*, 131–146. [CrossRef] [PubMed]

62. Woźniak, K.; Błasiak, J. Recognition and repair of DNA-cisplatin adducts. *Acta Biochim. Pol.* **2002**, *49*, 583–596. [PubMed]

63. Desoize, B. Metals and metal compounds in cancer treatment. *Anticancer Res.* **2004**, *24*, 1529–1544. [PubMed]

64. Chabner, B.A.; Amrein, P.C.; Druker, B.J.; Michaelson, M.D.; Mitsiades, C.S.; Goss, P.E.; Ryan, D.P.; Ramachandra, S.; Richardson, P.G.; Supko, J.G.; et al. Chemotherapy of neoplastic disease. In *Goodman & Gilman's the Pharmacological Basis of Therapeutics*, 11th ed.; Brunton, L.L., Lazo, J.S., Parker, K.L., Eds.; McGraw Hill: New York, NY, USA, 2006; pp. 1315–1403, ISBN 0-07-160891-5.

International Journal of
Molecular Sciences

MDPI

Article

Optical Properties of Silver-Mediated DNA from Molecular Dynamics and Time Dependent Density Functional Theory

Esko Makkonen [1], Patrick Rinke [1], Olga Lopez-Acevedo [1,2,*] and Xi Chen [1,*]

[1] Department of Applied Physics, Aalto University, P.O. Box 11100, 00076 Aalto, Finland;
    esko.makkonen@aalto.fi (E.M.); patrick.rinke@aalto.fi (P.R.)
[2] Grupo de Física Atómica y Molecular, Instituto de Física, Facultad de Ciencias Exactas y Naturales,
    Universidad de Antioquia UdeA, Calle 70 No. 52-21, 050010 Medellín, Colombia
*   Correspondence: olga.lopeza@udea.edu.co (O.L.-A.); xi.6.chen@aalto.fi (X.C.)

Received: 18 June 2018; Accepted: 20 July 2018; Published: 9 August 2018

Abstract: We report a combined quantum mechanics/molecular mechanics (QM/MM) molecular dynamics and time-dependent density functional (TDDFT) study of metal-mediated deoxyribonucleic acid (M-DNA) nanostructures. For the Ag^+-mediated guanine tetramer, we found the maug-cc-pvdz basis set to be sufficient for calculating electronic circular dichroism (ECD) spectra. Our calculations further show that the B3LYP, CAM-B3LYP, B3LYP*, and PBE exchange-correlation functionals are all able to predict negative peaks in the measured ECD spectra within a 20 nm range. However, a spurious positive peak is present in the CAM-B3LYP ECD spectra. We trace the origins of this spurious peak and find that is likely due to the sensitivity of silver atoms to the amount of Hartree–Fock exchange in the exchange-correlation functional. Our presented approach provides guidance for future computational investigations of other Ag^+-mediated DNA species.

Keywords: DNA; silver; ECD; QM/MM; TDDFT

1. Introduction

The nucleobases inside highly polymorphic deoxyribonucleic acid (DNA) molecules contain the essential genetic information for the creation and the functional properties of living organisms. In addition to hydrogen bonding, DNA is also capable of forming metallo base pairs in the presence of strongly bound metal ions. Interest in such metal-mediated DNA (M-DNA) nanostructures has rapidly grown due to their high potential for developing new materials for a wide range of biomedical and technological applications, such as bioimaging [1–4], metal ion detection [5–7], DNA sequencing [8], logic gates, and molecular devices [9]. Among the studied metal ions, Ag^+ is particularly interesting, because it has low toxicity to humans and binds exclusively to the base pairs rather than the sugar–phosphate backbone. C-Ag^+-C, G-Ag^+-G, T-Ag^+-T, A-Ag^+-T (where A = adenine, C = cytosine, G = guanine, and T = thymine) and Ag^+- mediated artificial DNA pairs have been reported [10–12]. In recent years, DNA-stabilized color-tunable fluorescent Ag clusters have also been synthesized, which contain both neutral Ag atoms and Ag^+ ions, and are stabilized by Ag^+-DNA interactions. They hold great potential for chemical and biochemical sensor applications and have attracted even more attention to the study of metal–DNA interactions [3,13].

Although the structures of some specific M-DNAs have been recently resolved via X-ray diffraction analysis [14–16], the atomic structures of M-DNAs are generally very difficult to obtain with X-ray crystallography, because M-DNA rarely crystallizes. Therefore, computational chemistry methods are important tools for deciphering the atomic structure of M-DNA and for interpreting the experimental spectroscopic data. In this context, electronic circular dichroism (ECD) spectroscopy is a versatile method to obtain structure information due to its high sensitivity to small structural changes in chiral optical active molecules, and a broad operational window in various solvent conditions. By comparing computed with measured ECD spectra, the structure of M-DNA can be inferred. This approach has been used for mapping the conformal properties of various DNA-based systems including M-DNAs [17–19]. Additionally, ECD provides a reliable benchmark method for validating computational approaches, due to its structural sensitivity.

However, modeling the ECD spectra of M-DNAs remains a difficult task. Since the prepared materials exist mostly in aqueous solution, the effect of the environment on the structures and properties have to be taken into account. This leads to an increased system size, which is prohibitive for a full quantum level calculation, and thus requires multi-scale strategies, i.e., hybrid quantum mechanics/molecular mechanics (QM/MM) methods. In order to compare with experiments, the structures for simulating ECD need to be chosen carefully from QM/MM simulations, bearing in mind that they exhibit the key structural properties of the system but are still computationally affordable. Last, as always with density functional theory (DFT), the used exchange-correlation functional has to be suitable for the system under study. It is well known that hybrid or long-range corrected hybrid functionals are usually more accurate for studying organic molecule, whereas the GGA functionals are commonly used for bulk metal and solid state systems. Currently, it is not fully conclusive how to choose the optimal exchange-correlation functional for M-DNA systems.

In this work, we chose to study Ag^+-mediated guanine tetramer, whose ECD has been reported recently [20] to systematically study how the basis sets, the exchange-correlation functionals, and the implicit water affect simulations of the ECD. This work will provide some useful knowledge to choose suitable computational methods for studying silver-DNA systems and other M-DNAs.

2. Results and Discussion

2.1. QM/MM Molecular Dynamics and the Atomic Structures of the Ag^+-Mediated Guanine Tetramer

To unveil the atomic structure of the Ag^+-mediated guanine duplex in water, we have previously performed QM/MM molecular dynamics calculations for the smallest Ag^+-mediated guanine duplex G_2-Ag_2^+-G_2, namely the Ag^+-mediated guanine tetramer [21]. Silver and DNA atoms were treated at the DFT level. Solvation was modeled with a 4 nm box of classical water molecules and the Amber force field [22]. The calculations were carried out with the CP2K code [23], and the Perdew–Burke–Ernzerhof (PBE) exchange and correlation potential [24]. Grimme's D3 dispersive corrections were used to account for long-range van der Waals interactions [25]. Before the QM/MM simulations, the system was first equilibrated at 300 K with a classical NVT run of 2 ns, using the Amber force field. Then the QM/MM simulations were performed for 21 ps with a time-step of 0.5 ps. In the QM/MM simulations, van de Waals interactions between the QM and MM regions were approximated by classical force fields. Electrostatic interactions between the two subsystems were accounted for using an embedding approach, where point charges located at the water molecules (MM region) can polarize the Ag^+-mediated DNA (QM region). By comparing the root mean square distance of the heavy atoms to the initial structure positions, we can conclude that the system reached an equilibrium regime after 16ps. Therefore, we used only the last 5 ps of the trajectory for analysis. Further details of these simulations can be found in Reference [21].

In this work we adopted the structures for the ECD analysis from our previous QM/MM study. We briefly recap the structural findings from Reference [21]. In brief, Ag^+ ions bind to the N atoms in the Hoogsteen region (Figure 1a) and the DNA forms a left-handed helix. The Ag–Ag distance oscillates around an average value of 3.45 Å with a standard deviation of 0.27 Å. A typical structure of an Ag^+-mediated tetramer is shown in Figure 1b.

(a)   (b)

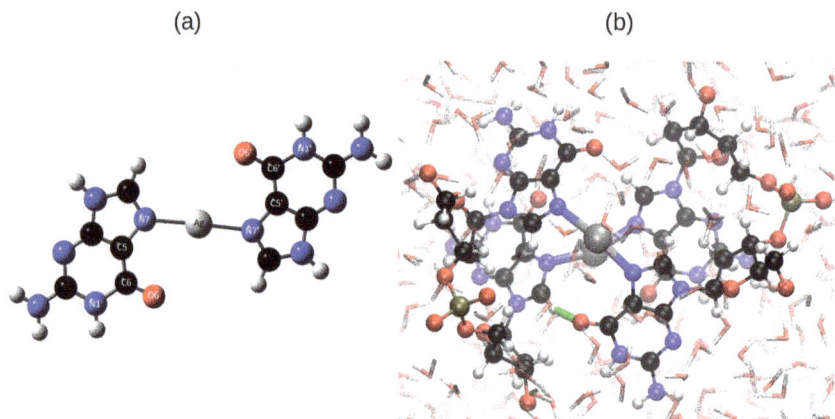

Figure 1. (**a**) Ag$^+$ binds to N7 and N7' in the Hoogsteen region; (**b**) One typical structure of G$_2$-Ag$_2^+$-G$_2$ from the final 0.5 ps of the hybrid quantum mechanics/molecular mechanics (QM/MM) simulation. The structure was taken from our previous work [21] and replotted here. The interplanar H-bond is highlighted with a green line. Atomic color code: Ag: silver, N: blue, C: black, O: red, P: Olive and H: white.

2.2. Time-Dependent Density Functional Theory Study for the Optical Properties of the Ag$^+$-Mediated Guanine Tetramer

Previously we have simulated the ECD of G$_2$-Ag$_2^+$-G$_2$ by linear-response time-dependent density functional theory (TDDFT) implemented in GPAW [26,27]. GPAW is a DFT code based on the projector-augmented wave (PAW) method and the wave functions are expanded on uniform real-space grids. We used the LB94 [28] exchange-correlation functional in that simulation. Although the simulated average ECD and the measured ECD of G$_2$-Ag$_2^+$-G$_2$ were in satisfactory agreement [21], it is worthwhile to check how the basis sets and the exchange-correlation functional affect the ECD simulations when we switch to the Gaussian16 code [29]. One particular question we are interested in is, can hybrid functionals improve the agreement between the calculated and the measured ECD?

In this work, we employed the Gaussian16 code [29] to simulate the ECD spectra. The details can be found in the Materials and Methods section.

2.2.1. Experimental ECD Spectra of Ag$^+$-Mediated Guanine Species

The experimental ECD spectra for three Ag$^+$-mediated guanine species obtained from Reference [20] are shown in Figure 2. The spectra of G$_6$-Ag$_6^+$-G$_6$ and G$_{20}$-Ag$_{20}^+$-G$_{20}$ exhibit similar features, but the G$_2$-Ag$_2^+$-G$_2$ spectrum differs. This is possibly due to aggregation induced by the Ag$^+$ ions, which prevents a comparison of the intensity between the predicted and the experimental ECD [20,30,31]. However, G$_2$-Ag$_2^+$-G$_2$ presents a good model system to study the ECD of Ag$^+$-mediated guanine duplexes, especially in terms of the negative peak locations and the absence of any positive peaks.

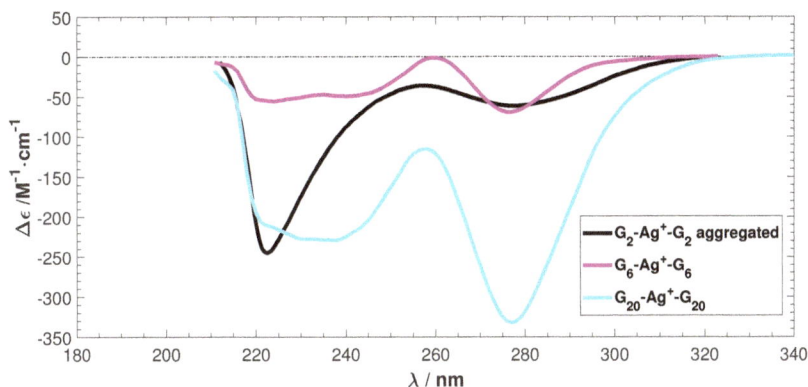

Figure 2. The experimental electronic circular dichroism (ECD) spectra of Ag^+ stabilized G_2, G_6, and G_{20}. $\Delta\epsilon$ is differential molar extinction coefficients $\Delta\epsilon = \epsilon_L - \epsilon_R$, λ is the wavelength. Data reproduced with permission from Swasey, S. and Gwinn, E. *New J. Phys.* **2016**, 18, 045008-045025. [20].

2.2.2. Tests for Basis Set Convergence

We first tested the basis set convergence on a single sampled structure. Two popular hybrid exchange-correlation functionals, B3LYP [32,33] and the Coulomb-attenuated B3LYP (CAM-B3LYP) [34], were used for the tests. A single compute node with 20 Haswell cores and 40 GB of shared memory was used. To assure convergence in the range of 180–360 nm, 300 and 150 excited singlet states were used in the B3LYP and CAM-B3LYP calculations, respectively.

Figure 3 shows the effects of employed basis sets for C, H, N, and O atoms and the related cost for calculating the ECD spectrum of a single sampled structure. For Ag, the LANL2DZ/ECP basis set was used in all calculations. Figure 3 shows that B3LYP is more sensitive to the basis sets when diffusion functions are added. For both functionals, basis set convergence is reached by the minimally augmented Dunning basis set, maug-cc-pvdz. It gives practically the same ECD spectra as the aug-cc-pvdz and 6-311++G**, which have been previously used for M-DNAs and pure base pair systems [19,35,36] with good performance. The capability of maug-cc-pvdz to match aug-cc-pvdz has been shown previously for other systems and similar to this study, doing so with much lower computational cost [37]. Based on the results presented in Figure 3, maug-cc-pvdz was chosen for the remainder of this work.

The two functionals give significantly different ECD spectra. CAM-B3LYP exhibits two peculiar positive peaks at 204 and 234 nm that are absent in the B3LYP spectrum. In the following, we will trace the origin of this difference and try to find the best functionals for studying the optical properties of Ag^+-DNA.

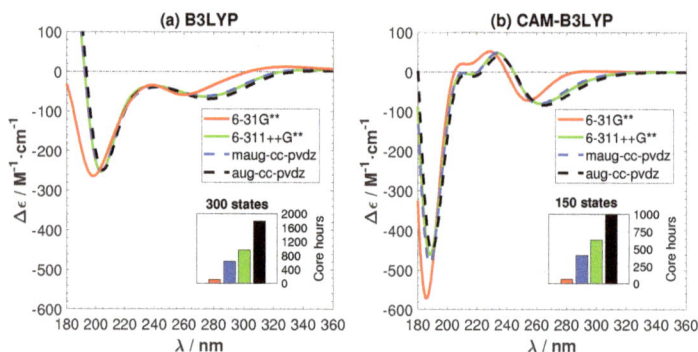

Figure 3. (main figures) The effects of basis sets and exchange correlation-functional on the ECD spectrum of a single snapshot structure and (inset figures) the related cpu time costs. (**a**) B3LYP; (**b**) CAM-B3LYP.

2.2.3. Calculated Average ECD and Comparison to the Experimental ECD

Previous work [21] has highlighted the importance of sampling molecular dynamics trajectories and comparing averaged ECD spectra to experiment. Typically ten sampled structures were found to be adequate for G_2-Ag_2^+-G_2 in water. Therefore, in this section, we will discuss the calculated average ECD spectra and compare them to the experimental ECD in Figure 2.

Figure 4 shows the calculated average and individual ECD with the B3LYP and CAM-B3LYP functionals, respectively. In order to show that ten samples are enough to capture the main features of the ECD, we have also added the 95% confidence interval of the average ECD in the figure. The negative peaks predicted by B3YLP lie around 200 and 270 nm, while the CAM-B3YLP peaks are around 190 and 260 nm. They are both shifted towards the short wavelength region compared to experiment (peaks around 220 and 280 nm). The B3LYP ECD is in good agreement with the experimental ECD spectrum. The main difference is a small shoulder (<20 $M^{-1}cm^{-1}$) between 220 and 240 nm. This shoulder becomes much more pronounced in CAM-B3YLP (about 50 $M^{-1}cm^{-1}$). Such a shoulder, or in fact any positive peak, is not present in the experimental ECD spectra. This poses the question of whether CAM-B3YLP predicts artifacts for our system. To make sure the peculiar positive feature in the CAM-B3LYP spectra is not due to the solution, we have calculated the ECD in an implicit solution environment. The positive peak remains. More details can be found in the Appendix A.

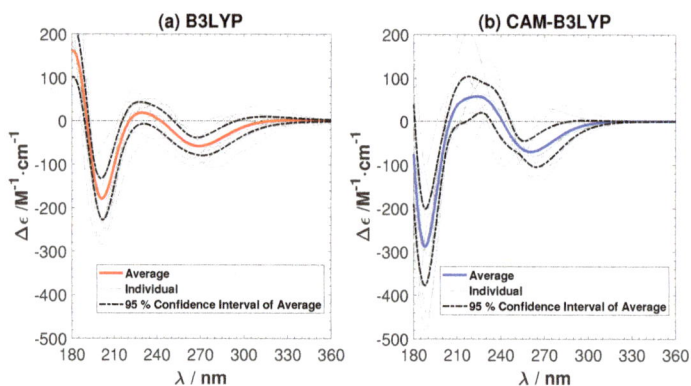

Figure 4. The average and individual ECD spectra of ten sampled structures with B3LYP (**a**) and CAM-B3LYP (**b**). Spectra are not shifted, but presented as calculated.

In the following, we will investigate different reasons for the difference between B3LYP and CAM-B3LYP and for CAM-B3LYP's erroneous behavior.

2.2.4. Amount of Hartree–Fock Exchange

The amount of Hartree–Fock (HF) exchange in hybrid functionals is known to be crucial in studying transition metal complexes [38,39], and lowering the amount of HF exchange in B3LYP from 20 % to 15 % has proven to be a valid modification to correctly obtain energetics for proton and electron transfer reactions in comparison with experimental values [38–41]. We have also tested this modified B3LYP, called B3LYP*, for the Ag^+-mediated guanine tetramer, along with a pure GGA functional PBE, which has no HF exchange.

The averaged ECD spectra from ten sampled B3LYP* and PBE structures are shown in Figure 5. The ECD of the individual samples and the 95% confidence interval of the average ECD are given in the same figure. The amount of HF exchange has a clear effect, since the positive shoulder has now vanished. Both B3LYP* and PBE successfully predict the two negative peaks in the experimental ECD. B3LYP* gives a better agreement with the intensity of the peaks in the experimental ECD of G_2-Ag_2^+-G_2, but PBE predicts the positions of the peaks more accurately. There are spurious peaks in the individual ECD spectra predicted by PBE in the long wavelength region. However they cancel each other out in the averaged ECD spectrum. Comparing the ECD spectra from B3YLP, CAM-B3YLP, B3YLP*, and PBE, we conclude that it is important to adjust the amount of Hartree–Fock exchange in exchange-correlation functional, in order to predict the ECD of G_2-Ag^+-G_2 accurately. The sensitivity to the amount of exact exchange might be a common issue for other M-DNA systems, especially for the Ag^+-mediated guanine-rich DNA systems.

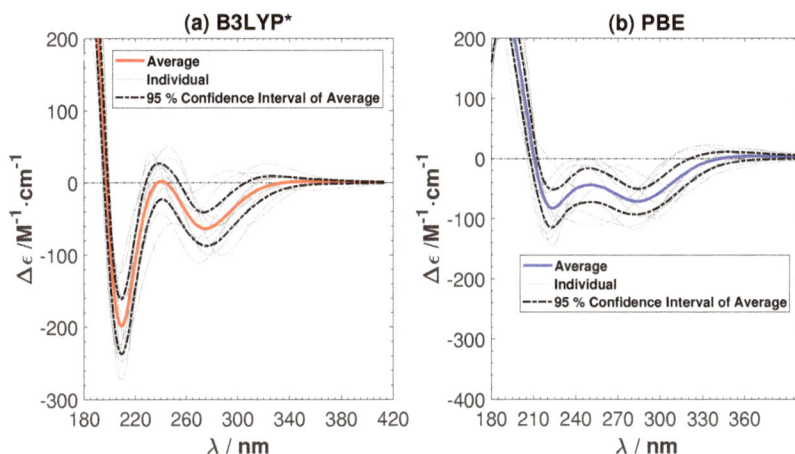

Figure 5. The average and individual ECD spectra of ten sampled structures with (**a**) the B3LYP* (15% Hartree–Fock (HF) exchange) and (**b**) the Perdew–Burke–Ernzerhof (PBE) functionals. No shiftings of the calculated spectra are done.

2.2.5. Effect of the Long-Range Correction in CAM-B3LYP

B3LYP adequately predicts ECD spectra for many organic molecules [42,43]. However, CAM-B3LYP is a functional built to overcome the deficiencies of B3LYP and to correctly describe charge transfer, local excitations, and Rydberg excitations [34]. It has been assessed for a broad set of small main group and organic molecules [44], and has been demonstrated to outperform B3LYP for chiral alkenes [45], chiral aromatic nitro compounds [46], and metal-porphyrin complexes [47]. It is very interesting to understand why for our Ag^+-guanine system, CAM-B3YLP gives less accurate ECD spectra than

B3LYP. For this purpose, we first tested the two functionals on two DNA systems without metal ions: one is a C_2-G_2 tetramer taken from a reported B-DNA structure (9BNA in the protein data bank), and another is a G_2-G_2 tetramer taken from the same sample used in Figure 3, in which the Ag atoms were deleted.

Figure 6 shows the experimental ECD of a B-DNA [48,49] (d(m5C-G-C-G-m5C-G)) and the calculated ECD spectra of the C_2-G_2 system using B3LYP and CAM-B3LYP. As for the Ag^+-mediated guanine tetramer we removed the sugar–phosphate backbone and optimized the replacement hydrogen atoms at the B3LYP/6-31G** level. B3LYP predicts the correct positions and intensity of the peaks in the experimental ECD. On the other hand, CAM-B3LYP also gives overall correct features, but with a shift of roughly 20 nm towards the lower wavelength region. A shift of the same order (roughly 20 nm) was also obtained between CAM-B3LYP and B3LYP for G_2-G_2 as shown in Figure 7. Although CAM-B3LYP predicts more intensive peaks with shifts to lower wavelengths, these two functionals give qualitatively the same ECD features for the two DNA structures without metal ions.

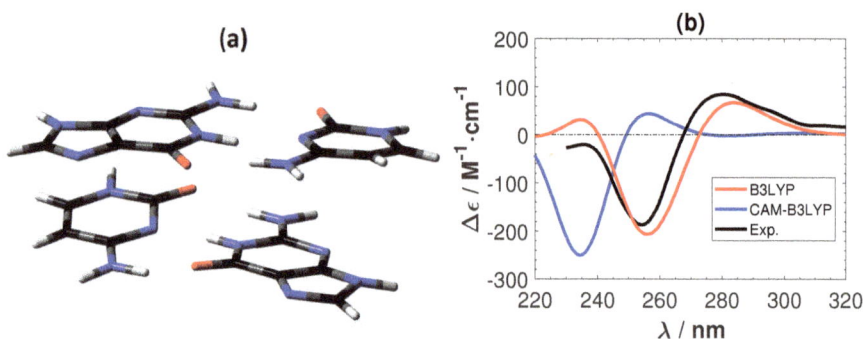

Figure 6. (a) The structure of a plain B-DNA tetramer. The color of atoms : O: red, C: black, N: blue and H: white. (b) The corresponding calculated and experimental [48,49] ECD spectra. No shifting of the calculated spectra are done.

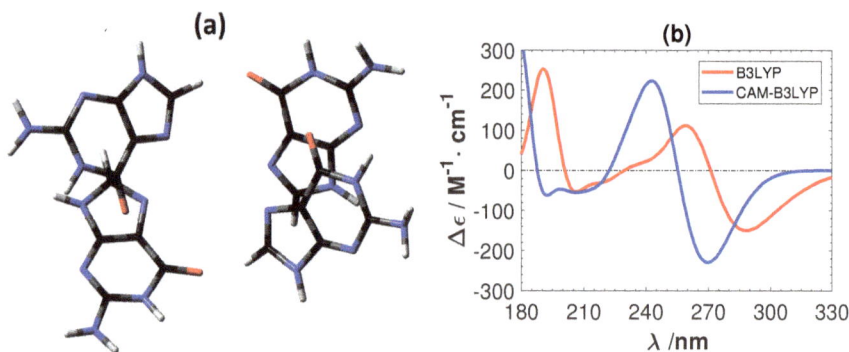

Figure 7. (a) The structure of a plain guanine tetramer. The color of atoms : O: red, C: black, N: blue and H: white. (b) The corresponding calculated ECD spectra. No shifting of the calculated spectra are done.

To further investigate the difference between B3LYP and CAM-B3LYP in our system, we calculated the ECD spectra of a sampled structure by gradually changing CAM-B3LYP to B3LYP. This was done by changing the short-range parameter α and the long-range parameter β in the definition of Coulomb-attenuation range r_{12} [34]:

$$\frac{1}{r_{12}} = \frac{1 - [\alpha + \beta \cdot \mathrm{erf}(\mu r_{12})]}{r_{12}} + \frac{\alpha + \beta \cdot \mathrm{erf}(\mu r_{12})}{r_{12}}. \tag{1}$$

The HF exchange has a weight of α at $r_{12} = 0$ and $\alpha + \beta$ at $r_{12} = \infty$. Correspondingly, DFT exchange is $1 - \alpha$ at $r_{12} = 0$ and $1 - (\alpha + \beta)$ at $r_{12} = \infty$. The default parameter values for CAM-B3LYP are $\alpha = 0.19, \beta = 0.46$, and $\mu = 0.33$. The parameter μ sets the midpoint behavior, and was kept at the default value of 0.33. With parameters $\alpha = 0.2$ and $\beta = 0$, B3LYP is obtained.

The results are shown in Figures 8 and 9. Figure 8 shows that we achieve a smooth transformation from CAM-B3LYP to B3LYP. The absolute intensity of the negative peak in the short wavelength region reduces and the ECD spectrum shifts towards the long wavelength direction. The positive features observed in CAM-B3LYP smoothly vanish. Figure 9 also shows the transformation, together with the involved rotatory strengths of the singlet excitations. The positive peak emerges from a collection of high intensity excitations, which are gradually shifting towards the higher wavelength region. From these tests, we can conclude that applying the long-range corrected hybrid function CAM-B3LYP to Ag atoms cause the unnatural features in the calculated ECD.

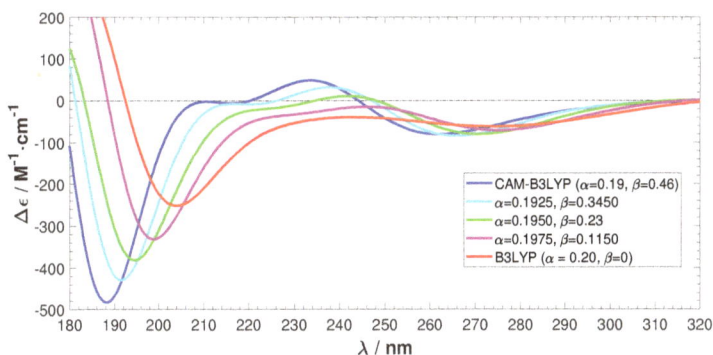

Figure 8. ECD spectra transformation from CAM-B3LYP/maug-cc-pvdz to B3LYP/maug-cc-pvdz.

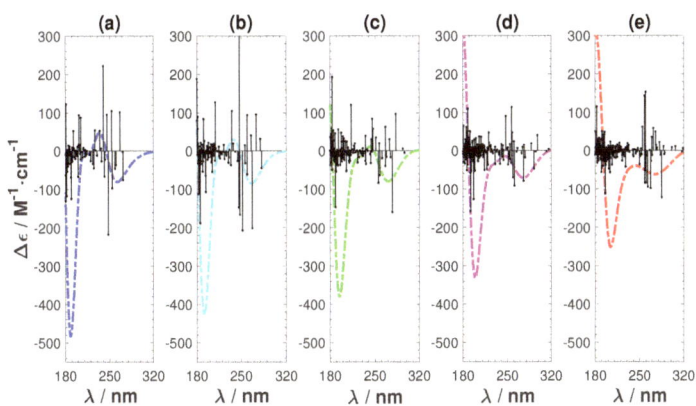

Figure 9. Transformation from CAM-B3LYP to B3LYP by changing the short- and long-range parameters. (**a**) CAM-B3LYP ($\alpha = 0.19, \beta = 0.46$); (**b**) $\alpha = 0.1925, \beta = 0.3450$; (**c**) $\alpha = 0.1950, \beta = 0.23$; (**d**) $\alpha = 0.1975, \beta = 0.1150$; and (**e**) B3LYP ($\alpha = 0.2, \beta = 0$).

Finally, to show the capability of our methodology to predict ECD spectra, our best predictions without any fitting are shown alongside experimental spectra in Figure 10. The two negative peaks are predicted within 20 nm.

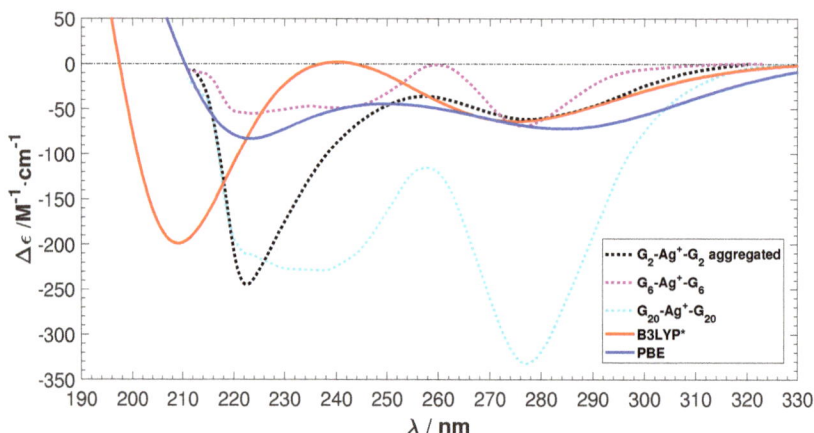

Figure 10. The experimental ECD [20] spectra shown along with our best prediction by B3LYP*. No shiftings of spectra are done.

3. Materials and Methods

In this work, we employed the Gaussian16 code [29] which uses Gaussian type local basis sets to simulate the ECD spectra. The sugar–phosphate backbones were replaced by hydrogen atoms to make the computational cost feasible. We argue that this is reasonable, because the sugar–phosphate backbone is not the main interest of our study and does not affect the optical properties in the region of interest (long distances from the center Ag^+-ions and transitions at high energies [50]). The replacement hydrogen atoms were then relaxed at the B3LYP/6-31G** level, while keeping the rest of the atoms frozen.

Singlet transitions were included in excited state calculation so that the high-energy wavelength region up to 170 nm was covered (to assure convergence up to 180 nm) for the Ag^+-mediated tetramer and guanine tetramer. The first 300, 500, 700, and 150 transitions were included for B3LYP, B3LYP*, PBE, and CAM-B3LYP calculations, respectively. For B-DNA, the high-energy wavelength region, up to 210 nm, was covered by the first 60 and 80 transitions for CAM-B3LYP and B3LYP, respectively.

The following relations were used in the ECD spectra simulations [51–53]:

$$\Delta\epsilon = 4a \sum_n R_n E_n \sigma_n(E) \tag{2}$$

$$a = \frac{4\pi N_A}{3\ln(10)10^3} \frac{2\pi}{hc} \tag{3}$$

$$\sigma_n(E) = \frac{1}{\sqrt{2\pi}\sigma} \exp\left(-\frac{1}{2\sigma^2}(E - E_n)^2\right), \tag{4}$$

where $\Delta\epsilon$ is molar circular dichroism (in $M^{-1}cm^{-1}$), N_A is Avogagro's constant (in mol^{-1}), h is Planck's constant (in Js), c is the speed of light (in cm s^{-1}), R_n is length-gauge rotatory strength (in 10^{-40} cgs), E is the energy of the incident light (in eV), E_n is the excitation energy to state n (in eV), and σ is the exponential half-width in Gaussian convolution (here 0.3 eV was used for the Ag^+-mediated tetramer and 0.2 eV for the B-DNA and plain guanine tetramers).

Int. J. Mol. Sci. **2018**, *19*, 2346

4. Conclusions

In summary, we have used TDDFT and QM/MM to study the ECD spectra of the Ag^+-mediated guanine tetramer. We have discussed how the basis sets and the exchange-correlation functional affect the computed ECD spectra. For the basis sets, convergence is achieved at the maug-cc-pvdz level. The tested functionals CAM-B3LYP, B3LYP, B3LYP* (15% HF), and PBE all successfully predict two negative peaks within a 20 nm error. B3LYP* and PBE give the best agreement of the measured ECD. Conversely, CAM-B3LYP produces a spurious positive feature, which is not observed in experiment. Our study highlights the importance of the exchange-correlation functional. In future ECD studies of metal enhanced DNA in water, we recommend to start with the PBE or B3LYP* functionals. Due to the much lower computational cost and satisfactory accuracy of PBE, it is a better choice for large M-DNA systems.

Author Contributions: O.L.-A., P.R., and X.C. conceived and designed the original simulations; E.M. performed all of the simulations under the supervision of P.R. and X.C.; E.M. and X.C. wrote the first version of the paper; All the authors contributed to the analysis of the results and to finalize the manuscript.

Acknowledgments: We acknowledge the computing resources by the CSC-IT Center for Science and the Aalto Science-IT project. This work was supported by the Academy of Finland, Projects 308647, 314298, 279240 and 312556.

Conflicts of Interest: The authors declare no conflict of interest.

Appendix A. Solvent Effect on the CAM-B3LYP Calculated ECD

To investigate if the peculiar positive feature in the CAM-B3LYP spectra is due to the solution, we calculate the ECD in a solvent environment. We modeled water implicitly by a polarizable continuum model (PCM) [54]. Here we used a linear-response procedure, where the solvent effect on the self-consistent field (SCF), density of solute, is computed by adding the required terms to the TDDFT equations [55]. Figure A1 shows the effects of the implicit solvent model on the average ECD of ten sampled structures. The positive peak remains.

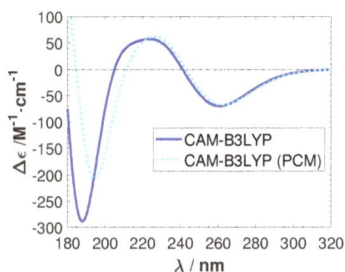

Figure A1. The average ECD spectra of ten sampled structures with CAM-B3LYP in the gas phase and in a polarizable continuum model (PCM) water environment. No calculated spectra were shifted.

References

1.   Copp, S.; Schultz, D.; Swasey, S.; Gwinn, E. Atomically Precise Arrays of Fluorescent Silver Clusters: A Modular Approach for Metal Cluster Photonics on DNA Nanostructures. *ACS Nano* **2008**, *10*, 2303–2310. [CrossRef] [PubMed]

2.   Yu, J.; Choi, S.; Richards, C.; Antoku, Y.; Dickson, R. Live cell surface labeling with fluorescent Ag nanocluster composites. *Photochem. Photobiol.* **2008**, *84*, 1435–1439. [CrossRef] [PubMed]

3.   Sharma, J.; Yeh, H.C.; Yoo, H.; Werner, J.; Martinez, J. A complementary palette of fluorescent silver nanoclusters. *Chem. Commun.* **2010**, *46*, 3280–3282. [CrossRef] [PubMed]

4.   Gwinn, E.; Schultz, D.; Copp, S.; Swasey, S. DNA-Protected Silver Clusters for Nanophotonics. *Nanomaterials* **2015**, *5*, 180–207. [CrossRef] [PubMed]

5.   Ono, A.; Togashi, H. Highly Selective Oligonucleotide-Based Sensor for Mercury(II) in Aqueous Solutions. *Angew. Chem. Int. Ed.* **2004**, *43*, 4300–4302. [CrossRef] [PubMed]

6. Xiang, Y.; Lu, Y. DNA as Sensors and Imaging Agents for Metal Ions. *Inorg. Chem.* **2015**, *53*, 1925–1942. [CrossRef] [PubMed]

7. Zhou, W.; Saran, R.; Liu, J. Metal Sensing by DNA. *Chem. Rev.* **2017**, *117*, 8272–8325. [CrossRef] [PubMed]

8. Obliosca, J.M.; Liu, C.; Batson, R.A.; Babin, M.C.; Werner, J.; Yeh, H.C. DNA/RNA Detection Using DNA-Templated Few Atom Silver Nanoclusters. *Biosensors* **2013**, *3*, 185–200. [CrossRef] [PubMed]

9. Krishnan, Y.; Simmel, F. Nucleic Acid Based Molecular Devices. *Angew. Chem. Int. Ed.* **2011**, *50*, 3124–3156. [CrossRef] [PubMed]

10. Ono, A.; Cao, S.; Togashi, H.; Tashiro, M.; Fujimoto, T.; Machinami, T.; Oda, S.; Miyake, Y.; Okamoto, I.; Tanaka, Y. Specific interactions between silver(I) ions and cytosine-cytosine pairs in DNA duplexes. *Chem. Commun.* **2008**, *0*, 4825–4827. [CrossRef] [PubMed]

11. Swasey, S.M.; Leal, L.E.; Lopez-Acevedo, O.; Pavlovich, J.; Gwinn, E.G. Silver(I) as DNA Glue: Ag^+-Mediated Guanine Pairing Revealed by Removing Watson-Crick Constraints. *Sci. Rep.* **2015**, *5*, 10163. [CrossRef] [PubMed]

12. Kondo, J.; Tada, Y.; Dairaku, T.; Hattori, Y.; Saneyoshi, H.; Ono, A.; Tanaka, Y. A metallo-DNA nanowire with uninterrupted one-dimensional silver array. *Nat. Chem.* **2017**, *9*, 956–960. [CrossRef] [PubMed]

13. Copp, S.; Schultz, D.; Swasey, S.; Pavlovich, J.; Debord, M.; Chiu, A.; Olsson, K.; Gwinn, E. Magic Numbers in DNA-Stabilized Fluorescent Silver Clusters Lead to Magic Colors. *J. Phys. Chem. Lett.* **2014**, *5*, 959–963. [CrossRef] [PubMed]

14. Kondo, J.; Yamada, T.; Hirose, C.; Okamoto, I.; Tanaka, Y.; Ono, A. Crystal Structure of Metallo DNA Duplex Containing Consecutive Watson-Crick-like T-Hg(II)-T Base Pairs. *Angew. Chem. Int. Ed.* **2014**, *53*, 2385–2388. [CrossRef] [PubMed]

15. Tanaka, Y.; Kondo, J.; Sychrovsky, V.; Sebera, J.; Dairaku, T.; Saneyoshi, H.; Urata, H.; Torigoe, H.; Ono, A. Structures, physicochemical properties, and applications of T-HgII-T, C-AgI-C, and other metallo-base-pairs. *Chem. Commun.* **2015**, *51*, 17343–17360. [CrossRef] [PubMed]

16. Kondo, J.; Sugawara, T.; Saneyoshi, H.; Ono, A. Crystal structure of a DNA duplex containing four Ag(I) ions in consecutive dinuclear Ag(I)-mediated base pairs: 4-thiothymine-2Ag(I)-4-thiothymine. *Chem. Commun.* **2017**, *53*, 11747–11750. [CrossRef] [PubMed]

17. Kypr, J.; Kejnovská, I.; Renčiuk, D.; Vorlíčková, M. Circular dichroism and conformational polymorphism of DNA. *Nucleic Acids Res.* **2009**, *37*, 1713–1725. [CrossRef] [PubMed]

18. Vorlíčková, M.; Kejnovská, I.; Bednářová, K.; Renčiuk, D.; Kypr, J. Circular Dichroism Spectroscopy of DNA: From Duplexes to Quadruplexes. *Chirality* **2012**, *24*, 691–698. [CrossRef] [PubMed]

19. Espinosa Leal, L.A.; Karpenko, A.; Swasey, S.; Gwinn, E.G.; Rojas-Cervellera, V.; Rovira, C.; Lopez-Acevedo, O. The Role of Hydrogen Bonds in the Stabilization of Silver-Mediated Cytosine Tetramers. *J. Phys. Chem. Lett.* **2015**, *6*, 4061–4066. [CrossRef] [PubMed]

20. Swasey, S.; Gwinn, E. Silver-mediated base pairings: Towards dynamic DNA nanostructures with enhanced chemical and thermal stability. *New J. Phys.* **2016**, *18*, 045008. [CrossRef]

21. Chen, X.; Makkonen, E.; Golze, D.; Lopez-Acevedo, O. Silver-Stabilized Guanine Duplexes: Structural and Optical Properties. *J. Phys. Chem. Lett.* **2018**, doi:10.1021/acs.jpclett.8b01908. [CrossRef]

22. Ponder, J.W.; Case, D.A. Force Fields for Protein Simulations. *Adv. Prot. Chem.* **2003**, *66*, 27–85.

23. Hutter, J.; Iannuzzi, M.; Schiffmann, F.; VandeVondele, J. cp2k: Atomistic simulations of condensed matter systems. *WIREs Comput. Mol. Sci.* **2014**, *4*, 15–25. [CrossRef]

24. Perdew, J.P.; Burke, K.; Ernzerhof, M. Generalized Gradient Approximation Made Simple. *Phys. Rev. Lett.* **1996**, *77*, 3865–3868. [CrossRef] [PubMed]

25. Grimme, S.; Antony, J.; Ehrlich, S.; Krieg, H. A consistent and accurate *ab initio* parametrization of density functional dispersion correction (DFT-D) for the 94 elements H-Pu. *J. Chem. Phys.* **2010**, *132*, 154104. [CrossRef] [PubMed]

26. Enkovaara, J.; Rostgaard, C.; Mortensen, J.; Chen, J.; Dułak, M.; Ferrighi, L.; Gavnholt, J.; Glinsvad, C.; Haikola, V.; Hansen, H.; et al. Electronic structure calculations with GPAW: A real-space implementation of the projector augmented-wave method. *J. Phys. Condens. Matter* **2010**, *22*, 253202. [CrossRef] [PubMed]

27. Walter, M.; Häkkinen, H.; Lehtovaara, L.; Puska, M.; Enkovaara, J.; Rostgaard, C.; Mortensen, J. Time-dependent density-functional theory in the projector augmented-wave method. *J. Chem. Phys.* **2008**, *128*, 244101. [CrossRef] [PubMed]

28. Van Leeuwen, R.; Baerends, E.J. Exchange-correlation potential with correct asymptotic behavior. *Phys. Rev. A* **1994**, *49*, 2421–2431. [CrossRef] [PubMed]

29. Frisch, M.J.; Trucks, G.W.; Schlegel, H.B.; Scuseria, G.E.; Robb, M.A.; Cheeseman, J.R.; Scalmani, G.; Barone, V.; Petersson, G.A.; Nakatsuji, H.; et al. *Gaussian 16, Revision A.03*; Gaussian, Inc.: Wallingford, CT, USA, 2014.

30. Valery, A.; Petr, B. Circular dichroism enhancement in large DNA aggregates simulated by a generalized oscillator model. *J. Comput. Chem.* **2008**, *29*, 2693–2703.

31. Wolf, B.; Berman, S.; Hanlon, S. Structural transitions of calf thymus DNA in concentrated lithium chloride solutions. *Biochemistry* **1977**, *16*, 3655–3662. [CrossRef] [PubMed]

32. Becke, A.D. Density-functional thermochemistry. III. The role of exact exchange. *J. Chem. Phys.* **1993**, *98*, 5648–5652. [CrossRef]

33. Cohen, A.J.; Handy, N.C. Dynamic correlation. *Mol. Phys.* **2001**, *99*, 607–615. [CrossRef]

34. Yanai, T.; Tew, D.; Handy, N. A new hybrid exchange-correlation functional using the Coulomb-attenuating method (CAM-B3LYP). *Chem. Phys. Lett.* **2004**, *393*, 51–57. [CrossRef]

35. Üngördü, A.; Tezer, N. The solvent (water) and metal effects on HOMO-LUMO gaps of guanine base pair: A computational study. *J. Mol. Graph. Model.* **2017**, *74*, 265–272. [CrossRef] [PubMed]

36. Di Meo, F.; Pedersen, M.N.; Rubio-Magnieto, J.; Surin, M.; Linares, M.; Norman, P. DNA Electronic Circular Dichroism on the Inter-Base Pair Scale: An Experimental–Theoretical Case Study of the AT Homo-Oligonucleotide. *J. Phys. Chem. Lett.* **2015**, *6*, 355–359. [CrossRef] [PubMed]

37. Papajak, E.; Zheng, J.; Xu, X.; Leverentz, H.R.; Truhlar, D.G. Perspectives on Basis Sets Beautiful: Seasonal Plantings of Diffuse Basis Functions. *J. Chem. Theory Comput.* **2011**, *7*, 3027–3034. [CrossRef] [PubMed]

38. Lundberg, M.; Siegbahn, P.E. Agreement between experiment and hybrid DFT calculations for OH bond dissociation enthalpies in manganese complexes. *J. Comput. Chem.* **2005**, *26*, 661–667. [CrossRef] [PubMed]

39. Reiher, M.; Salomon, O.; Artur Hess, B. Reparameterization of hybrid functionals based on energy differences of states of different multiplicity. *Theor. Chem. Acc.* **2001**, *107*, 48–55. [CrossRef]

40. Salomon, O.; Reiher, M.; Hess, B.A. Assertion and validation of the performance of the B3LYP* functional for the first transition metal row and the G2 test set. *J. Chem. Phys.* **2002**, *117*, 4729–4737. [CrossRef]

41. Siegbahn, P.E.M. The performance of hybrid DFT for mechanisms involving transition metal complexes in enzymes. *J. Biol. Inorg. Chem.* **2006**, *11*, 695–701. [CrossRef] [PubMed]

42. Ding, Y.; Li, X.C.; Ferreira, D. Theoretical Calculation of Electronic Circular Dichroism of a Hexahydroxydiphenoyl-Containing Flavanone Glycoside. *J. Nat. Prod.* **2009**, *72*, 327–335. [CrossRef] [PubMed]

43. Nugroho, A.; Morita, H. Circular dichroism calculation for natural products. *J. Nat. Med.* **2014**, *68*, 1–10. [CrossRef] [PubMed]

44. Peach, M.; Helgaker, T.; Salek, P.; Keal, T.; Lutnaes, O.; Tozer, D.; Handy, N. Assessment of a Coulomb-attenuated exchange-correlation energy functional. *Phys. Chem. Chem. Phys.* **2006**, *8*, 558–562. [CrossRef] [PubMed]

45. Jorge, F.; Jorge, S.; Suave, R. Dichroism of Chiral Alkenes: B3LYP and CAM-B3LYP Calculations. *Chirality* **2015**, *27*, 23–31. [CrossRef] [PubMed]

46. Komjáti, B.; Urai, Á.; Hosztafi, S.; Kökösi, J.; Kováts, B.; Nagy, J.; Horváth, P. Systematic study on the TD-DFT calculated electronic circular dichroism spectra of chiral aromatic nitro compounds: A comparison of B3LYP and CAM-B3LYP. *Spectrochim. Acta A Mol. Biomol. Spectrosc.* **2016**, *155*, 95–102. [CrossRef] [PubMed]

47. Bruhn, T.; Witterauf, F.; Götz, D.; Grimmer, C.; Würtemberger, M.; Radius, U.; Bringmann, G. C,C- and N,C-Coupled Dimers of 2-Aminotetraphenylporphyrins: Regiocontrolled Synthesis, Spectroscopic Properties, and Quantum-Chemical Calculations. *Chem. Eur. J.* **2014**, *20*, 3998–4006. [CrossRef] [PubMed]

48. Tran-Dinh, S.; Taboury, J.; Neumann, J.M.; Huynh-Dinh, T.; Genissel, B.; Langlois d'Estaintot, B.; Igolen, J. Proton NMR and circular dichroism studies of the B and Z conformations of the self-complementary deoxyhexanucleotide d(m^5C-G-C-G-m^5C-G): Mechanism of the Z-B-coil transitions. *Biochemistry* **1984**, *23*, 1362–1371. [CrossRef] [PubMed]

49. Miyahara, T.; Nakatsuji, H.; Sugiyama, H. Helical Structure and Circular Dichroism Spectra of DNA: A Theoretical Study. *J. Phys. Chem. A* **2013**, *117*, 42–55. [CrossRef] [PubMed]

50. Li, J.H.; Chai, J.D.; Guo, G.Y.; Hayashi, M. Significant role of the DNA backbone in mediating the transition origin of electronic excitations of B-DNA - implication from long range corrected TDDFT and quantified NTO analysis. *Phys. Chem. Chem. Phys.* **2012**, *14*, 244101. [CrossRef] [PubMed]

51. Schellman, J.A. Circular dichroism and optical rotation. *Chem. Rev.* **1975**, *75*, 323–331. [CrossRef]
52. Crawford, T.D. Ab initio calculation of molecular chiroptical properties. *Theor. Chem. Acc.* **2006**, *115*, 227–245. [CrossRef]
53. Swasey, S.M.; Karimova, N.; Aikens, C.M.; Schultz, D.E.; Simon, A.J.; Gwinn, E.G. Chiral Electronic Transitions in Fluorescent Silver Clusters Stabilized by DNA. *ACS Nano* **2014**, *8*, 6883–6892. [CrossRef] [PubMed]
54. Tomasi, J.; Mennucci, B.; Cammi, R. Quantum Mechanical Continuum Solvation Models. *Chem. Rev.* **2005**, *105*, 2999–3094. [CrossRef] [PubMed]
55. Cossi, M.; Barone, V. Time-dependent density functional theory for molecules in liquid solutions. *J. Chem. Phys.* **2001**, *115*, 4708–4717. [CrossRef]

International Journal of
Molecular Sciences

MDPI

Article

Binding of Organometallic Ruthenium Anticancer Complexes to DNA: Thermodynamic Base and Sequence Selectivity

Suyan Liu [1,2], Aihua Liang [1], Kui Wu [2,3,*], Wenjuan Zeng [2,4], Qun Luo [2,4] and Fuyi Wang [2,4,*]

[1]   Institute of Chinese Materia Medica, China Academy of Chinese Medical Sciences, Beijing 100700, China;
      syliu@icmm.ac.cn (S.L.); ahliang@icmm.ac.cn (A.L.)
[2]   Beijing National Laboratory for Molecular Sciences, National Centre for Mass Spectrometry in Beijing,
      CAS Key Laboratory of Analytical Chemistry for Living Biosystems, CAS Research/Education Centre for
      Excellence in Molecular Sciences, Institute of Chemistry, Chinese Academy of Sciences,
      Beijing 100190, China; zengwj2014@iccas.ac.cn (W.Z.); qunluo@iccas.ac.cn (Q.L.)
[3]   School of Chemistry and Chemical Engineering, Wuhan University of Science and Technology,
      Wuhan 430081, China
[4]   University of Chinese Academy of Sciences, Beijing 100049, China
*   Correspondence: wukui@wust.edu.cn (K.W.); fuyi.wang@iccas.ac.cn (F.W.); Tel.: +86-10-6252-9069 (F.W.)

Received: 14 June 2018; Accepted: 4 July 2018; Published: 23 July 2018

Abstract: Organometallic ruthenium(II) complexes [(η^6-arene)Ru(en)Cl][PF$_6$] (arene = benzene (**1**), *p*-cymene (**2**), indane (**3**), and biphenyl (**4**); en = ethylenediamine) are promising anticancer drug candidates both in vitro and in vivo. In this paper, the interactions between ruthenium(II) complexes and 15-mer single- and double-stranded oligodeoxynucleotides (ODNs) were thermodynamically investigated using high performance liquid chromatography (HPLC) and electrospray ionization mass spectroscopy (ESI-MS). All of the complexes bind preferentially to G$_8$ on the single strand 5′-CTCTCTT$_7$G$_8$T$_9$CTTCTC-3′ (**I**), with complex **4** containing the most hydrophobic ligand as the most reactive one. To the analogs of **I** (changing T$_7$ and/or T$_9$ to A and/or C), complex **4** shows a decreasing affinity to the G$_8$ site in the following order: -AG$_8$T- (K: 5.74 × 10^4 M^{-1}) > -CG$_8$C- > -TG$_8$A- > -AG$_8$A- > -AG$_8$C- > -TG$_8$T- (**I**) ≈ -CG$_8$A- (K: 2.81 × 10^4 M^{-1}). In the complementary strand of **I**, the G bases in the middle region are favored for ruthenation over guanine (G) bases in the end of oligodeoxynucleotides (ODNs). These results indicate that both the flanking bases (or base sequences) and the arene ligands play important roles in determining the binding preference, and the base- and sequence-selectivity, of ruthenium complex in binding to the ODNs.

Keywords: organometallic ruthenium complexes; anticancer; oligodeoxynucleotide; base/sequence selectivity; thermodynamics; LC-MS

1. Introduction

Organometallic ruthenium(II) arene complexes [(η^6-arene)Ru(YZ)(X)]$^{n+}$ have interesting anticancer properties both in vitro and in vivo, including cytotoxic activity towards cisplatin-resistant cell lines [1–9]. The arene ligand occupies three coordination sites in this type of pseudo-octahedral complexes and stabilizes Ru in its +2 oxidation state [10]. These mono-functional complexes appear to have a novel mechanism of action, differing from those of bi-functional cisplatin [11] and of the Ru(III) anticancer complexes, for example, (ImH)[*trans*-RuCl$_4$Im(Me$_2$SO)] (NAMI-A, Im = imidazole) and (IndH)[*trans*-RuCl$_4$(Ind)$_2$] (KP1019, Ind = indazole), which have entered into clinical trials [12–15]. The cytotoxicity of the chloro ethylenediamine complexes [(η^6-arene)Ru(en)Cl][PF$_6$] (arene = benzene, *p*-cymene, biphenyl, tetrahydroanthracene, etc.; en = ethylenediamine) increases with the size of coordinated arene. DNA has been shown to be a potential target for these Ru(II) arene complexes, most of which bind selectively to N7 of guanine [16–19]. The biphenyl Ru complex shows a decreasing activity towards mononucleotides in the following order: 5′-GMP(N7) > 5′-TMP(N3) >> 5′-CMP(N3) > 5′-AMP (N7/N1), being more discriminatory between guanine (G) and adenine (A) bases than cisplatin, which exhibits affinity to adenine second to guanine [11]. The formation of H-bonds between en-NH$_2$ groups of ruthenium complexes and C6O of guanine was found to associate with this selective coordination [16,17]. The guanine bases in oligonucleotides of different sequences and in calf thymus DNA have been demonstrated to be the only sites ruthenated by the Ru(II) arene complexes [18,20–23]. Importantly, complexes containing π-rich arene ligands, like biphenyl or tetrahydroanthracene, showed combined coordination to guanine N7 and non-covalent interactions between the arene ligands and DNA bases, including arene intercalation and minor groove binding [16,17,20,21].

We have previously demonstrated that the thymine (T) bases in 15-mer single-stranded oligodeoxynucleotides (ODNs) can kinetically compete with G for binding to ruthenium arene complexes, but such T-bound mono-ruthenated ODN complexes were not thermodynamically stable and finally transformed to stable G-bound adducts, including mono-G-bound ODNs and minor G,T-bound di-ruthenated adducts [24]. Interestingly, the T-bases in G-quadruplex DNA are both kinetically and thermodynamically competitive with G-bases for ruthenation by ruthenium arene complexes because G-N7 is involved in H-bonding in G-quadruplex DNA [25].

The chloride ligand in the mono-functional compounds [(η^6-arene)Ru(en)(Cl)]$^+$ is readily hydrolyzed to give the more reactive aqua species [(η^6-arene)Ru(en)(H$_2$O)]$^{2+}$. However, reaction of the aqua ruthenium complexes with DNA is retarded at high pH, suggesting that Ru–OH$_2$ bonds are more reactive towards DNA than Ru–OH bonds [24,26,27]. Such behavior appears parallel to that of Pt(II) diam(m)ine anticancer complexes [28]. The bi-functional complex cisplatin undergoes a two-step aquation to form 1,2-intrastrand cross-linked DNA adducts, which is thought to be the main lesions on DNA that triggers apoptosis signaling initiated by high mobility group box 1 (HMGB1) protein recognition [11]. These two steps of hydrolysis, especially the second step, can be rate-limiting for interactions of cisplatin with DNA bases [28]. For instance, the formation of 5′-GpA intrastrand crosslinks by cisplatin is more preferred than that of 5′-ApG in d(ApGpA) and d(TpApGpApT) [29]. The rate of hydrolysis of the chloride ligand in the mono-aquated cisplatin was shown to be the key step that controlled the final bi-functional intrastrand adduct profiles when cisplatin bound to DNA with various X-purine-purine-Y motif [30]. However, aquation cannot be the only factor that controls the rate and selectivity of DNA binding of cisplatin. The diaquated form of cisplatin, [Pt(NH$_3$)$_2$(OH$_2$)$_2$]$^{2+}$, displays no preference for the 3′- or 5′-guanine in a TpGpGpT sequence of a hairpin duplex oligonucleotide [31] or a single-stranded oligonucleotide [32]. Surprisingly, the 5′-guanine in TpGpGpC motif of a single-stranded oligonucleotide is preferred by a factor of two, and even by a factor of twelve over the 3′-guanine in a duplex oligonucleotide [33]. These indicate that DNA structure (or/and sequence) is a primary factor for determining the binding preference of bifunctional cisplatin to DNA. For the mono-functional Ru(II) arene complexes, binding to DNA may not be as complicated as that of cisplatin, but the varied flanking sequences next to the binding sites (e.g., guanine bases) may significantly affect the selective ruthenium binding, finally causing distinct response to the damaged DNA. However, the DNA binding properties, in particular sequence selectivity, of Ru(II) arene complexes are poorly understood.

In this present work, the interactions of [(η^6-arene)Ru(en)Cl][PF$_6$] (arene = benzene (**1**), *p*-cymene (**2**), indane (**3**), and biphenyl (**4**)) with 15-mer single- or double-stranded ODNs have been thermodynamically investigated using high performance liquid chromatography (HPLC) and LC-mass spectroscopy (MS). Based on the thermodynamic binding constants, we found the arene ligands of ruthenium complexes and adjacent bases of guanine have great influence on the binding affinity of ruthenium to guanine. We elucidated the base-selectivity and sequence-selectivity on the DNA binding of the ruthenium arene complexes, and hope to provide novel insights into the molecular mechanism of action of organometallic ruthenium anticancer complexes.

2. Results

2.1. Reactions of Organometallic Ruthenium(II) Complexes with One-G-centered Single-Stranded ODNs

Chart 1 shows the structures of the Ru(II) arene complexes and sequences of ODNs used in this work. Firstly, each ruthenium complex was incubated with 1 mol equiv strand **I** in 50 mM triethylammonium acetate buffer (TEAA) buffer solution (pH 7) at 310 K for 24 h. The reaction mixtures were then separated and analyzed by HPLC with ultraviolet (UV) detection at 260 nm (Figure 1), and each fraction was identified by electrospray ionization mass spectroscopy (ESI-MS). The observed negatively-charged ions corresponding to ruthenated **I** are listed in Table 1, and the mass spectra are shown in Figure S1 in the supplementary materials. The reaction of complex **1** with strand **I** gave rise to only one mono-ruthenated product, as indicated by the triply-charged ion $[\textbf{I}\text{-}\{(\eta^6\text{-benzene})\text{Ru(en)}\}]^{3-}$ at m/z 1556.56 (Figure S1, supplementary materials). However, minor amount of di-ruthenated adducts were detected for the reaction mixtures of complexes **2–4** with 1 mol equiv strand **I** apart from the respective mono-ruthenated adducts at larger amount compared with the mono-ruthenated **I** by complex **1** (Figure 1, Figure S1 in supplementary materials). Moreover, even in the presence of 2-fold complex **1**, only ca. 67% of strand **I** was ruthenated (Figure 2) to produce mono-ruthenated ODN complex, while at the same molar ratio (Ru/**I** = 2.0), over 90% of strand **I** was ruthenated by complex **2**, **3**, or **4**, affording the mono-ruthenated ODN as the main product with a significant amount of di-ruthenated ODN complex (Figure 2).

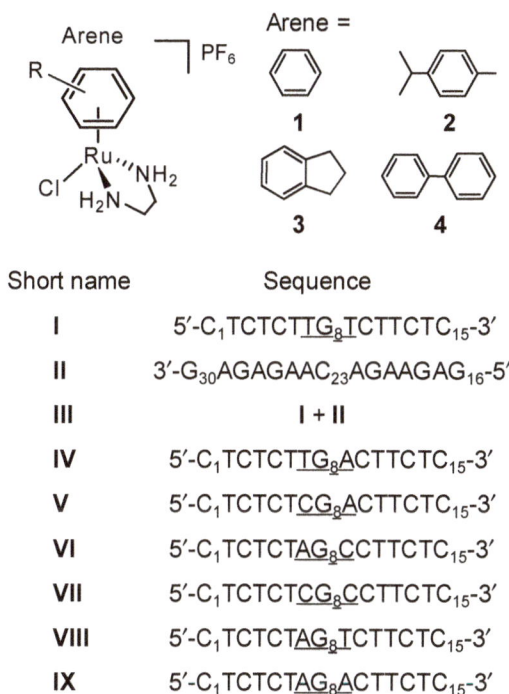

Short name	Sequence
I	5′-C$_1$TCTCTTG$_8$TCTTCTC$_{15}$-3′
II	3′-G$_{30}$AGAGAAC$_{23}$AGAAGAG$_{16}$-5′
III	I + II
IV	5′-C$_1$TCTCTTG$_8$ACTTCTC$_{15}$-3′
V	5′-C$_1$TCTCTCG$_8$ACTTCTC$_{15}$-3′
VI	5′-C$_1$TCTCTAG$_8$CCTTCTC$_{15}$-3′
VII	5′-C$_1$TCTCTCG$_8$CCTTCTC$_{15}$-3′
VIII	5′-C$_1$TCTCTAG$_8$TCTTCTC$_{15}$-3′
IX	5′-C$_1$TCTCTAG$_8$ACTTCTC$_{15}$-3′

Chart 1. Structures of ruthenium arene anticancer complexes and sequences of oligodeoxynucleotides (ODNs) used in this work. ODNs **IV–IX** are the analogues of strand **I** with only variations at the neighboring bases of the central guanine (underlined).

In order to make sure that the reactions of complexes **1–4** with strand **I** reached equilibrium within 24 h, we analyzed the reaction mixtures of complex **4** with strand **I** for various times. As shown in

Figure 1B, no pronounced changes in the product ratios were observed after 24 h of incubation even in the presence of 5 mM NaCl, which mimics the concentration of Cl$^-$ in nuclear as Cl$^-$ was thought to retard the hydrolysis of ruthenium arene complexes [26], subsequently slowing down their reaction with ODNs. These results indicate that the reactions of the ruthenium complexes with single-strand ODNs reached equilibrium within 24 h.

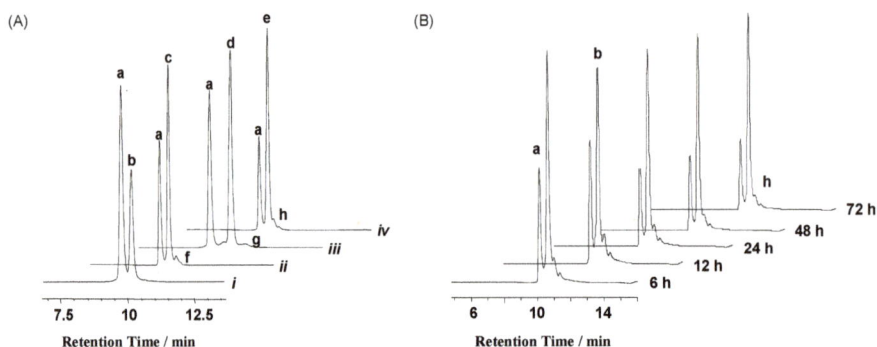

Figure 1. High performance liquid chromatography (HPLC) chromatograms with ultraviolet (UV) detection at 260 nm for reactions of ruthenium complexes with single strand **I**. (**A**) Complex **1** (*i*), **2** (*ii*), **3** (*iii*), or **4** (*iv*) with single strand **I** (0.1 mM, Ru/**I** = 1.0) in 50 mM triethylammonium acetate buffer (TEAA) (pH 7) at 310 K for 24 h; (**B**) complex **1** with **I** (0.1 mM, Ru/**I** = 1.0) in 50 mM TEAA (pH 7) and 5 mM NaCl at 310 K for different times. Peak assignments: a, unruthenated strand **I**; b, c, d, and e, mono-ruthenated **I**; f, g, and h, di-ruthenated **I**. For mass spectra, see Figure S1 in Supplementary Materials.

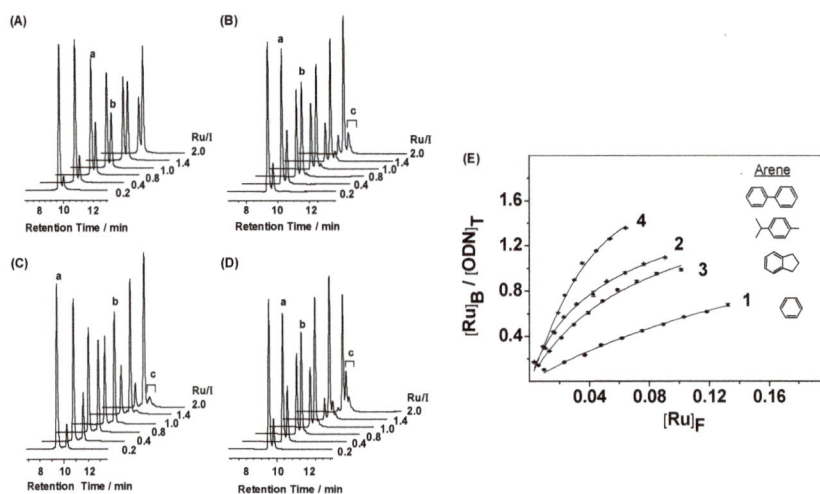

Figure 2. (**A–D**) HPLC chromatograms for reaction mixtures of complex **1** (**A**), **2** (**B**), **3** (**C**), or **4** (**D**) with single strand **I** (0.1 mM) at various molar ratios in 50 mM TEAA at 310 K for 24 h. Peak assignments: a, unruthenated strand **I**; b, mono-ruthenated **I**; c, di-ruthenated **I**. (**E**) A direct plot (dots) of moles of ODN-bound **1**, **2**, **3**, or **4** ([Ru]$_B$) as a function of the concentration of free ruthenium complexes ([Ru]$_F$). Computer-fitting (lines) of the experimental data to the ligand–receptor binding Equation (8) (see details in Supplementary Materials) gave rise to the equilibrium constants for the two-site binding reactions listed in Table 2.

Table 1. Negatively-charged ions observed by electrospray ionization mass spectroscopy (ESI-MS) for the high performance liquid chromatography (HPLC) fractions (Figure 1) from the reaction mixtures of complexes **1–4** with 1 mol equiv of single-strand **I**. For mass spectra, see Figure S1.

Ru Complex	*m/z*: Observed (Calculated)		
	$[I]^{3-}$	$[I + \{Ru\}^{1}]^{3-}$	$[I + \{Ru\}_2]^{3-}$
1	a: 1477.22 (1477.24)	b: 1556.56 (1556.58)	— [2]
2	a: 1477.22 (1477.24)	c: 1575.25 (1575.27)	f: 1672.96 (1672.95)
3	a: 1477.22 (1477.24)	d: 1569.91 (1569.92)	g: 1662.65 (1662.60)
4	a: 1477.22 (1477.24)	e: 1581.89 (1581.92)	h: 1686.66 (1686.60)

[1] $\{Ru\} = \{(\eta^6\text{-arene})Ru(en)\}^{2+}$, arene = benzene (**1**), *p*-cymene (**2**), indane (**3**), or biphenyl (**4**); [2] not detectable.

Table 2. Equilibrium binding constants (*K*) for the reactions of complexes **1–4** with strand **I**.

Ru Complex	K_1 (10^4 M^{-1})	K_2 (10^4 M^{-1})	K_1/K_2
1	0.79 ± 0.04	0.18 ± 0.03	4.39
2	3.94 ± 0.24	0.51 ± 0.05	7.72
3	2.52 ± 0.13	0.43 ± 0.04	5.86
4	2.92 ± 0.46	2.75 ± 0.47	1.06

We have previously shown that the G_8 base in strand **I** is the main binding site for the ruthenium biphenyl complex **4**, and that T_7 and T_{11} in the same strand are the kinetically favored, yet thermodynamically unfavored, ruthenation sites by complex **4** [24]. Taking the structural similarity of the four ruthenium arene complexes into account, we assume that the mono-ruthenated adducts formed by the reactions of strand **I** with complexes **1–4** are G_8-bound ODN complexes, and the minor di-ruthenated adducts are G_8,T_x-bound (x = 7 or 11) ruthenated **I** [24].

Next, the reaction mixtures of the four ruthenium complexes with strand **I** at various molar ratios of Ru to **I** were analyzed by HPLC so as to determine the equilibrium binding constants of the two-site receptor/ligand reaction. The results are shown in Figure 2A,D. On the basis of the HPLC peak areas with UV detection at 260 nm (the coordination of Ru(II) arene complexes to ODNs had little effect on their extinction coefficients at 260 nm), the extent of saturation (B in Equation (8) shown in Materials and Method) of the ODN binding sites was plotted as a function of the amount of unbound Ru complex (L) [25]. The resulting curves were computer-fitted to Equation (8) (Figure 2E), giving rise to the equilibrium constants listed in Table 2.

It can be seen that among the four ruthenium complexes, the benzene complex **1** had the lowest equilibrium constants for both binding steps to strand **I**, whereas the *p*-cymene complex **2** exhibits the highest binding affinity to strand **I** for the first step, and the biphenyl complex **4** is the most highly active to bind to strand **I** for the second step. It is notable that for complexes **2** and **3**, the *K* values for the first step binding are much higher than those for the second step, but complex **4** has a similar affinity for both of the binding steps. In other words, complexes **2** and **3** are more discriminative between the G and T in the single-stranded ODN than complex **4**.

To investigate the sequence selectivity of Ru(II) arene complexes binding to DNA, six analogues of single-stranded ODN **I** with sequence variants only at the adjacent bases to G_8 (Chart 1) were selected to react with complex **4** at molar ratios of Ru/ODN ranging from 0.2 to 2 under the same conditions as described above. Analysis of the reaction mixtures by HPLC (Figure 2D, Figures S2A–S7A in supplementary materials) shows a similar binding profile for complex **4** to all the seven ODNs, that is, at Ru/ODN = 1.0, about 70% of the ODN was ruthenated, and at Ru/ODN = 2.0, less than 5% ODN remained intact. However, based on the thermodynamic G_8,T_i-diruthenated model (Scheme 1), the equilibrium constants (Table S1) resulting from the computer-fits of the titration data to Equation (8) showed a pronounced difference (Figure 3). The equilibrium constants for the first (mono-ruthenation) step of binding decreased in the following order: -AG$_8$T- (**VIII**) > -CG$_8$C- (**VII**) > -TG$_8$A- (**IV**) > -AG$_8$A-

(IX) > -AG$_8$C- > (VI) -TG$_8$T- (I) $\approx$ -CG$_8$A- (V), whereas the equilibrium constants for the second step of binding, which formed the di-ruthenated ODNs, decreased in the following order: -TG$_8$T- (I) $\approx$ -CG$_8$A- (IV) > -AG$_8$T- (VIII) > -CG$_8$C- (VII) > -AG$_8$C- (VI) $\approx$ -TG$_8$A- (IV) > -AG$_8$A- (IX). Except for the -TG$_8$T- and -CG$_8$A- sequences, the equilibrium constants for the first step of binding of complex **4** to the other five ODNs are almost two-fold higher than those of the second step of binding, implying significant discrimination between the first ruthenation site (G$_8$) and the second ruthenation site at T$_x$ (Figure 3) [24].

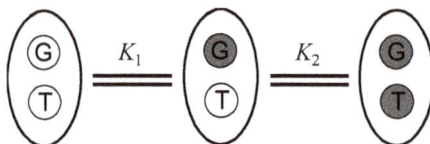

Scheme 1. Schematic representative model for the binding reactions of organometallic ruthenium anticancer complexes to oligodeoxynucleotides (ODNs) **I** and **IV–IX** at the guanine (G) and thymine (T) sites based on the divalent receptor system. Bare circle with G/T means unoccupied site and black circle with G/T means occupied site.

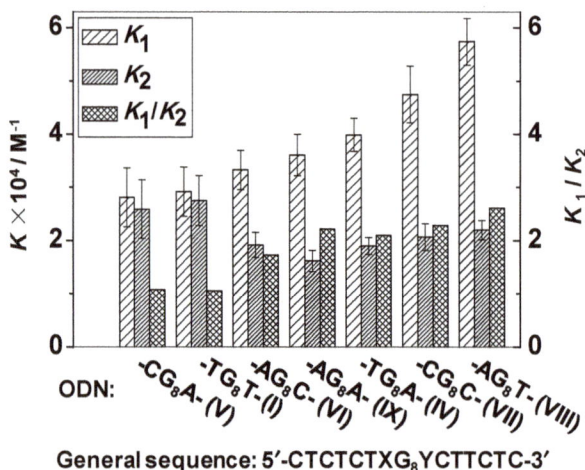

General sequence: 5'-CTCTCTXG$_8$YCTTCTC-3'

Figure 3. Sequence selectivity of organometallic ruthenium complex **4** binding to single-strand ODNs **I** and **IV–IX**. K_1 and K_2, of which the values are listed in Table S2 in supplementary materials, are the binding constants of complex **4** to each ODN at the first and the second step of binding, respectively.

2.2. Reactions of Organometallic Ruthenium(II) Complexes with Single-Strand II

The ruthenium complex **1**, **2**, **3**, or **4** was individually incubated with single-strand **II**, the complementary strand of **I**, at a molar ratio of Ru/**II** = 1.0 or 3.0 in 50 mM TEAA buffer solution at 310 K for 24 h. The reaction mixtures were then analyzed by HPLC and ESI-MS, and the observed negatively-charged ions for ruthenated **II** are listed in Table 3. The HPLC chromatograms with TIC (total ion count) detection and mass spectra of each HPLC fraction are shown in Figure 4 and Figure S8 in supplementary materials, respectively. At a molar ratio of Ru/**II** = 1.0, a significant amount of mono- and di-ruthenated **II** resulted from the reactions of **II** with complex **2**, **3**, or **4**, but the reaction of the benzene complex **1** with **II** afforded only mono-ruthenated **II**. Even in the presence of three-fold excess of complex **1**, nearly half of strand **II** remained intact after 24 h of reaction. In contrast, the reaction of strand **II** with three-fold excess of complex **2**, **3**, or **4** resulted in ruthenation of ca. 85% **II**, giving a pronounced amount of triply-ruthenated **II**. For **3** and **4**, a small amount of tetra-ruthenated **II**, as indicated by the triply-charged ions at m/z 1938.42 and 1986.32, respectively, were also observed (Table 3, Figure S8).

Table 3. Negative ions observed by ESI-MS for the HPLC fractions shown in Figure 4 (for mass spectra, see Figure S8 in supplementary materials).

Ru Complex	*m/z*: Observed (Calculated)			
	[**II** + {Ru} [1]]$^{3-}$	[**II** + {Ru}$_2$]$^{3-}$	[**II** + {Ru}$_3$]$^{3-}$	[**II** + {Ru}$_4$]$^{3-}$
1	b: 1647.31 (1647.28)	c: 1726.69 (1726.62)	c: 1805.81 (1805.62)	— [2]
2	d: 1666.32 (1666.30)	e: 1764.34 (1764.33)	m: 1862.10 (1862.02)	— [2]
3	f: 1660.96 (1660.96)	g: 1753.37 (1753.30)	n: 1846.01 (1845.99)	n: 1938.42 (1938.34)
4	h: 1672.96 (1672.96)	k: 1777.33 (1777.30)	p: 1882.04 (1881.99)	p: 1986.32 (1986.33)

[1] {Ru} = {(η^6-arene)Ru(en)}$^{2+}$ (arene = benzene (**1**), *p*-cymene (**2**), indane (**3**) or biphenyl (**4**); [2] Not detected.

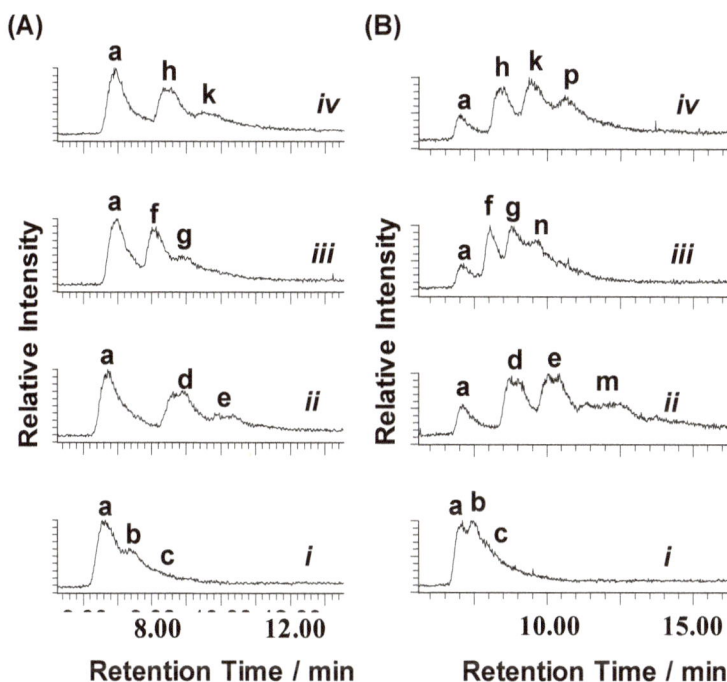

Figure 4. Total ion count (TIC) chromatograms for reactions of complex **1** (*i*), **2** (*ii*), **3** (*iii*), or **4** (*iv*) with single strand **II** (0.1 mM) at the molar ratio of (**A**) Ru/**II** = 1.0 or (**B**) Ru/**II** = 3.0 in 50 mM TEAA (pH 7) at 310 K for 24 h. Peak assignments: a, unruthenated strand **II**; b, d, f, and h, mono-ruthenated **II**; c, e, g, and k, di-ruthenated **II**; m, n, and p, tri-ruthenated **II**. Notably the fraction c in (**B**) contains a little tri-ruthenated **II** by **1** and fractions n and p contains a small amount of tetra-ruthenated **II** by **3** or **4** as shown in Figure S8 in supplementary materials.

To identify the preferential binding sites of the tested ruthenium complexes in strand **II**, a three-fold excess of complex **1** or **2** was incubated with strand **II** at 310 K for 24 h. The adducts were digested by snake venom phosphodiesterase (SVP) after removing unbound ruthenium through centrifugation, and then analyzed by HPLC-ESI-MS in negative mode. The observed ions corresponding to ruthenated ODN fragments are listed in Table S2 in supplementary materials, and the mass spectra are shown in Figure S9 in the supplementary materials. Firstly, the results showed that the SVP digestion stopped at G_{21}, as evidenced by the detection of doubly-charged ions at m/z 1050.53 and 1078.24, which are assignable to the mono-ruthenated ODN fragments $[F_{21}\text{-}\mathbf{1}']^{2-}$ and $[F_{21}\text{-}\mathbf{2}']]^{2-}$ ($F_{21} = 3'\text{-}G_{21}A_{20}A_{19}G_{18}A_{17}G_{16}\text{-}5'$, $\mathbf{1}' = \{(\eta^6\text{-ben})Ru(en)\}$, $\mathbf{2}' = \{(\eta^6\text{-}p\text{-cym})Ru(en)\}$), respectively. These are parallel to the reactions of ODN **II** with complexes **3–4** [19]. Secondly, doubly-charged ions at m/z 1225.29 assignable to the di-ruthenated ODN fragment $[F_{21}\text{-}\mathbf{2}'_2]^{2-}$ were observed, indicating that the fragment $3'\text{-}G_{21}A_{20}A_{19}G_{18}A_{17}G_{16}\text{-}5'$ contains two binding sites for complex **2**, most likely G_{21} and G_{18} as for complex **3** reported previously [19]. Thirdly, the triply-charged ions at m/z 1335.63, which correspond to the ruthenated fragments $[F_{26}\text{-}\mathbf{2}'_2]^{3-}$ ($F_{26} = 3'\text{-}G_{26}A_{25}A_{24}C_{23}A_{22}G_{21}A_{20}A_{19}G_{18}A_{17}G_{16}\text{-}5'$), were detected, providing evidence for the formation of di-ruthenated **II** at G_{21} and G_{26} by complex **2** as observed for the reaction of the same strand with complex **4** [19]. These results suggest that G_{21} in the middle region of single strand **II** is the common preferential binding site for complexes **1–4**. Additionally, G_{18} and G_{26} are the secondary binding sites for complexes **3** and **4**, respectively, and both G_{21} and G_{26} are the secondary binding sites for complex **2**.

2.3. Reactions of Organometallic Ruthenium(II) Complexes with Duplex III

Next, mixtures of complex **1** or **4** with duplex **III** (= **I** + **II**) at Ru/**III** = 1.0 or 6.0 in 50 mM TEAA buffer (pH 7) containing 100 mM $NaClO_4$ were incubated at 310 K for 48 h, and then analyzed by HPLC followed by ESI-MS analysis under negative-ion mode. The chromatograms for the reaction mixture of duplex **III** with complex **1** or **4** are shown in Figure 5A, and Figures S10A and S11A, and the corresponding mass spectra for HPLC fractions are shown in Figure 5B, and Figures S10B and S11B, respectively. The observed ions for the reaction mixtures at the molar ratio of Ru/**III** = 6.0 are listed in Table 4. It can be seen that at a low reaction molar ratio (Ru/**III** = 1.0) (Figure 5B), no ruthenated **I** by complex **1** or **4** was detected, but a little amount of mono-ruthenated **II** by complex **1** ($[\mathbf{II}\text{-}\mathbf{1}']^{4-}$) and both mono- and di-ruthenated **II** adducts $[\mathbf{II}\text{-}\mathbf{4}']^{4-}$ and $[\mathbf{II}\text{-}\mathbf{4}'_2]^{4-}$ by complex **4** were obviously identified (Figure 5B). Even in the presence of six-fold excess of complex **1**, only a small amount of strand **I** was ruthenated, accompanied with the observation of a pronounced amount of mono-ruthenated **II** and a minor amount of di-ruthenated **II** adducts (Figure S10B). In contrast, for the binding of complex **4** to duplex **III** at **4**/**III** = 6.0, a series of multi-ruthenated **I** and **II** adducts were detected by MS following HPLC separation. The last HPLC fraction (peak e in Figure S11A) was identified as a mixture of tri-ruthenated **I**, and tetra-, penta-, and hexa-ruthenated **II** by **4** (Figure S11B). Meanwhile, little free **II** (peak a, Figure S11A) remained, while a large amount of unreacted strand **I** (peak b, Figure S11A) was detected in the reaction mixture of complex **4** to duplex **III** at **4**/**III** = 6.0. It is noticed that under the given chromatographic conditions, the HPLC separation denatured the duplex, as well as its ruthenated adducts, into single-stranded ODNs, but a small amount of ruthenated duplex **III** was still detected in the fraction c and d in the forms of di-, tri-, tetra-, and penta-ruthenated duplex **III** (Figure S11B in supplementary materials).

Circular dichroism (CD) spectroscopy was introduced to elucidate the effect of the Ru arene complexes on the conformation of duplex ODN **III**. A CD spectrum provides unique signals in diagnosing changes in DNA conformation during drug-DNA interactions, for example, the maximum positive adsorption at ~275 nm attributed to DNA base stacking, and the maximum negative adsorption at ~245 nm assigned to the right-handed helicity of duplex DNA, may change in both intensity and wavelength subject to reactions of DNA with drug molecules [34]. CD spectra for 15-mer duplex **III** (in 50 mM TEAA buffer containing 100 mM $NaClO_4$) in the absence and presence of increasing amounts of Ru arene complexes **1**, **2**, or **4** are compared as shown in Figure 5C and Figure S12 in supplementary materials, with the applied molar ratios of Ru/**III** = 1.0, 3.0, and 6.0, respectively. Cisplatin was also introduced to interact with the duplex DNA **III** as a reference. The 15-mer duplex used in our experiments has a typical B-DNA conformation in the 100 mM $NaClO_4$ solution with the maximum positive absorption (base stacking) at ~277 nm and the maximum negative absorption (right-handed helicity) at ~242 nm. The increasing concentration of the ruthenium complexes made more Ru bind to the duplex DNA, leading to an obvious decrease in the intensity of both positive and negative CD signals. Complex **4** resulted in the largest decrease in the signal intensity and caused the two absorption bands to both approach the baseline gradually (Figure S12C). Another interesting variation was that the maximal positive absorption wavelength had a little red-shift from 277 nm to 282 nm because of the increasing level of ruthenation by complex **4**, while the other two Ru complexes caused no obvious band shift in the CD spectrum of the duplex ODN. At the same molar ratio (Ru/III = Pt/III = 6.0), the intensity decrease of positive signal caused by the ruthenium complexes and cisplatin was in the following order: **1** < **2** < cisplatin < **4**, while for negative signal, the order was as follows: **1** < cisplatin < **2** < **4** (Figure 5C).

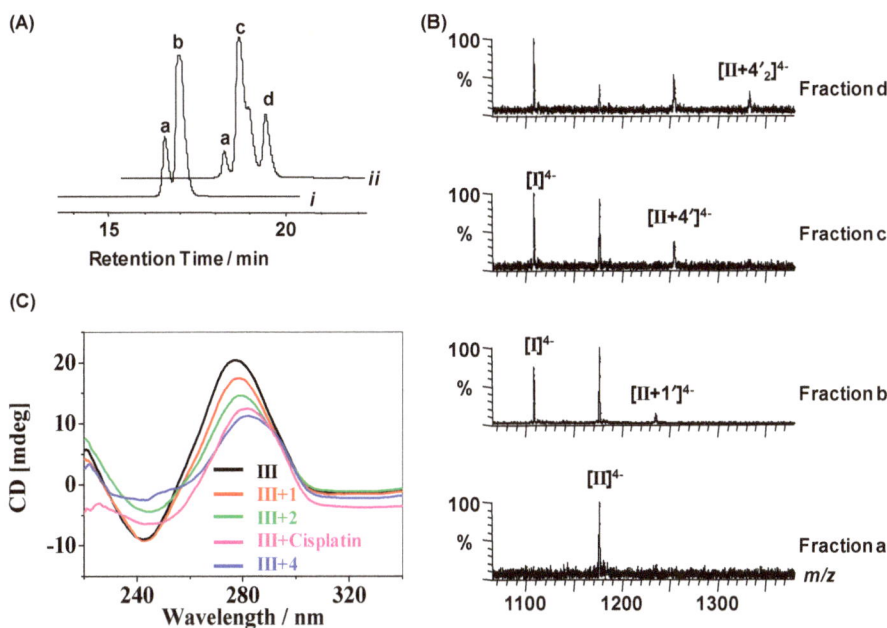

Figure 5. (**A**) HPLC chromatograms with ultraviolet (UV) detection at 260 nm for the reaction mixture of duplex **III** with complex **1** (*i*) or complex **4** (*ii*) (Ru/III = 1.0) in 50 mM TEAA buffer (pH 7) and 100 mM NaClO₄ incubated at 310 K for 48 h; (**B**) mass spectra for HPLC fractions shown in (**A**); (**C**) circular dichroism (CD) spectra of free duplex **III** (black) and the reaction mixtures of duplex **III** with complex **1** (red), **2** (green), **4** (blue) and cisplatin (magenta) incubated at 310 K for 24 h at a molar ratio of Ru/III or Pt/III = 6.0.

Table 4. Negatively-charged ions observed by ESI-MS coupled to denatured HPLC for reaction mixture of Ru complex **1** or **4** with duplex **III** (Ru/III = 6.0, **III** = **I** + **II**). For chromatograms and mass spectra, see Figures S10 and S11 in supplementary materials.

Ru Complex	Observed (Calculated) *m/z*	Observed Ions
	1107.89 (1107.91)	$[I]^{4-}$
	1167.44 (1167.43)	$[I\text{-}1'\ 1]^{4-}$
1	1175.92 (1175.96)	$[II]^{4-}$
	1235.44 (1235.46)	$[II\text{-}1']^{4-}$
	1294.69 (1294.71)	$[II\text{-}1'_2]^{4-}$
	1342.93 (1342.96)	$[II\text{-}1'_3]^{4-}$

Table 4. *Cont.*

Ru Complex	Observed (Calculated) *m/z*	Observed Ions
	1186.40 (1186.44)	$[I\text{-}4'_1]^{4-}$
	1264.67 (1264.70)	$[I\text{-}4'_2]^{4-}$
	1343.18 (1343.21)	$[I\text{-}4'_3]^{4-}$
	1477.57 (1477.57)	$[I]^{3-}$
	1582.23 (1582.25)	$[I\text{-}4']^{3-}$
	1686.69 (1686.60)	$[I\text{-}4'_2]^{3-}$
	1791.38 (1791.28)	$[I\text{-}4'_3]^{3-}$
	1254.19 (1254.22)	$[II\text{-}4']^{4-}$
4	1332.76 (1332.73)	$[II\text{-}4'_2]^{4-}$
	1411.23 (1411.24)	$[II\text{-}4'_3]^{4-}$
	1489.46 (1489.50)	$[II\text{-}4'_4]^{4-}$
	1567.73 (1567.76)	$[II\text{-}4'_5]^{4-}$
	1646.30 (1646.27)	$[II\text{-}4'_6]^{4-}$
	1568.36 (1568.28)	$[II]^{3-}$
	1672.66 (1672.63)	$[II\text{-}4']^{3-}$
	1777.35 (1777.30)	$[II\text{-}4'_2]^{3-}$
	1882.07(1881.99)	$[II\text{-}4'_3]^{3-}$
	1986.45(1986.33)	$[II\text{-}4'_4]^{3-}$
	2090.80(2090.68)	$[II\text{-}4'_5]^{3-}$

[1] $1' = \{(\eta^6\text{-benzene})Ru(en)\}^{2+}$; $4' = \{(\eta^6\text{-biphenyl})Ru(en)\}^{2+}$.

3. Discussion

Through recognizing different sequences of nucleic acids, the repressor and activator proteins can regulate the gene expressions. The selective manipulation of gene expression can be achieved by using small molecules that can target DNA to selectively activate or repress gene expression have a high potential for being anticancer therapeutics [35–39]. Therefore, it is of great importance to study the base and sequence selectivity of anticancer metallodrugs and candidates as DNA binders.

The most studied organometallic ruthenium(II) complexes are $[(\eta^6\text{-arene})Ru(en)(Cl)]^+$, where arene = biphenyl (bip), tetrahydroanthracene (tha), dihydroanthracene (dha), *p*-cymene (cym), or benzene (ben) have been previously reported to bind preferentially to N7 of guanosine and to N3 of thymidine, but weakly to N3 of cytidine, and little to adenosine . Such base-selectivity appears to be enhanced by strong intra-molecular H-bonding between the en NH_2 groups and exocyclic oxygens. However, reacting with the single-stranded ODNs d(ATACATG$_7$G$_8$TACATA) (**X**), d(TATG$_{25}$TACCATG$_{18}$TAT) (**XI**) or the duplex **XII** (= **X** + **XI**), the biphenyl Ru(II) complex was demonstrated to bind only to N7 of guanine bases (G$_7$, G$_8$, G$_{18}$, and G$_{25}$), but not to thymine residues in either of the single- or double-stranded ODNs [20]. We have recently shown that thymine bases in single-strand **I** and its analogs are kinetically competitive with guanine bases for binding to organometallic ruthenium(II) complexes [24]. The T-bound mono-ruthenated adducts are not stable and could dissociate or transfer to G-bound adducts or G$_8$,T$_x$-diruthenated adducts [25]. In the present work, our thermodynamic studies further prove that the di-ruthenated adducts formed by the reactions of complexes **2–4** with single strand **I** and **IV–IX** bearing a single G base were still detectable even after 24 h of reaction at 310 K. The binding constants of Ru complexes **1–4** to both G$_8$ and T$_x$ in the single-stranded ODNs increase with the size of the arene ligands, consistent with previous reports [16]. The similar binding preference of Ru to G over T was also observed for the reactions of the ruthenium complexes with the duplex ODN **III**. In these cases, the thymine ruthenation occurred only when the most of guanine bases in the duplex **III** were ruthenated. This is just the opposite with the base selectivity of Zn(II)–acridinylcyclen complexes, which bound to G only after most of T was occupied in GpT, d(GTGTCGCC), or in duplex d(CGCTAGCG)$_2$ in neutral or basic solution [40]. However, in acidic solution or nonpolar solvent, the attack of Zn to T was restricted because of the protonation of T-N3 [41]. We have also found that the competition resulting from the protonation of T-N3 reduces the binding affinity of the ruthenium arene complexes to T-N3 in single-stranded ODNs [24]. At pH 4.8, no T-bound adducts were detected in the reaction mixtures of the biphenyl complex **4** with strand **I** at Ru/**I** = 1.0.

The intercalation between the aromatic ligands of Ru arene complexes and the purine ring of DNA was one of driving forces for the (arene)Ru-G-N7/T-N$_3$ coordination, especially for these with arene ligands with more than one ring, like biphenyl and tetrahydroanthracene [16,20,23,42–45]. The intercalating ligands in coordinately saturated octahedral Ru^{II}, Os^{II}, and Rh^{III} complexes showed similar function in increasing their DNA-binding affinity, and conferred them with capacity recognizing DNA bases via shape selection [46]. The strong π–π stacking of acridine–thymine rings was also found to enhance the affinity of Zn(II)–acridinylcyclen complexes to N3 of dT [47]. However, for some metal complexes, e.g., *trans*-[Pt(quinoline)(NH$_3$)-(9EtG)Cl]$^+$, the rigid quinoline ring retards the binding of Pt to N7 of 5′-GMP [48]. On the other hand, when the extended arene ligand rings in transition metal complexes intercalate into DNA bases, it may extend the phosphate spacing along the helix axis; as a consequence, lengthening and unwinding DNA helices [49,50]. Such intercalation distorts the structure of DNA helices, and makes the hidden sites, for example, T-N3, which participated in H-bonding with adenine base in duplex DNA, exposed for metal attacking. The reaction of complex **4** with duplex **III** produced pronounced amounts of G, T-bound di-/tri- ruthenated strand **I**, but only G-bound mono-ruthenated strand **I** formed by reaction of complex **1** with the same duplex ODN, strongly supporting this deduction [16,51].

Although both cisplatin and the Ru arene complexes can bind to DNA, and unwind and loosen the DNA helices, the changes in the helical structure of the bound DNA are different. For example, cisplatin unwinds DNA duplex to about 25° [52], while ruthenium complexes only unwind at most 14° [23]. The main reason is that cisplatin mainly forms intrastrand crosslinks, which bend the helices toward the platinated sites, while ruthenium complexes form mono-functional adducts accompanied by the intercalation of arene ligands between nucelobases, bending DNA duplexes toward the direction opposite to the ruthenated site. Therefore, the changes in the helicity of DNA due to Ru binding are dependent on the size of arene ligands of the Ru complexes as evidenced by CD spectra shown in Figure 5C.

Furthermore, our thermodynamic data demonstrate that the arene ligands of the ruthenium complexes regulate not only base selectivity, but also sequence selectivity of these complexes to DNA. For instance, there are six G residues in the single strand **II**. Our studies show that G$_{21}$ in the middle region is the common binding site for all the four ruthenium complexes, and that G$_{18}$ in the 5′-region and G$_{26}$ in the 3′-region are the secondary binding site for complexes **3** and **4**, respectively, while both G$_{18}$ and G$_{26}$ are the secondary binding site for complex **2**. These results imply that the ruthenium arene complexes preferentially bind to guanine bases in the middle region of the ODN strand.

For the reactions of polyanionic DNA and the cationic [(η^6-arene)Ru(en)(H$_2$O)]$^{2+}$, which is the reactive species of the Ru arene complexes produced by hydrolysis in aqueous solution, electrostatic interactions and H-bonding are the initial recognition forces between Ru complexes and DNA prior to Ru–DNA coordination [16]. Such binding is noncovalent and reversible. It may significantly influence the site preference of DNA-metal bindings, which are kinetically controlled [53]. It is proposed that the preferential binding of cisplatin to guanines is predominantly a result of electrostatic attraction of the platinum toward the most nucleophilic sites [54,55], for example, the nitrogen N7 atoms of purines as they are the most electron-dense and accessible sites in DNA for electrophilic attack by platinum, or by ruthenium in this work. Moreover, the G-N7 sites are exposed in the major groove of the double helix, and are not involved in base-pair hydrogen-bonding [56,57], which makes G-N7 sites more favored for metalation. However, variations at the neighboring bases of G may cause a change in the microenvironment that in turn modulates metal coordination and alters the affinity of Ru atom to guanine. Indeed, our thermodynamic studies show that the binding constants of complex **4** to the centered G base in seven 15-mer single-stranded ODNs containing a variation in the flanking bases of the G site are significantly different. The binding constant of complex **4** to G in -A$_7$G$_8$T$_9$-

sequence is nearly two-fold higher than that to G in -C$_7$G$_8$A$_9$- sequence. An interesting result was that for the single-G sequences with adenine being the flanking base in one side, the affinity of Ru to G decreased with the variation of the other side base in an order as follows: T > A > C, whether A was at the 3′- or 5′-side of the central G site. This is in line with the reactivity order of the biphenyl organometallic ruthenium complex to mononucleotides where Ru binding to A and C is much weaker than binding to G and T, implying that the binding of the ruthenium complex to an adjacent T base may be synergetic to the thermodynamically favored G-binding [24,25]. It has been previously reported that the flanking bases play an important role in the sequence selectivity of the bindings of *cis*-platinum anticancer drugs to DNA [58]. One of the early studies on this topic demonstrated that the reaction of cisplatin (*cis*-[PtCl$_2$(NH$_3$)$_2$]) with salmon sperm DNA gave rise to chelating adducts with Pt–GG (65%) and Pt–AG (25%), but no Pt–GA adducts (0%) additional to Pt–G mono-functional adduct (10%) [59]. However, the mechanism of crosslinking did not necessarily follow the sequence selectivity of the initial coordination. The fact that the 5′-monoplatinated adducts are formed more rapidly than the 3′-monoplatinated adducts may reflect the inherently greater reactivity of the 5′-G compared with the 3′-G [60]. To explain the differences, Kozelka and Chottard et al. proposed a model according to which the neighboring bases have an influence on the sequence selectively of Pt–G coordination by changing the electronegativity of the N7 site, and presenting different steric hindrances when platinum complex approaches the G-N7 site [61]. Other non-metallic DNA-binding drugs, like aflatoxin [62] and nitrogen mustards [63], also showed discriminated reactivity to G-N7 upon the local sequence context.

4. Materials and Methods

4.1. Materials

[(η^6-arene)Ru(en)Cl][PF$_6$] (arene = benzene (ben) (**1**), *p*-cymene (cym) (**2**), indane (ind) (**3**), and biphenyl (bip) (**4**); en = ethylenediamine; Chart 1) were synthesized as described in the literature [2,17,64,65]. Triethylammonium acetate buffer (TEAA, 1 M) was purchased from AppliChem (Darmstadt, Germany), and acetonitrile (HPLC grade) from Tedia (Fairfield, OH, USA). HPLC-purified 15-mer oligodeoxynucleotides (ODNs, Chart 1) were obtained from TaKaRa (Dalian, China), and the concentration was determined by UV at 260 nm absorption according to the instruction of manufacturer. Snake venom phosphodiesterase (SVP) was purchased from Orientoxin (Shandong, China). Microcon YM-3 ultrafilters were purchased from Millipore (Billerica, MA, USA). Aqueous solutions were prepared using MilliQ water (MilliQ Reagent Water System, Molsheim, France).

4.2. Sample Preparation

Stock solutions of Ru(II) arene complexes (5 mM) were prepared by dissolving individual complex in deionized water and diluted as required before use. The ODNs were dissolved in water to give 2 mM stock solutions. The reaction mixtures containing 0.1 mM single-stranded ODNs in 50 mM TEAA (pH 7) or 0.4 mM duplex **III** in 50 mM TEAA buffer containing 100 mM NaClO$_4$ (pH 7) and various concentrations of Ru complexes were incubated at 310 K in the dark for 24 h, unless otherwise stated, and then separated or analyzed by HPLC or/and HPLC-ESI-MS. The general procedure to identify the binding sites of Ru complexes on ODNs was similar to that reported in our previous works by using exonuclease digestions combined with HPLC-ESI-MS [19] and the top-down MS method [24,66].

4.3. High Performance Liquid Chromatography (HPLC)

An Agilent 1200 series HPLC system was applied coupled with a quaternary pump, a 20-μL Rheodyne sample injector, and a UV-Vis diode-array-detector (DAD) detector. All the HPLC data was processed by the Chemstation data processing system. Water containing 20 mM TEAA and acetonitrile containing 20 mM TEAA were used as mobile phases A and B, respectively. HPLC assays for reaction mixtures of Ru(II) arene complexes with different single-stranded ODNs were carried out on a Varian Pursuit XRs C18 reversed-phase column (100 × 2.0 mm, 3 μm, 0.2 mL·min^{-1}, Varian, Inc., Palo Alto,

CA, USA). The gradient for separating reaction mixtures of complexes **2** and **4** with strand **I** (Chart 1) was as follows: 8% solvent B from 0 to 3 min, increasing to 60% at 10 min, 80% from 11 min to 15 min, and resetting to 8% at 16 min. The same gradient was also applied to separate the reaction mixtures of complex **4** with other six single-stranded ODNs. For reaction mixtures of complexes **1** and **3** with strand **I**, the applied gradient was as follows: from 8% to 18% during first 10 min, 80% from 11 min to 15 min, and resetting to 8% at 16 min. The gradient was also applied to separate the reaction mixtures of complexes **1** and **4** with strand **II** (Chart 1). The same column was used to separate the products from enzymatic digestions of ruthenated ODNs by exonucleases prior to MS analysis with a gradient as follows: 1% to 5% solvent B from 0 to 5 min, 5% from 5 to 8 min, then increasing to 80% and keeping it from 15 to 21 min, and finally resetting to 1% at 22 min. An Agela C8 column (50×2.1 mm, 0.2 mL min^{-1}, Agela Technology, Tianjin, China) was used to separate the reaction mixtures of ruthenium complexes with duplex **III** (Chart 1) prior to MS analysis with a gradient as follows: 1% solvent B for 3 min, 5% to 10% from 4 to 9 min and kept at 10% for 5 min, then increased to 80% at 15 min and retained for 4 min, and finally reset to 1% at 20 min.

4.4. Electrospray Ionization Mass Spectroscopy (ESI-MS)

A Micromass Q-TOF (Waters Corp., Manchester, UK) coupled with an Agilent 1200 system was applied under negative-ion mode for the online HPLC-ESI-MS assays. The HPLC conditions were the same as described above, with a flow rate of 0.2 mL·min^{-1} and a splitting ratio of 1/3 into mass spectrometer. All the MS data was analyzed and post-processed by a Masslynx (ver. 4.0, Wasters Corp., Manchester, UK) data processing system. The MS conditions were similar as in our previously reported paper [19] and the typical conditions are as follows: spray voltage 2.8~3.8 kV; cone voltage 55~70 V; desolvation temperature 393 K; source temperature 373 K; cone gas flow rate 50 L·h^{-1}; desolvation gas flow rate 500 L·h^{-1}; and collision energy 10 V. Mass spectra were acquired in the range of 200–3000 m/z with mass accuracy of all measurements within 0.01 m/z unit, which were calibrated versus a NaI calibration file. All the m/z values are the mass-to-charge ratios of the most abundant isotopomer for observed ions.

4.5. Circular Dichroism (CD) Spectroscopy

For CD spectroscopy analysis, the reaction mixtures, containing 15 μL 0.1 mM duplex **III** in 100 mM NaClO$_4$ and 50 mM TEAA (pH 7), and 15 μL various concentrations of Ru complexes in 50 mM TEAA, were incubated at 310 K for 24 h in dark, and then diluted to 5 μM using 50 mM TEAA buffer prior to CD analysis. CD spectra were recorded by using a Jasco J-810 spectropolarimeter in the range 220–340 nm in 0.5 nm increments with an average time of 1 s, the cell path-length was 1 cm. Each spectrum was the average of ten scans and corrected by the blank buffer solution.

4.6. Determination of Equilibrium Binding Constants of Organometallic Ruthenium Complexes to ODNs

The peak areas of ruthenated ODNs separated by HPLC with UV detection at 260 nm were used to determine the equilibrium binding constants of Ru(II) arene complexes to single-stranded ODNs. Klotz's methods (affinities from a stoichiometric perspective) were applied to construct the reaction model and determine the stoichiometric equilibrium constants (Scheme 1) [67]. The G- and T-bound two-site divalent model was applied as there are only G- and T-binding sites of ruthenium complexes on each single strand ODN identified as described previously [24]. Although there are more than one T binding sites on single strands **I** and **IV–VI**, it is difficult to fully separate the different G$_8$, T$_1$-bound ODN adducts by HPLC [24]. Therefore, for clarity, the HPLC peak areas corresponding to di-ruthenated ODN adducts with different T-sites were summed to calculate the content of di-ruthenated species.

For the divalent receptor system, the uptake of ligand L (Ru complexes) by receptor R (ODNs) can be described by two steps:

$$R + L = RL_1 \tag{1}$$

$$RL_1 + L = RL_2 \tag{2}$$

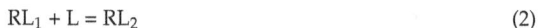

with the following equilibrium constants:

$$K_1 = [RL_1]/[R][L] \tag{3}$$

$$K_2 = [RL_2]/[RL_1][L] \tag{4}$$

Thus,

$$[RL_1] = K_1[R][L] \tag{5}$$

$$[RL_2] = K_2[RL_1][L] = K_1 K_2[R][L]^2 \tag{6}$$

Therefore, the extent of site occupation (B) can be defined as follows:

$$B = \text{mole of bound ligand/mole of total receptor} = ([RL_1] + 2[RL_2])/([R] + [RL_1] + [RL_2]) \tag{7}$$

Replacing $[RL_1]$ and $[RL_2]$ by Equations (5) and (6) into Equation (7), we obtain the stoichiometric binding equation:

$$B = (K_1[L] + 2K_1 K_2[L]^2)/(1 + K_1[L] + K_1 K_2[L]^2) \tag{8}$$

where K_1 and K_2 are the binding constants for the first and the second step of reaction, respectively. The B $---$ [L] data obtained by HPLC analysis were fitted computationally using Equation (8) to give rise to the stoichiometric equilibrium constants K_1 and K_2.

5. Conclusions

In summary, we have thermodynamically studied the reactions of the organometallic ruthenium(II) anticancer complexes [(η^6-arene)Ru(en)Cl][PF$_6$] (arene = benzene (**1**), *p*-cymene (**2**), indane (**3**), biphenyl (**4**); en = ethylenediamine) with the 15-mer one-G-containing single-stranded ODN **I**; its complementary strand **II** and the duplex **III** (= **I** + **II**); and a series of analogue strands of **I** by HPLC, ESI-MS, and CD spectroscopy. Different arene ligands regulate the reactivity of the ruthenium complexes towards G and T of strand **I**, with complex **2** being the most discriminatory one between G and T bases, and complex **4** the most reactive one. In the complementary single-strand **II**, G$_{21}$ is the common preferential binding site for complexes **1–4**; G$_{18}$ and G$_{26}$ are the secondary binding sites for complexes **3** and **4**, respectively; and both G$_{21}$ and G$_{26}$ are the secondary binding sites for complex **2**. This implies that the G bases in the middle region are more favored for ruthenation than G bases in the end of ODNs. When reacted with different single-G-containing ODNs, complex **4** shows a decreasing affinity to the G$_8$ site in the following order: -AG$_8$T- (*K*: 5.74 × 10^4 M^{-1}) > -CG$_8$C- > -TG$_8$A- > -AG$_8$A- > -AG$_8$C- > -TG$_8$T- (**I**) ≈ -CG$_8$A- (*K*: 2.81 × 10^4 M^{-1}), indicating that adjacent bases play an important role in the sequence selectivity of the bindings of the organometallic ruthenium(II) complexes to DNA, perhaps related to the arene intercalation and the steric hindrance of adjacent bases. Among the tested ruthenium complexes, complex **4** caused the most severe distortion to DNA duplex conformation as a result of the intercalation of biphenyl ligand into bases. These findings provide a molecular basis for better understanding in the base and sequence selectivity of the binding of ruthenium arene anticancer complexes to DNA, and is helpful to design and development of novel ruthenium-based anticancer complexes.

Supplementary Materials: Supplementary materials can be found at http://www.mdpi.com/1422-0067/19/7/2137/s1.

Author Contributions: F.W. and K.W. designed the experiments; S.L., A.L., K.W. and W.Z. conducted experiments; all authors analyzed and interpreted data. S.L., K.W. and F.W. carried out writing—original draft; F.W. and K.W. edited the manuscript. All authors have reviewed the manuscript and agreed to publish the results.

Acknowledgments: We thank NSFC (Grant Nos. 21575145, 21621062, 21505141) for support.

Conflicts of Interest: The authors declare no conflict of interest.

References

1. Aird, R.E.; Cummings, J.; Ritchie, A.A.; Muir, M.; Morris, R.E.; Chen, H.; Sadler, P.J.; Jodrell, D.I. In vitro and in vivo activity and cross resistance profiles of novel ruthenium(II) organometallic arene complexes in human ovarian cancer. *Br. J. Cancer* **2002**, *86*, 1652–1657. [CrossRef] [PubMed]

2. Morris, R.E.; Aird, R.E.; Murdoch, P.D.; Chen, H.M.; Cummings, J.; Hughes, N.D.; Parsons, S.; Parkin, A.; Boyd, G.; Jodrell, D.I.; et al. Inhibition of cancer cell growth by ruthenium(II) arene complexes. *J. Med. Chem.* **2001**, *44*, 3616–3621. [CrossRef] [PubMed]

3. Yan, Y.K.; Melchart, M.; Habtemariam, A.; Sadler, P.J. Organometallic chemistry, biology and medicine: Ruthenium arene anticancer complexes. *Chem. Commun.* **2005**, 4764–4776. [CrossRef] [PubMed]

4. Sava, G.; Bergamo, A.; Dyson, P.J. Metal-based antitumour drugs in the post-genomic era: What comes next? *Dalton Trans.* **2011**, *40*, 9069–9075. [CrossRef] [PubMed]

5. Hartinger, C.G.; Dyson, P.J. Bioorganometallic chemistry-from teaching paradigms to medicinal applications. *Chem. Soc. Rev.* **2009**, *38*, 391–401. [CrossRef] [PubMed]

6. Meggers, E. Targeting proteins with metal complexes. *Chem. Commun.* **2009**, 1001–1010. [CrossRef] [PubMed]

7. Hambley, T.W. Developing new metal-based therapeutics: Challenges and opportunities. *Dalton Trans.* **2007**, 4929–4937. [CrossRef] [PubMed]

8. Mjos, K.D.; Orvig, C. Metallodrugs in Medicinal Inorganic Chemistry. *Chem. Rev.* **2014**, *114*, 4540–4563. [CrossRef] [PubMed]

9. Suss-Fink, G. Arene ruthenium complexes as anticancer agents. *Dalton Trans.* **2010**, *39*, 1673–1688. [CrossRef] [PubMed]

10. Bennett, M.A.; Byrnes, M.J.; Kovacik, I. The fragment bis(acetylacetonato)ruthenium: A meeting-point of coordination and organometallic chemistry. *J. Organomet. Chem.* **2004**, *689*, 4463–4474. [CrossRef]

11. Jamieson, E.R.; Lippard, S.J. Structure, recognition, and processing of cisplatin-DNA adducts. *Chem. Rev.* **1999**, *99*, 2467–2498. [CrossRef] [PubMed]

12. Sava, G.; Bergamo, A. Ruthenium-based compounds and tumour growth control. *Int. J. Oncol.* **2000**, *17*, 353–365. [CrossRef] [PubMed]

13. Galanski, M.; Arion, V.B.; Jakupec, M.A.; Keppler, B.K. Recent developments in the field of tumor-inhibiting metal complexes. *Curr. Pharm. Des.* **2003**, *9*, 2078–2089. [CrossRef] [PubMed]

14. Rademaker-Lakhai, J.M.; van den Bongard, D.; Pluim, D.; Beijnen, J.H.; Schellens, J.H.M. A phase I and pharmacological study with imidazolium-trans-DMSO-imidazole-tetrachlororuthenate, a novel ruthenium anticancer agent. *Clin. Cancer Res.* **2004**, *10*, 3717–3727. [CrossRef] [PubMed]

15. Jakupec, M.A.; Galanski, M.; Arion, V.B.; Hartinger, C.G.; Keppler, B.K. Antitumour metal compounds: More than theme and variations. *Dalton Trans.* **2008**, 183–194. [CrossRef] [PubMed]

16. Chen, H.M.; Parkinson, J.A.; Morris, R.E.; Sadler, P.J. Highly selective binding of organometallic ruthenium ethylenediamine complexes to nucleic acids: Novel recognition mechanisms. *J. Am. Chem. Soc.* **2003**, *125*, 173–186. [CrossRef] [PubMed]

17. Chen, H.M.; Parkinson, J.A.; Parsons, S.; Coxall, R.A.; Gould, R.O.; Sadler, P.J. Organometallic ruthenium(II) diamine anticancer complexes: Arene-nucleobase stacking and stereospecific hydrogen-bonding in guanine adducts. *J. Am. Chem. Soc.* **2002**, *124*, 3064–3082. [CrossRef] [PubMed]

18. Novakova, O.; Kasparkova, J.; Bursova, V.; Hofr, C.; Vojtiskova, M.; Chen, H.M.; Sadler, P.J.; Brabec, V. Conformation of DNA modified by monofunctional Ru(II) arene complexes: Recognition by DNA binding proteins and repair. Relationship to cytotoxicity. *Chem. Biol.* **2005**, *12*, 121–129. [CrossRef] [PubMed]

19. Wu, K.; Luo, Q.; Hu, W.B.; Li, X.C.; Wang, F.Y.; Xiong, S.X.; Sadler, P.J. Mechanism of interstrand migration of organoruthenium anticancer complexes within a DNA duplex. *Metallomics* **2012**, *4*, 139–148. [CrossRef] [PubMed]

20. Liu, H.K.; Berners-Price, S.J.; Wang, F.Y.; Parkinson, J.A.; Xu, J.J.; Bella, J.; Sadler, P.J. Diversity in guanine-selective DNA binding modes for an organometallic ruthenium arene complex. *Angew. Chem. Int. Ed.* **2006**, *45*, 8153–8156. [CrossRef] [PubMed]

21. Liu, H.K.; Wang, F.Y.; Parkinson, J.A.; Bella, J.; Sadler, P.J. Ruthenation of duplex and single-stranded d(CGGCCG) by organometallic anticancer complexes. *Chem. Eur. J.* **2006**, *12*, 6151–6165. [CrossRef] [PubMed]

22. Wang, F.Y.; Bella, J.; Parkinson, J.A.; Sadler, P.J. Competitive reactions of a ruthenium arene anticancer complex with histidine, cytochrome c and an oligonucleotide. *J. Biol. Inorg. Chem.* **2005**, *10*, 147–155. [CrossRef] [PubMed]

23. Novakova, O.; Chen, H.M.; Vrana, O.; Rodger, A.; Sadler, P.J.; Brabec, V. DNA interactions of monofunctional organometallic ruthenium(II) antitumor complexes in cell-free media. *Biochemistry* **2003**, *42*, 11544–11554. [CrossRef] [PubMed]

24. Wu, K.; Hu, W.B.; Luo, Q.; Li, X.C.; Xiong, S.X.; Sadler, P.J.; Wang, F.Y. Competitive Binding Sites of a Ruthenium Arene Anticancer Complex on Oligonucleotides Studied by Mass Spectrometry: Ladder-Sequencing versus Top-Down. *J. Am. Soc. Mass Spectrom.* **2013**, *24*, 410–420. [CrossRef] [PubMed]

25. Wu, K.; Liu, S.Y.; Luo, Q.; Hu, W.B.; Li, X.C.; Wang, F.Y.; Zheng, R.H.; Cui, J.; Sadler, P.J.; Xiang, J.F.; et al. Thymines in Single-Stranded Oligonucleotides and G-Quadruplex DNA Are Competitive with Guanines for Binding to an Organoruthenium Anticancer Complex. *Inorg. Chem.* **2013**, *52*, 11332–11342. [CrossRef] [PubMed]

26. Wang, F.; Chen, H.M.; Parsons, S.; Oswald, L.D.H.; Davidson, J.E.; Sadler, P.J. Kinetics of aquation and anation of Ruthenium(II) arene anticancer complexes, acidity and X-ray structures of aqua adducts. *Chem. Eur. J.* **2003**, *9*, 5810–5820. [CrossRef] [PubMed]

27. Wang, F.Y.; Habtemariam, A.; van der Geer, E.P.L.; Fernandez, R.; Melchart, M.; Deeth, R.J.; Aird, R.; Guichard, S.; Fabbiani, F.P.A.; Lozano-Casal, P.; et al. Controlling ligand substitution reactions of organometallic complexes: Tuning cancer cell cytotoxicity. *Proc. Natl. Acad. Sci. USA* **2005**, *102*, 18269–18274. [CrossRef] [PubMed]

28. Guo, Z.J.; Sadler, P.J. Medicinal inorganic chemistry. *Adv. Inorg. Chem.* **2000**, *49*, 183–306.

29. Dijt, F.J.; Chottard, J.C.; Girault, J.P.; Reedijk, J. Formation and Structure of Reaction-Products of Cis-Ptcl$_2$(Nh$_3$)$_2$ with D(Apg) and or D(Gpa) in Dinucleotide, Trinucleotide and Penta-Nucleotide—Preference for Gpa Chelation over Apg Chelation. *Eur. J. Biochem.* **1989**, *179*, 333–344. [CrossRef]

30. Davies, M.S.; Berners-Price, S.J.; Hambley, T.W. Rates of platination of AG and GA containing double-stranded oligonucleotides: Insights into why cisplatin binds to GG and AG but not GA sequences in DNA. *J. Am. Chem. Soc.* **1998**, *120*, 11380–11390. [CrossRef]

31. Legendre, F.; Kozelka, J.; Chottard, J.C. GG versus AG platination: A kinetic study on hairpin-stabilized duplex oligonucleotides. *Inorg. Chem.* **1998**, *37*, 3964–3967. [CrossRef] [PubMed]

32. Reeder, F.; Guo, Z.J.; Murdoch, P.D.; Corazza, A.; Hambley, T.W.; Berners-Price, S.J.; Chottard, J.C.; Sadler, P.J. Platination of a GG site on single-stranded and double-stranded forms of a 14-base oligonucleotide with diaqua cisplatin followed by NMR and HPLC—Influence of the platinum ligands and base sequence on 5′-G versus 3′-G platination selectivity. *Eur. J. Biochem.* **1997**, *249*, 370–382. [CrossRef] [PubMed]

33. Reeder, F.; Gonnet, F.; Kozelka, J.; Chottard, J.C. Reactions of the double-stranded oligonucleotide d(TTGGCCAA)$_2$ with *cis*-[Pt(NH$_3$)$_2$(H$_2$O)$_{(2)}$]$^{2+}$ and [Pt(NH$_3$)$_3$(H$_2$O)]$^{2+}$. *Chem.-A Eur. J.* **1996**, *2*, 1068–1076. [CrossRef]

34. Garbett, N.C.; Ragazzon, P.A.; Chaires, J.B. Circular dichroism to determine binding mode and affinity of ligand-DNA interactions. *Nat. Protoc.* **2007**, *2*, 3166–3172. [CrossRef] [PubMed]

35. Boger, D.L.; Fink, B.E.; Brunette, S.R.; Tse, W.C.; Hedrick, M.P. A simple, high-resolution method for establishing DNA binding affinity and sequence selectivity. *J. Am. Chem. Soc.* **2001**, *123*, 5878–5891. [CrossRef] [PubMed]

36. Neidle, S. Recent developments in triple-helix regulation of gene expression. *Anticancer Drug Des.* **1997**, *12*, 433–442. [PubMed]

37. Choo, Y.; Sanchezgarcia, I.; Klug, A. In-Vivo Repression by a Site-Specific DNA-Binding Protein Designed against an Oncogenic Sequence. *Nature* **1994**, *372*, 642–645. [CrossRef] [PubMed]

38. Mrksich, M.; Parks, M.E.; Dervan, P.B. Hairpin Peptide Motif—A New Class of Oligopeptides for Sequence-Specific Recognition in the Minor-Groove of Double-Helical DNA. *J. Am. Chem. Soc.* **1994**, *116*, 7983–7988. [CrossRef]

39. Werstuck, G.; Green, M.R. Controlling gene expression in living cells through small molecule-RNA interactions. *Science* **1998**, *282*, 296–298. [CrossRef] [PubMed]

40. Kimura, E.; Kitamura, H.; Ohtani, K.; Koike, T. Elaboration of selective and efficient recognition of thymine base in dinucleotides (TpT, ApT, CpT, and GpT), single-stranded d(GTGACGCC), and double-stranded d(CGCTAGCC)$_2$ by Zn^{2+}-acridinylcyclen (acridinylcyclen = (9-acridinyl)methyl-1,4,7,10-tetraazacyclododecane). *J. Am. Chem. Soc.* **2000**, *122*, 4668–4677.

41. Shionoya, M.; Kimura, E.; Shiro, M. A New Ternary Zinc(II) Complex with [12]Anen(4) (=1,4,7,10-Tetraazacyclododecane) and Azt (=3′-Azido-3′-Deoxythymidine)—Highly Selective Recognition of Thymidine and Its Related Nucleosides by a Zinc(II) Macrocyclic Tetraamine Complex with Novel Complementary Associations. *J. Am. Chem. Soc.* **1993**, *115*, 6730–6737.

42. Chen, H.M.; Parkinson, J.A.; Novakova, O.; Bella, J.; Wang, F.Y.; Dawson, A.; Gould, R.; Parsons, S.; Brabec, V.; Sadler, P.J. Induced-fit recognition of DNA by organometallic complexes with dynamic stereogenic centers. *Proc. Natl. Acad. Sci. USA* **2003**, *100*, 14623–14628. [CrossRef] [PubMed]

43. Liu, H.K.; Parkinson, J.A.; Bella, J.; Wang, F.Y.; Sadler, P.J. Penetrative DNA intercalation and G-base selectivity of an organometallic tetrahydroanthracene Ru^{II} anticancer complex. *Chem. Sci.* **2010**, *1*, 258–270. [CrossRef]

44. Liu, H.K.; Sadler, P.J. Metal Complexes as DNA Intercalators. *Acc. Chem. Res.* **2011**, *44*, 349–359. [CrossRef] [PubMed]

45. Ren, J.S.; Jenkins, T.C.; Chaires, J.B. Energetics of DNA intercalation reactions. *Biochemistry* **2000**, *39*, 8439–8447. [CrossRef] [PubMed]

46. Erkkila, K.E.; Odom, D.T.; Barton, J.K. Recognition and reaction of metallointercalators with DNA. *Chem. Rev.* **1999**, *99*, 2777–2795. [CrossRef] [PubMed]

47. Shionoya, M.; Ikeda, T.; Kimura, E.; Shiro, M. Novel Multipoint Molecular Recognition of Nucleobases by a New Zinc(II) Complex of Acridine-Pendant Cyclen (Cyclen=1,4,7,10-Tetraazacyclododecane). *J. Am. Chem. Soc.* **1994**, *116*, 3848–3859. [CrossRef]

48. Bierbach, U.; Farrell, N. Modulation of nucleotide binding of *trans*-platinum(II) complexes by planar ligands. A combined proton NMR and molecular mechanics study. *Inorg. Chem.* **1997**, *36*, 3657–3665. [CrossRef] [PubMed]

49. Bjorndal, M.T.; Fygenson, D.K. DNA melting in the presence of fluorescent intercalating oxazole yellow dyes measured with a gel-based assay. *Biopolymers* **2002**, *65*, 40–44. [CrossRef] [PubMed]

50. Maeda, Y.; Nunomura, K.; Ohtsubo, E. Differential scanning calorimetrics study of the effect of intercalators and other kinds of DNA-binding drugs on the stepwise melting of plasmid DNA. *J. Mol. Biol.* **1990**, *215*, 321–329. [CrossRef]

51. Rodger, A.; Norden, B. *Circular Dichroism and Linear Dichroism*; Oxford University Press: Oxford, UK; New York, NY, USA; Tokyo, Japan, 1997.

52. Gelasco, A.; Lippard, S.J. NMR solution structure of a DNA dodecamer duplex containing a cis-diammineplatinum(II) d(GpG) intrastrand cross-link, the major adduct of the anticancer drug cisplatin. *Biochemistry* **1998**, *37*, 9230–9239. [CrossRef] [PubMed]

53. Keenea, F.R.; Smith, J.A.; Collins, J.G. Metal complexes as structure-selective binding agents for nucleic acids. *Coord. Chem. Rev.* **2009**, *253*, 2021–2035. [CrossRef]

54. Elmroth, S.K.C.; Lippard, S.J. Surface and Electrostatic Contributions to DNA-Promoted Reactions of Platinum(II) Complexes with Short Oligonucleotides—A Kinetic-Study. *Inorg. Chem.* **1995**, *34*, 5234–5243. [CrossRef]

55. Elmroth, S.K.C.; Lippard, S.J. Platinum binding to d(GpG) target sequences and phosphorothioate linkages in DNA occurs more rapidly with increasing oligonucleotide length. *J. Am. Chem. Soc.* **1994**, *116*, 3633–3634. [CrossRef]

56. Pizarro, A.M.; Sadler, P.J. Unusual DNA binding modes for metal anticancer complexes. *Biochimie* **2009**, *91*, 1198–1211. [CrossRef] [PubMed]

57. Saenger, W. Principles of Nucleic Acid Structure. In *Springer Advanced Texts in Chemistry*; Cantor, C.R., Ed.; Springer: New York, NY, USA, 1984; pp. 201–219.

58. Lippert, B. *Cisplatin—Chemistry and Biochemistry of a Leading Anticancer Drug*; Wiley-VCH: Weinheim, Germany, 1999.

59. Fichtinger-Schepman, A.M.J.; Vanderveer, J.L.; Denhartog, J.H.J.; Lohman, P.H.M.; Reedijk, J. Adducts of the Antitumor Drug *Cis*-Diamminedichloroplatinum(II) with DNA—Formation, Identification, and Quantitation. *Biochemistry* **1985**, *24*, 707–713. [CrossRef] [PubMed]

60. Kozelka, J. Molecular origin of the sequence-dependent kinetics of reactions between cisplatin derivatives and DNA. *Inorg. Chim. Acta* **2009**, *362*, 651–668. [CrossRef]

61. Monjardet-Bas, V.; Elizondo-Riojas, M.A.; Chottard, J.C.; Kozelka, J. A combined effect of molecular electrostatic potential and N7 accessibility explains sequence-dependent binding of cis-[Pt(NH$_3$)$_2$(H$_2$O)$_2$]$^{2+}$ to DNA duplexes. *Angew. Chem. Int. Ed.* **2002**, *41*, 2998–3001. [CrossRef]

62. Benasutti, M.; Ejadi, S.; Whitlow, M.D.; Loechler, E.L. Mapping the binding site of aflatoxin B1 in DNA: Systematic analysis of the reactivity of aflatoxin B1 with guanines in different DNA sequences. *Biochemistry* **1988**, *27*, 472–481. [CrossRef] [PubMed]

63. Mattes, W.B.; Hartley, J.A.; Kohn, K.W. DNA sequence selectivity of guanine–N7 alkylation by nitrogen mustards. *Nucleic Acids Res.* **1986**, *14*, 2971–2987. [CrossRef] [PubMed]

64. Zelonka, R.A.; Baird, M.C. Benzene Complexes of Ruthenium(II). *Can. J. Chem.* **1972**, *50*, 3063–3072. [CrossRef]

65. Bennett, M.A.; Smith, A.K. Arene Ruthenium(II) Complexes Formed by Dehydrogenation of Cyclohexadienes with Ruthenium(III) Trichloride. *J. Chem. Soc. Dalton Trans.* **1974**, 233–241. [CrossRef]

66. Liu, S.Y.; Wu, K.; Zheng, W.; Zhao, Y.; Luo, Q.; Xiong, S.X.; Wang, F.Y. Identification and discrimination of binding sites of an organoruthenium anticancer complex to single-stranded oligonucleotides by mass spectrometry. *Analyst* **2014**, *139*, 4491–4496. [CrossRef] [PubMed]

67. Klotz, I.M. *Ligand Receptor Energetics: A Guide for the Perplexed*; John Wiley & Sons, Inc.: New York, NY, USA, 1997.

International Journal of
Molecular Sciences

MDPI

Article

Poly Organotin Acetates against DNA with Possible Implementation on Human Breast Cancer

George K. Latsis [1], Christina N. Banti [1,*], Nikolaos Kourkoumelis [2,*],
Constantina Papatriantafyllopoulou [3], Nikos Panagiotou [3], Anastasios Tasiopoulos [3],
Alexios Douvalis [4], Angelos G. Kalampounias [5], Thomas Bakas [4] and Sotiris K. Hadjikakou [1,*]

[1] Section of Inorganic and Analytical Chemistry, Department of Chemistry, University of Ioannina, 45110 Ioannina, Greece; glatsis@icloud.com
[2] Medical Physics Laboratory, Medical School, University of Ioannina, 45110 Ioannina, Greece
[3] Department of Chemistry, University of Cyprus, 1678 Nicosia, Cyprus; constantina.papatriantafyllopo@nuigalway.ie (C.P.); panagiotou.nikos@ucy.ac.cy (N.P.); atasio@ucy.ac.cy (A.T.)
[4] Mössbauer Spectroscopy and Physics of Material Laboratory, Department of Physics, University of Ioannina, 45110 Ioannina, Greece; adouval@cc.uoi.gr (A.D.); tbakas@uoi.gr (T.B.)
[5] Physical Chemistry Laboratory, Department of Chemistry, University of Ioannina, 45110 Ioannina, Greece; akalamp@cc.uoi.gr
* Correspondence: cbanti@cc.uoi.gr (C.N.B.); nkourkou@uoi.gr (N.K.); shadjika@uoi.gr (S.K.H.); Tel.: +30-2651-008374 (S.K.H.)

Received: 16 June 2018; Accepted: 12 July 2018; Published: 14 July 2018

Abstract: Two known tin-based polymers of formula $\{[R_3Sn(CH_3COO)]_n\}$ where R = *n*-Bu– (**1**) and R = Ph– (**2**),were evaluated for their in vitro biological properties. The compounds were characterized via their physical properties and FT-IR, ^{119}Sn Mössbauer, and ^{1}H NMR spectroscopic data. The molecular structures were confirmed by single-crystal X-Ray diffraction crystallography. The geometry around the tin(IV) ion is trigonal bi-pyramidal. Variations in O–Sn–O···Sn′ torsion angles lead to zig-zag and helical supramolecular assemblies for **1** and **2**, respectively. The in vitro cell viability against human breast adenocarcinoma cancer cell lines: MCF-7 positive to estrogens receptors (ERs) and MDA-MB-231 negative to ERs upon their incubation with **1** and **2** was investigated. Their toxicity has been studied against normal human fetal lung fibroblast cells (MRC-5). Compounds **1** and **2** exhibit 134 and 223-fold respectively stronger antiproliferative activity against MDA-MB-231 than cisplatin. The type of the cell death caused by **1** or **2** was also determined using flow cytometry assay. The binding affinity of **1** and **2** towards the CT-DNA was suspected from the differentiation of the viscosity which occurred in the solution containing increasing amounts of **1** and **2**. Changes in fluorescent emission light of Ethidium bromide (EB) in the presence of DNA confirmed the intercalation mode of interactions into DNA of both complexes **1** and **2** which have been ascertained from viscosity measurements. The corresponding apparent binding constants (K_{app}) of **1** and **2** towards CT-DNA calculated through fluorescence spectra are 4.9×10^4 (**1**) and 7.3×10^4 (**2**) M^{-1} respectively. Finally, the type of DNA binding interactions with **1** and **2** was confirmed by docking studies.

Keywords: biological inorganic chemistry; acetic acid; organotins; bio-polymer; anti-cancer activity; cell cycle

1. Introduction

Platinum-based compounds are at the focal point of research on potent anticancer drugs since the discovery of the anticancer potential of cisplatin, back during 70's [1–4]. However, platinum-based cancer treatments are being dominated by serious side effects [5]. Moreover, cancer cells resistance against platinum based drugs is developed in a short while [6]. Nowadays, organometallic compounds, with a different pharmacological profile than that of platinum one, are developed and tested against various types of cancer cells [7].

During the last decades the anticancer activity of organotin compounds (OTCs) has been well studied [8–29]. Therefore, the investigation of new OTCs of low toxicity and improved anticancer activity are known to induce apoptosis in several cancer cell lines [17–19]. Their activity can be coupled to the lipophilicity of alkyl or aryl groups attached to the tin atoms [21]. In addition, due to their lipophilicity, OTCs are able to permeate membranes and reach the cell nucleus, where the dissociable ligands yield intermediate molecules capable of binding DNA [16,26].

The advantages of the use of polymeric drugs as anticancer agents have been described earlier [30]. This is because: (i) the polymers overcome cellular resistance mechanisms; (ii) the polymers could be used as carriers in high-dose chemotherapy; (iii) polymers are filtered out by the kidneys more slowly than small compounds increasing the body retention time; (iv) the size and structure of the polymer provide more binding sites to cellular targets; (v) the polymers can be act as hybrid drugs incorporating multiple anticancer agents against cells through different mechanisms; and (vi) the polymers accumulate in solid tumors more than in normal tissues [30].

In the course of our studies on the design and synthesis of new metallodrugs [13–29], the known $\{[R_3Sn(CH_3COO)]_n\}$ where R = *n*-Bu– (**1**) and R = Ph– (**2**) compounds were isolated from the reaction between acetic acid with tributyltin or triphenyltin oxides. The compounds were characterized via their physical properties and their FT-IR, ^{119}Sn Mössbauer, and ^{1}H NMR spectroscopic data, while their structures were verified by single-crystal X-Ray diffraction crystallography. The enhancement on the biological activity against tumor cells of the polymeric **1** and **2** is studied in relation to their polymeric intermolecular architecture (helical and zig-zag). The presence of acetic acid is also, expected to adjust the lipophilicity of the metallodrugs. The in vitro cell viability against MCF-7 (estrogen receptor (ER) positive) and MDA-MB-231 (estrogen receptor (ER) negative) was evaluated. Their genotoxicity has been studied against normal human fetal lung fibroblast cells (MRC-5). The type of the cell death caused by **1** and **2** was studied by flow cytometry assay. Finally, conclusions on Structure Activity Relationship are derived, in the light of the results obtained for OCT's from our group up to now.

2. Results and Discussion

2.1. General Aspects

Complexes **1** and **2** were synthesized as pale white powders by refluxing a benzene solution of tri-aryl-tin oxide and acetic acid (glacial) in a 1:1 molar ratio (Scheme 1), using a Dean–Stark water trap.

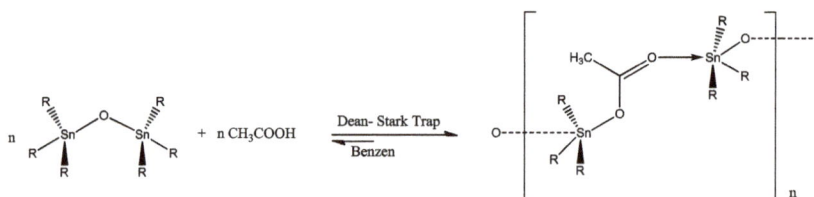

Scheme 1. Preparation route of **1** and **2**.

The formulae of **1** and **2** were first deduced by melting point and spectroscopic data. Crystals of the complexes **1** and **2** are stable in air. Complexes **1** and **2** are soluble in DMSO, DMF, toluene, ethanol, acetone, and diethyl-ether.

2.2. Solid State

2.2.1. Vibrational Spectroscopy

The $\nu_{as}(COO^-)$ vibrations are observed at 1584 (**1**) and 1574 (**2**) cm^{-1} respectively, while the bands at 1418 (**1**) and 1416 (**2**) cm^{-1} are assigned to $\nu_s(COO^-)$ (Figure S1). The $\Delta\nu$ [$\nu_{as}(COO^-) - \nu_s(COO^-)$] value is 166 (**1**) and 158 (**2**) cm^{-1}, respectively. Monodentate coordination of the carboxylic group results in significantly higher difference values $\Delta\nu$ than those observed for the ionic compounds of the ligand [20], while when the ligand chelates, the $\Delta\nu$ is considerably smaller than that observed for its ionic compounds. For asymmetric bidentate coordination, the values are in the range of monodentate one [20]. When the –COO– group bridges metal ions, the $\Delta\nu$ values are higher than that of the chelating mode and nearly the same as that observed for ionic compounds [20]. In the case of sodium acetate, the $\Delta\nu[\nu_{as}(COO-)-\nu_s(COO-)]$ value is 170 cm^{-1} [31]. Since the $\Delta\nu$ values in **1** and **2** (166 (**1**) and 158 (**2**) cm^{-1}) is in the range of the corresponding one of sodium acetate (170 cm^{-1}) the bridging coordination mode is concluded for the carboxylic group in **1** and **2** (Figure S1). Bands at 493 (**1**) and 457 (**2**) cm^{-1} in the spectra of **1** and **2** are assigned to the ν(Sn–O) bond vibrations [20], while the corresponding bands at 672, 614 (**1**) and 730, 698 (**2**) cm^{-1} are assigned to the antisymmetric and symmetric vibrations of Sn–C bonds [29].

2.2.2. ^{119}Sn Mössbauer Spectroscopy

^{119}Sn Mössbauer spectra at 80 K are shown in Figure 1.

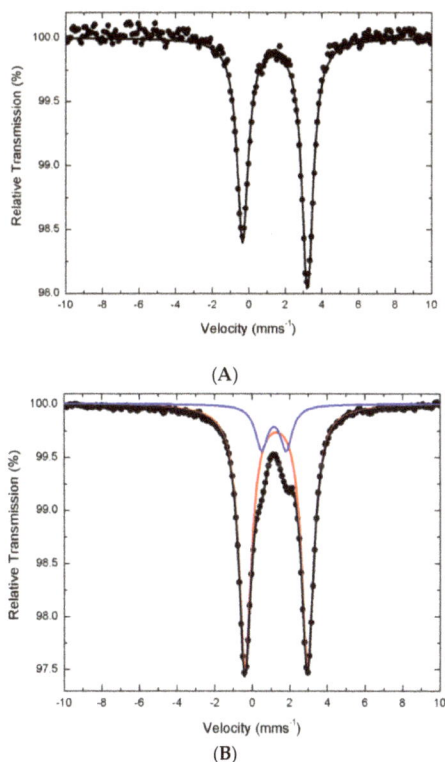

(A)

(B)

Figure 1. ^{119}Sn Mössbauer spectra of **1** (**A**) and **2** (**B**) at 80 K.

The spectrum of **1** consists of one asymmetric Lorentzian doublet. The absorption line intensity asymmetry could be attributed to the recoilless fraction (f) asymmetry, which could be a consequence of vibrational-bond anisotropic involving the Sn ions [32]. Preferred orientation of the crystallites in the powdered sample cannot also be excluded in order to justify this asymmetry. The occurrence of one Lorentzian double, indicates either the existence of one type of Sn atom in **1** or one structural isomer [25,29]. The corresponding spectrum of **2** consists by two symmetric Lorentzian doublets. The occurrence of two symmetric Lorentzians, however, indicates two kinds of tin centers under different environment in **2** with 85–15% molar ratio [25,29]. The values of the Isomer Shifts (I.S.) of +1.44 (**1**), 1.28 (**2A**) and 1.15 (**2B**) mm·s^{-1} corresponds to the (4+) oxidation state [25,29]. The quadrupole splitting parameter (ΔEq) values are 3.57 (**1**), 3.35 (**2A**) and 1.33 (**2B**) mm·s^{-1}. Therefore, trigonal-bibyramidal (tbp) geometry should be concluded for tin(IV) ions in **1** and **2A** as in the case of tbp R$_3$Sn(IV) (eg-R$_3$ = alkyl) geometry where the ΔEq values lie between 2.50–4.00 mm·s^{-1} [25,29]. On the contrary, tetrahedral (tet) geometry R$_3$Sn(IV) results in to ΔEq values of 1.30–3.00 mm·s^{-1} [25,29], tet conformation should be attributed to the tin(IV) atoms in **2B**.

2.2.3. Crystal and Molecular Structures of [Bu$_3$SnCH$_3$COO]$_n$ (**1**) and [Ph$_3$SnCH$_3$COO]$_n$ (**2**)

Crystals suitable for X-ray analysis were obtained by slow evaporation of diethyl-ether solutions of **1** and **2**. Their formula was confirmed here by single crystal X-ray diffraction analysis at ambient conditions. The structure of **1** is identical to that already reported by M. Adeel Saeed et. al. [33]. Thus **1** crystallizes in P2$_1$/c space group, a = 10.1845(3), b = 20.2542(7), c = 16.2466(6) Å, β = 94.739(3)°, V = 3339.87 Å^3; while the reported one crystallizes in P2$_1$/c space group, a = 10.386(4), b = 20.924(3), c = 16.584(6) Å, β = 92.87(2)°, V = 3599(2) Å^3 [33]. Although, **2** crystallizes in Pn space group with a = 16.7427(6), b = 10.0426(2), c = 25.5119(8) Å, α = 89.999(2), β = 100.936(3), γ = 89.998(2)°, V = 4211.68 Å^3, the already reported crystallizes in P2$_1$/c, space group, a = 8.969(4), b = 10.146(5), c = 19.540(7) Å, β = 93.70(4)°, V = 1774.5 Å^3 [34] suggesting a polymorphism between **2** and the published one. However, the extended disorder on the density observed in **2** prevent its accurate refinement allowing only qualitative conclusions to be drawn. Although the obtained X-ray data do not allow discussion about bond lengths and angles, they are of sufficient quality to determine the connectivity and the packing of **2**. The molecular diagrams of **1** and **2** are shown in Figure 2.

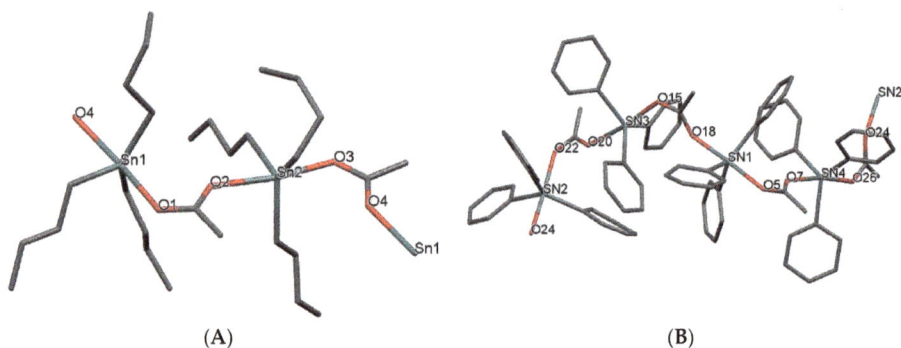

(A)   (B)

Figure 2. Molecular diagrams of **1** (**A**) and **2** (**B**).

Three C atoms from the alkyl groups and two O atoms from two de-protonated CH$_3$COOH molecules form the trigonal bipyramidal arrangement around the Sn ions in **1**. The average C–O bond lengths found in **1** and **2** (1.260 ± 0.020 Å), indicates a bond order of 1.5 e. This delocalization of the electron density in the –COO$^-$ group suggests a charge distribution shown in Scheme 2.

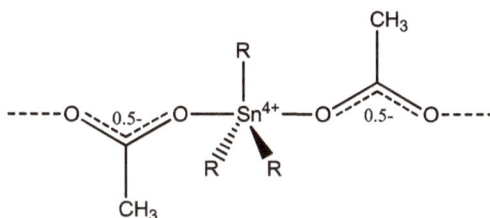

Scheme 2. Charge distribution.

2.3. Solution Studies

^{1}H-NMR Spectroscopy

The ^{1}H-NMR spectrum of free acetic acid in DMSO-d_6 is dominated by a single resonance signal at 1.91 (s, H) ppm for the methyl protons, which is shifted upon its coordination to the tin(IV) ion at 1.776 (s, H) ppm in **1** and at 1.758 (s, H) ppm in **2** (Figure S2). In the case of **1** four additional signals are observed at 1.52 ppm (t, H^a(–^{a}CH$_2$–Sn)), 1.27 ppm (m, H^b(–^{b}CH$_2$–CH$_2$–Sn)), at 1.00 ppm (m, H^c((–^{c}CH$_2$–CH$_2$–CH$_2$–Sn)), and 0.84 ppm (t, H^d ((^{d}CH$_3$–CH$_2$–CH$_2$–CH$_2$–Sn)) of the butyl substituent bind on tin(IV) ions (Figure S2). These four signals (1.52–0.84 ppm) have been replaced by the signals at 7.79–7.45 ppm in the spectrum of **2** (Figure S2), which were attributed into the aromatic protons of phenyl substituent of organotin moieties. Since the cells were incubated for 48 h the stability of **1** and **2** is checked for this period with ^{1}H-NMR spectroscopy. No changes were observed between the initial spectra of freshly prepared solutions and the corresponding spectra when measured after 48 h confirming the retention of the structures in solution (Figure S2).

2.4. Biological Tests

2.4.1. Anti-Proliferative Activity

Organotin compounds **1** and **2** were tested for their in vitro cytotoxic activity against human breast adenocarcinoma cell lines, MCF-7 and MDA-MB-231, by the mean of sulforhodamine B (SRB) assay [16,17]. The cells were incubated for 48 h with **1** and **2**. Since ERs are expressed in 65% of human breast cancer, (a hormone dependent malignancy) the MCF-7 and MDA-MB-231 cells were used in order to ascertain the influence of the ER's in the mechanism of action of **1** and **2** [16,17]. MCF-7 cells serve as a valuable model system to elucidate pathways of hormone response and resistance. Especially, the MCF-7 cells were used for studying estrogen response both in vitro and in vivo [35]. MDA-MB-231 human breast cancer cells, on the other hand, are used as a model of ER-negative breast cancers [36].

The IC$_{50}$ values of **1** and **2** against MCF-7 cells lie in the range of nM and they are 0.25 ± 0.02 and 0.21 ± 0.01 μM respectively, while their corresponding IC$_{50}$ values against MDA-MB-231 cells are 0.20 ± 0.01 and 0.12 ± 0.01 μM. The activity of **1** follows reverse order to the corresponding one of **2** against these cell lines suggesting no interference of the estrogen receptors to their mechanism. By taking into account the IC$_{50}$ value of cisplatin against MCF-7 and MDA-MB-231 cells (5.5 ± 0.4 and 26.7 ± 1.1 μM respectively), both **1** and **2** exhibit extremely cytotoxic activity against these cell lines. These values indicate 22 and 26-fold higher activity of **1** and **2** against MCF-7 cells than cisplatin and 134 and 223-fold against MDA-MB-231 cells. Despite their strong activity against tumor cells, **1** and **2** also exhibit high toxic activity against MRC-5 cells with IC$_{50}$ values of 0.22 ± 0.01 (**1**) and 0.11 ± 0.01 (**2**) μM, respectively. The IC$_{50}$ values against MCF-7, MDA-MB-231, and MRC-5 cells of organotin compounds studied from our group, are summarized in Table 1. The following conclusions are made: (i) tri-organotin derivatives are more active than the corresponding di-organtin ones; (ii) organotin derivatives of carboxylic acids are generally more active than those of other types of ligands; (iii) organotin compounds are also highly toxic, even more toxic than cispaltin; (iv) however the selectivity index, which is defined as the IC_{50} value against MRC-5 towards the corresponding value against MCF-7, and is an indicator of the therapeutic potency of a compound (the higher the value is, the better potency), shows that **1** and **2** are more potent therapeutics (TPI values of 0.88 (**1**) and 0.52 (**2**)) than cisplatin (TPI = 0.20). (v) The tri-*n*-butyl tin compound of thiobarbituric acid (*n*-Bu)$_3$Sn(*o*-HTBA)(H$_2$O) exhibits the higher TPI value of 1.6 (Table 1).

Table 1. Bioactivity data recorded for **1** and **2** in comparison with those of other reported organotin compounds.

Compound	IC$_{50}$ (μM)				
	MCF-7	MDA-MB-231	MRC-5	TPI *	Ref.
1	0.25 ± 0.02	0.20 ± 0.01	0.22 ± 0.01	0.88	[present]
2	0.21 ± 0.01	0.12 ± 0.01	0.11 ± 0.01	0.52	[present]
{[Ph$_3$Sn]$_2$(mna)·[(CH$_3$)$_2$CO]}	0.030		>0.200		[18]
[Me$_2$Sn(Sal)$_2$]	0.142 ± 0.043		0.0975 ± 0.00015	0.69	[19]
[(*n*-Bu)$_2$Sn(Sal)$_2$]	0.108 ± 0.0026		0.1041 ± 0.0002	0.96	[19]
[(*n*-Bu)$_3$Sn(Sal)]	0.724 ± 0.0054		0.0981 ± 0.0001	0.14	[19]
[Ph$_3$Sn(Sal)]	0.121 ± 0.0037		0.0945 ± 0.000.2	0.78	[19]
[(*n*-Bu)$_3$Sn(pHbza)]	0.325 ± 0.0023		0.0784 ± 0.0002	0.24	[19]
{[Ph$_3$Sn(*o*-HTBA)]}$_n$	0.103	0.203	0.130	1.26	[17]
(*n*-Bu)$_3$Sn(*o*-HTBA)(H$_2$O)	0.068	0.106	0.108	1.59	[17]
[(tert-Bu–)$_2$(HO–Ph)]$_2$SnCl$_2$	3.12 ± 0.38				[14]
[(tert-Bu–)$_2$(HO–Ph)]$_2$Sn(PMT)$_2$	7.86 ± 0.87				[14]
[(tert-Bu–)$_2$(HO–Ph)]$_2$Sn(MPMT)$_2$	0.58 ± 0.1				[14]
{[(tert-Bu–)$_2$(HO–Ph)]$_2$SnCl(PYT)}	>30				[14]
[(tert-Bu–)$_2$(HO–Ph)]$_2$SnCl(MBZT)}	>30				[14]
Ph$_3$SnCl	0.130	0.166	0.141	1.08	[16]
[Ph$_3$SnOH]$_n$	0.070	0.165	0.090	1.29	[16]
[(Ph$_2$Sn)$_4$Cl$_2$O$_2$(OH)$_2$]	>10	>10	>10		[16]
Me$_2$Sn((tert-Bu–)$_2$(HO–Ph–S)$_2$	19.20 ± 1.70		19.50 ± 1.40	1.02	[15]
Et$_2$Sn(((tert-Bu–)$_2$(HO–Ph–S))$_2$	6.20 ± 0.80		7.30 ± 0.60	1.18	[15]
(*n*-Bu)$_2$Sn–(((tert-Bu–)$_2$(HO–Ph–S))$_2$	0.40 ± 0.06		0.61 ± 0.07	1.53	[15]
Ph$_2$Sn(((tert-Bu–)$_2$(HO–Ph–S))$_2$	6.20 ± 0.80		12.40 ± 1.40	2.00	[15]
[(tert-Bu–)$_2$(HO–Ph)]$_2$Sn(((tert-Bu–)$_2$(HO–Ph–S))$_2$	>30	>30	>30		[15]
Me$_3$Sn((tert-Bu–)$_2$(HO–Ph–S))	4.90 ± 0.50		3.36 ± 0.13	0.69	[15]
Ph$_3$Sn(((tert-Bu–)$_2$(HO–Ph–S))	0.25 ± 0.03		0.22 ± 0.01	0.88	[15]
Cisplatin	5.5 ± 0.4	26.7 ± 1.1	1.1 ± 0.2	0.20	[37]

* TPI = IC$_{50}$(MRC-5)/IC$_{50}$(MCF-7), mna = 2-mercapto-nicotinic acid, salH = salicylic acid, pHbzaH = *p*-Hydroxyl-benzoic acid, H$_2$TBA = 2-thiobarbituric acid, PMTH = 2-mercapto-pyrimidine, MPMTH = 2-mercapto-4-methyl-pyrimidine, PYTH = 2-mercapto-pyridine, MBZTH = 2-mercapto-benzothiazole.

2.4.2. Evaluation of Genotoxicity by Micronucleus Assay In Vitro

Micronucleus assay is a reliable and an accessible technique to evaluate the appearance of genetic damage on a cell. The detection of micronucleus (MN) indicates mutagenic, genotoxic, or teratogenic effects [37]. In the presence of exogenous genotoxic factors, the MN is formed due to the metaphase–anaphase transition of the mitotic cycle. The possible induction of micronucleus frequencies was evaluated when MRC-5 cells were treated by **1** and **2** at the concentrations of their IC$_{50}$ values. The micronucleus frequency in the MRC-5 cell culture without treatment is 0.91 ± 0.02%, while it is 1.0 ± 0.1% upon treatment with DMSO. However, the micronucleus frequencies are slightly increased when the cells are incubated with **1** and **2** to 2.10 ± 0.04% (**1**) and 2.19 ± 0.03% (**2**) (Figure S3). The compounds **1** and **2** show a slightly increase in micronucleus frequency in contrast to the control or to cisplatin (1.6%) at its IC$_{50}$ value of 26 μM [37]. Cisplatin is used as a reference control. Doxorubicin has also been used as reference control from other groups. MRC-5 cells show slight increasing MN's when they are treated with **1** and **2** than the case of cisplatin indicating higher genotoxocity indeed. However, the treatment of MRC-5 cells with 0.18 μM or 0.014 μM of doxorubicin increases the percent of micronucleus at (94.7 ± 20.0)% and (17.00 ± 1.73)%, respectively, in contrast to the control ones (15.0 ± 2.64)% [38]. Thus, despite its higher genotoxicity of **1** and **2** towards MRC-5 cells than cisplatin both exhibit significant lower MN percentage than doxorubicin a medication against tumors in humans in used. It is therefore considered that **1** and **2** are nongenotoxic substances.

2.4.3. Cell Cycle Studies

Internucleosomal DNA fragmentation has been described as one of the main characteristics of the apoptotic process and can be identified by a sub-G_1 peak on DNA frequency histograms [37]. Therefore, the apoptotic type of the cell deaths caused by the regulation of cell cycle progression because of **1** and **2** can be evaluated by flow cytometric analysis since these cells give a sub-G_1 peak [37]. The percentage of cells in the phases of the cell cycle was analyzed after 48 h exposure of MCF-7 cells with **1** and **2** at their IC_{50} values.

The effect on the cell cycle which is illustrated in Figure 3, as the number of cells towards DNA content in sub-G_1, G_0/G_1, S, and G_2/M phases is caused by **1** and **2**. The untreated cells are spread in 6.1% sub-G1 phase, 46.5% in G_0/G_1, 18.3% in S, and 28.9% in G_2/M phases. After incubation of MCF-7 cells with **1** and **2**, a significant increase in the number of apoptotic cells in sub-G_1 phase (14.4% (**1**), 24.1% (**2**), respectively) was observed towards the control group (6.1%). In the case of **1**, the cells in G_0/G_1 phase are reduced to 37.9%, on the contrary, with 46.5% for the untreated cells while in the case of **2**, the corresponding value decrease to 35.8%. However, the percentage of MCF-7 cells in S phase, was increased to 25.1% (**1**), 22.9% (**2**). Finally, the percentage of MCF-7 cells in G_2/M phase was reduced to 22.1% (**1**) and 17.1% (**2**), respectively. In the DMSO-treated cells, the distribution in phases sub-G_1 (6.5%), G_0/G_1 (42.7%), S (20.6%), and G_2/M (29.5%) were similar to the corresponding ones of control cells'. All the data obtained in cell cycle studies of **1** and **2** are summarized in Table 2.

Table 2. Cell cycle studies data of **1** and **2**.

Description	Phases of cell cycle			
	Sub-G1	G_0/G_1	S	G_2/M
Untreated cells	6.1%	46.5%	18.3%	28.9%
Treated cells with DMSO	6.5	42.7	20.6	29.5
1	14.4	37.9%	25.1	22.1
2	24.1	35.8%.	22.9	17.1

In conclusion, **1** and **2** stimulate S-phase cell cycle arrest, thus suppressing cell proliferation by inhibiting DNA synthesis, in accordance to other anticancer agents (resveratrol, mitomycin C, and hydroxyurea) [37]. Likewise, cisplatin causes cell cycle arrest at S and G_2/M phases and the percentage of MCF-7 cells in sub-G_1 phase is increased, exhibiting an increasing number of apoptotic cells [37]. To summarize, metal drugs of this type induce cell cycle arrest either in G_0/G_1 or in S phase (like cisplatin), resulting in MCF-7 cell growth inhibition. Therefore, the reduced cell growth caused by **1** and **2** is attributed to the apoptotic type of cell death, in accordance to the way of action of cisplatin [37].

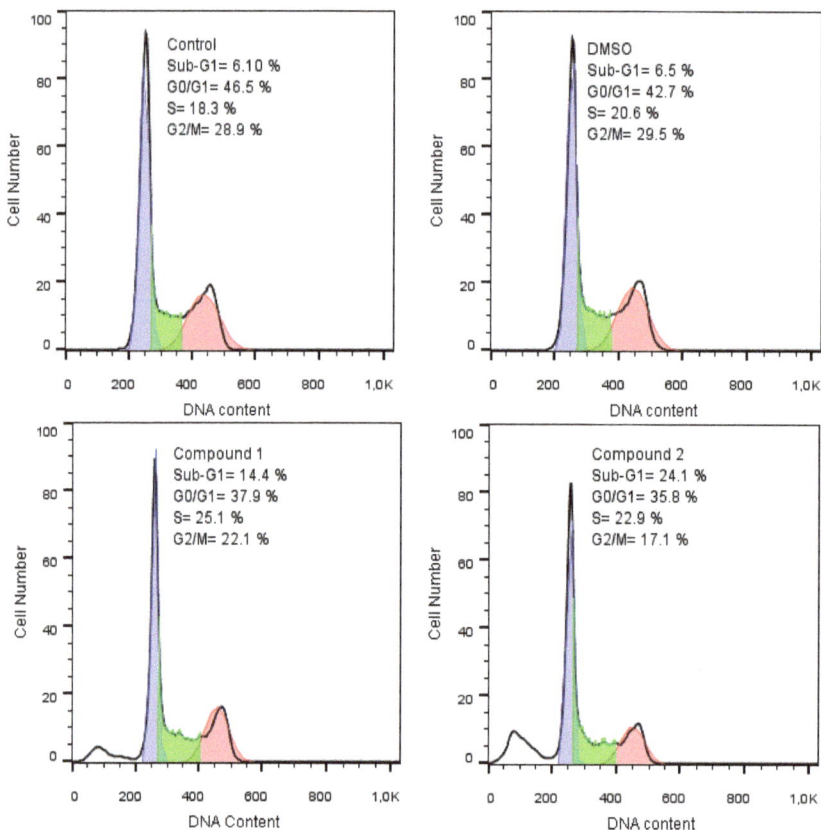

Figure 3. Number of MCF-7 cells in sub-G_1, G_0/G_1, S, and G_2/M phases, upon their treatment with **1** and **2**. The meaning of color labeling is white= Sub-G1, blue= G0/G1, green= S, pink= G2/M.

2.4.4. Detection of the Loss of the Mitochondrial Membrane Permeabilization (MMP Assay)

The releasing of cytochrome c in the cytosol through the loss of mitochondrial membrane permeability activates the mitochondrion cell apoptosis pathway [37]. The induction of loss in mitochondrial membrane permeability in tumor cells is one of the main accomplishments of targeted chemotherapy. The MMP assay is based on the cationic hydrophobic mitochondrial potential dye which accumulates in normal mitochondria. When cells are treated with a metallo-agent, the mitochondrial membrane permeability collapses, and the fluorescence emission of the dye decreases simultaneously.

The MCF-7 cells were treated with **1** and **2** at their IC_{50} values, for 48 h, and the fluorescence of the MMP assay dye decreased by 6.71% (**1**) and 6.52% (**2**), respectively. When the MCF-7 cells are treated with cisplatin at its IC_{50} values (5.5 μM) the fluorescence of the MMP assay dye decreased by 54.9% [37]. Therefore, the MMP assay should not support mitochondrial membrane permeability loss. Thus, **1** and **2** cause cell death by a different mechanism. Since apoptosis has been observed from cell cycle studies, it should be activated by a different mechanism than mitochondrion pathways.

2.4.5. DNA Binding Studies

DNA is the main target of the successful chemotherapeutics, the interaction of **1** and **2** towards CT-DNA was investigated by viscosity measurements and fluorescence spectroscopic studies.

(a) Viscosity measurements: DNA length changes upon its incubation with anticancer agents affect strongly affecting the viscosity of its solution. Thus (i) if the agent intercalates in the DNA strands, this results in its lengthening and a viscosity increase; (ii) if the agent interacts electrostatically with the DNA, no effect on DNA length is caused and therefore no significant change in viscosity is observed; (iii) however, in case the DNA strands are cleaved by an agent, the length of the DNA decreases, and also, the viscosity decreases significantly (iv) bending of the DNA helix caused by the agent reduces the viscosity. Therefore, viscosity exhibits high sensitivity to changes in the DNA and it is used for the study of the binding modes of an agent towards DNA [39]. The solution of CT-DNA (10 mM) is incubated with increasing amounts of **1** and **2** so that the [compound]/[DNA] molar ratio reaches $r = 0.27$. The relative viscosity of the solution which contains the agent/DNA/buffer, towards the corresponding one which contains DNA/buffer, increased for both compounds (Figure 4), suggesting an intercalation mode of interaction between **1** and **2** and CT-DNA.

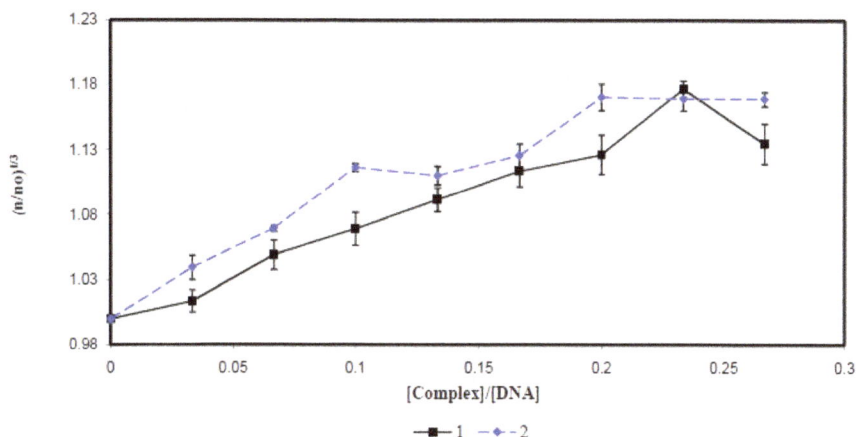

Figure 4. Effect of increasing concentrations of **1** and **2** on the relative viscosity of CT-DNA at 25 °C. ([DNA] = 10 mM, r = [compound]/[DNA], n is the viscosity of DNA in the presence of **1** or **2** and n_o is the viscosity of DNA alone).

(b) Fluorescence Spectroscopic Studies: In order to verify the intercalation mode of **1** and **2** towards CT-DNA, fluorescence spectroscopic studies were carried out. In the fluorescence spectroscopic studies the dye ethidium bromide (EB) is used. In the presence of DNA, EB emits, due to its strong intercalation between the adjacent DNA base pairs [37]. The displacement of EB during titration with the agent suggests an intercalative binding mode. The emission data of the solutions of EB with CT-DNA at 610 nm, with increasing concentrations of **1** and **2** (0–250 µM) upon their excitation at 532 nm, were recorded (Figure 5). The decreasing percent in fluorescence upon increasing of concentration of **1** and **2** at 610 nm was 31.5% (**1**) and 30.1% (**2**), respectively, confirming that both compounds can interact with DNA by the intercalation mode. The corresponding apparent binding constants (K_{app}) of **1** and **2** towards CT-DNA calculated through fluorescence spectra are $(4.9 \pm 0.5) \times 10^4$ (**1**) and $(7.3 \pm 1.3) \times 10^4$ (**2**) M^{-1}, respectively, indicating stronger binding affinity of the triphenyltin than the tri-*n*-butyltin acetates.

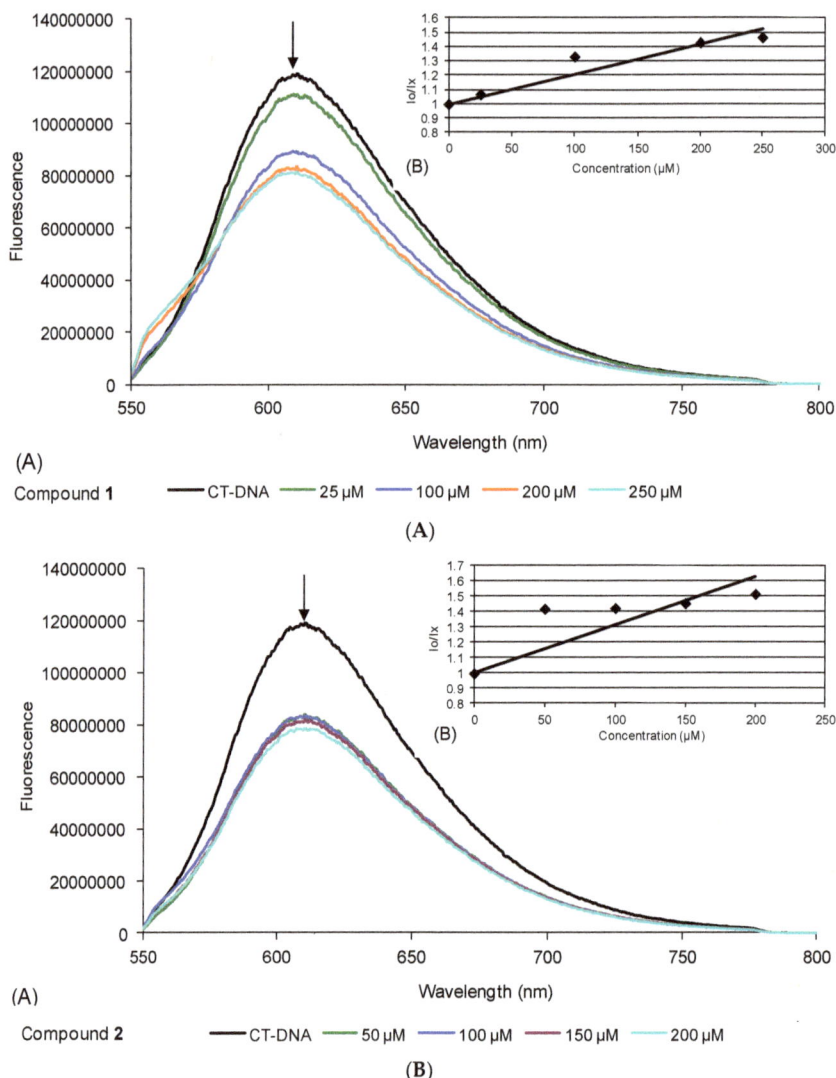

Figure 5. Emission spectrum of EB bound to DNA (peak around 610 nm) decreases in order of the concentration of the complex (**1** (**A**) and **2** (**B**)). The arrows show the intensity changing upon increasing complex concentration. Inset shows the plots of emission intensity Io/I vs. [complex].

(c) Computational studies—molecular docking: In order to verify the type of interaction between DNA with **1** or **2** molecular docking calculations were performed. Small aromatic molecules can typically bind to DNA by intercalation. Generally, either a planar molecule or a fragment is inserted between two adjacent DNA base pairs to form a hydrophobic pocket in the DNA structure. This non-covalent interaction is usually stabilized by π-interactions; at the same time, additional interactions with the groves are possible. Statistically, the CG intercalation site is preferable for intercalation [40]. Upon ligand binding, the DNA structure is slightly elongated and accommodates a small gap between two consecutive base pairs for the intercalators. Consequently, the latter can act as potent antitumor drugs and mutagens, inhibiting DNA replication and transcription. Targeting DNA sequences is a challenging task for docking software, especially when intercalation sites are unavailable, because gap openings cannot be simulated with rigid receptors. Moreover, it has been shown that popular docking software fails to predict the intercalation into canonical B-DNA when analogous solved structures are lacking [41]. For these reasons, the DNA target chosen for our study was 1DSC (PDB: www.rcsb.org) which is an octamer d(GAAGCTTC)$_2$ complexed with actinomycin D [42]. Actinomycin D is a potent antibiotic with high antibacterial and antitumor activity. It intercalates to DNA by localizing a phenoxazone ring at a GpC sequence while the two cyclic polypeptides of the drug bind to the DNA minor groove. 1DSC has been already used successfully for docking of metal complexes [43] and organic compounds [44] acting as DNA intercalators. The geometry of compounds **1** and **2** is trigonal bipyramidal (TBP) in the solid state. However, upon solvation in aqueous media their structure are converted to tetrahedral by cleavage of the Sn-O$_{(acetate)}$ dative bond. Tetrahedral metal complexes, unlike those with TBP geometry, can effectively bind to DNA by intercalation, although they normally cannot penetrate as deep as square planar complexes [45]. The optimized structures of the two organotin compounds [R$_3$Sn(CH$_3$COO)], (R = *n*-Bu (**1**), Ph– (**2**)) were used as ligands for DNA docking using AutodockVina. Molecular docking evaluates affinity potentials through a precalculated grid finding favorable binding positions for a flexible ligand towards a rigid macromolecular target. Complexes **1** and **2** were successfully docked into the B-DNA duplex and the lowest energy poses are depicted in Figure 6. The tri-phenyl derivative can adopt two different conformations intercalating in the GC region through either one pnenyl-ring (Figure 6(**2a**)) or the carboxylic moiety (Figure 6(**2b**)) with computed binding free energies −5.5 and −5.1 kcal/mol, respectively. The planarity of the phenyl ring favors π-π stacking interactions with the DNA base pairs, while additional hydrogen bonding interactions with the carboxylate O (G4:N2-Lig O, 2.97 Å and Lig O-C5O4, 3.1 Å) and with the pi-orbitals of the phenyl ring (G12:N2-Lig pi, 2.96 Å and G4:N2-Lig pi, 3.6 Å). On the other hand, the minimum energy conformation of the tri-*n*-butyl-derivative shows a semi-intercalation into the same GC region (Figure 6(**1**)) with lower binding affinity (−3.7 kcal/mol). The structure is also stabilized by hydrogen bonding through the carboxylic moiety (G4:N2-Lig O, 3.13 Å). Possibly, the bulk and agile butyl groups significantly increase the steric hindrance and prevent the intercalation. These results support the experimental results showing that the phenyl derivative exhibits higher antitumor activity possibly through its intercalative mode of action (Table 1).

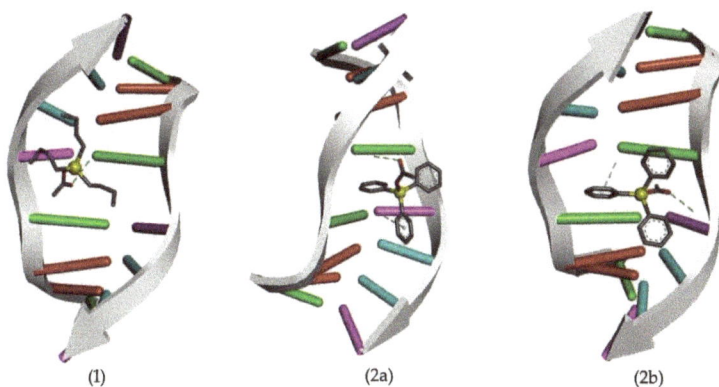

Figure 6. DNA docking and H-bonding interactions between B-DNA and compounds **1** and (**2a, 2b**).

3. Experimental

3.1. Materials and Instruments

All solvents used were of reagent grade, (Aldrich, Merck, Darmstadt, Germany) and they were used with no further purification. Infrared spectra in the region of 4000–370 cm^{-1} were obtained in KBr pellets with a Jasco FT-IR-6200 spectrometer. The ^{1}H-NMR spectra were recorded on a Bruker AC 250, 400 MHFT-NMR instrument in DMSO-d_6. Chemical shifts are given in ppm using ^{1}H-TMS as internal reference. Elemental analysis for C, H, N, and S were carried out with a Carlo Erba EA MODEL 1108 (Waltham, MA, USA). The ^{119}Sn Mössbauer spectra were collected at sample temperature of 80 K using a constant acceleration spectrometer equipped with a Ca^{119m}SnO$_3$ source kept at room temperature. The isomer shift values of the components used to fit the spectra are given relative to SnO$_2$ at room temperature. The ^{119}Sn Mössbauer spectra were recorded with Constant acceleration WissEl-Wissenschaftliche Elektronik GmbH spectrometer (Starnberg, Germany).

3.2. Synthesis and Crystallization of {[(n-Bu)$_3$Sn(CH$_3$COO)]$_n$} (1) and {[Ph$_3$Sn(CH$_3$COO)]$_n$} (2)

Although the synthesis of these compounds is already known [33,34] we briefly described the procedure follows here. 0.5 mmol of tri-*n*-butyltin oxide (C$_{24}$H$_{54}$OSn$_2$, 0.298 g) for **1**, or triphenyltin(IV) hydrooxide (C$_{24}$H$_{16}$OSn, 0.183 g) for **2**, were diluted with 0.5 mmol acetic acid, in 20 mL benzene in a 100-mL spherical flask. The flask was fitted with a Dean–Stark moisture trap and the reaction mixture was refluxed for 3 h. The solution was filtered and the clear filtrate was concentrated to dryness. Crystals of **1** and **2**, suitable for X-ray analysis, were formed by slow evaporation of a diethyl ether solution.

1: Yield: 40%; m.p: 75–76 °C; (C$_{14}$H$_{30}$O$_2$Sn)$_n$·(MW = 349.04); elemental analysis: found C = 48.23, H = 6.65%; calcd: C = 48.18, H = 8.66%. MID-IR (cm^{-1}) (KBr): 3045 w, 2330 w, 2295 s, 1959 w, 1821 w, 1659 s, 1643 w, 1573 w, 1555 s, 1428 s, 1334 vs, 1261 vs, 1077 vs, 1025 w, 997 w, 729 w, 696 w, 665 w, 609 w, 499 w, 455 w. ^{1}H NMR (ppm) in DMSO-d_6: 1.776 (s), 1.578–1.501 (q), 1.327–1.236 (q), 1.043–1.001 (t), 0.880–0.843 (t).

2: Yield: 40%; m.p: 110–115 °C; (C$_{20}$H$_{18}$O$_2$Sn)$_n$ (MW = 409.02); elemental analysis: found C = 58.91, H = 4.40%; calcd: C = 58.73, H = 4.43%. MID-IR (cm^{-1}) (KBr): 2400 w, 1574 w, 1556 s, 1415 w, 1384 s, 1015 s, 867 w, 671 w, 612 s. ^{1}H NMR (ppm) in DMSO-d_6: 7.850–7.707 (q), 7.452–7.390 (q), 1.758 (single), 1.114–1.079 (t).

3.3. X-Ray Structure Determination

Single crystal X-ray diffraction data for **1** and **2** were collected on an Oxford-Diffraction Supernova diffractometer, equipped with a CCD area detector utilizing Cu Kα (λ = (1.5418 Å)) radiation. A suitable crystal was mounted on a Hampton cryoloop with Paratone-N oil and transferred to a goniostat where it was cooled for data collection. Empirical absorption corrections (multiscan based on symmetry-related measurements) were applied using CrysAlis RED software [46]. The structures were solved by direct methods using SIR2004 [47] and refined on F^2 using full-matrix least-squares with SHELXL-2014/7 [48]. Software packages used were as follows: CrysAlis CCD for data collection [46], CrysAlis RED for cell refinement and data reduction [46], WINGX for geometric calculations [49]. The non-H atoms were treated anisotropically, whereas the aromatic H atoms were placed in calculated, ideal positions and refined as riding on their respective carbon atoms. The single crystals of compound **2** exhibited a fairly poor diffraction pattern. As a result moderate quality X-ray data were collected which did not lead to a publishable crystal structure. For this reason the X-ray data of **2** are not quoted here and have not been deposited in Cambridge Structural Database.

Supplementary data (1_bu3sn2o_asp_checkcif_new.pdf and 1_bu3sn2o_asp_FINAL.cif) are available from CCDC, 12 Union Road, Cambridge CB2 1EZ, UK, (e-mail:deposit@ccdc.cam.ac.uk), on request, quoting the deposition numbers CCDC-1846946 (**1**).

1: $(C_{14}H_{30}O_2Sn)_n$: MW = 349.04, Monoclinic, space group P21/c, $a = 10.1845(3)$, $b = 20.2542(7)$, $c = 16.2466(6)$ Å, $\beta = 94.739(3)°$, $V = 3339.9(2)$ Å^3, $Z = 4$, $T = 100$ K, ρ(calc) = 1.388 g·cm^{-3}, $\mu = 1.522$ mm^{-1}, $F(000) = 1440$. 22782 reflections measured, 5864 unique ($R_{int} = 0.041$), 5131 with $I > 2\sigma(I)$. The final $R_1 = 0.0265$ (for 5131 reflections with $I > 2\sigma(I)$) and $wR_2(F^2) = 0.0678$ (all data), $S = 1.09$.

3.4. Biological Tests

Biological experiments were carried in dimethyl sulfoxide Dulbecco's Modified Eagle's Medium solutions (DMEM) DMSO/DMEM (0.02–0.2% v/v) for the complexes **1–2**. Stock solutions of the complexes **1–2**, (0.01 M) in DMSO were freshly prepared and diluted in with cell culture medium to the desired concentrations (0.05–0.4 μM). Results are expressed in terms of IC$_{50}$ values, which is the concentration of drug required to inhibit cell growth by 50% compared to control, after 48 h incubation of the complexes towards cell lines. Since there is no "universal" correct time point to find the IC$_{50}$ value of a given compound. Generally the inhibitory concentration of the 50% of the cells is determined at once at the time of the cells (of interest) doubling time, and twice the time of the doubling time. In cell culture, the vast majority of adherent cell lines display a doubling time between 18 and 24 h. In our research unit, we prefer to use 48 h since 24 h is too short to determine a reliable IC$_{50}$ concentration. The cell viability was determined by SRB assay as previously described [37] and it is briefly described here: Cells were plated (100 μL per well) in 96-well flat-bottom microplates at various cell inoculation densities (MCF-7, MDA-MB-231 and MRC-5: 6000, 6000 and 2000 cells/well respectively). Cells were incubated for 24 h at 37 °C and they were exposed to tested agents for 48 h afterwards, followed by the addition of an equal volume (100 μL) of complete culture medium only in the well containing the controls, or twice the final drug concentrations diluted in complete culture medium in the wells where the compounds are tested. Drug activity was measured by means of a SRB colorimetric assay giving the percent of the survival cells towards the control (untreated cells) absorbance. The culture medium was aspirated before fixation and 50 μL of 10% cold trichloroacetic acid (TCA) were gently added to the wells. Microplates were left for 30 min at 4 °C, washed five times with deionized water and left to dry at room temperature for at least 24 h. Subsequently, 70 μL of 0.4% (w/v) sulforhodamine B (Sigma, Darmstadt, Germany) in 1% acetic acid solution was added to each well and left at room temperature for 20 min. SRB was removed and the plates were washed five times with 1% acetic acid before air-drying. Bound SRB was solubilized with 200 μL of 10 mM un-buffered Tris-base solution. Absorbance was read in a 96-well plate reader at 540 nm.

Evaluation of genotoxicity by micronucleus assay, cell cycle studies, detection of the loss of the Mitochondrial Membrane Permeabilization (MMP assay), and fluorescence spectral studies were performed as previously reported [37]. However, they are all quoted here in brief: (a) Micronucleus: MRC-5 cells were seeded (at a density of 2×10^4 cells/well) in glass cover slips which were afterwards placed in six-well plates, with 3 mL of cell culture medium and incubate for 24 h. MRC-5 cells exposed with **1** and **2** in IC$_{50}$ values for a period of 48 h. After the exposure of **1** and **2**, the cover slips were washed three times with PBS and with a hypotonic solution (75 Mm KCl) for 10 min at room temperature. The hypotonized cells were fixed by at least three changes of 1/3 acetic acid/methanol. The cover slips were also washed with cold methanol containing 1% acetic acid. The cover slips were stained with acridine orange (5 μgr/mL) for 15 min at 37 °C. After, the cover slips were rinsed three times with PBS to remove any excess acridine orange stain. The number of micronucleated cells per 1000 cells was determined. (b) Cell cycle: MCF-7 cells were seeded at a density of 10^5 cells/well in six-well plates at 37 °C for 24 h. Cells were treated with **1** and **2** at the indicated IC$_{50}$ values for 48 h. The cells were then trypsinized and washed twice with phosphate-buffered saline (PBS) and separated by centrifugation. With the addition of 1 mL of cold 70% ethanol, the cells were incubated overnight at −20 °C. For analysis, the cells were centrifuged and transferred into PBS, incubated with RNase (0.2 mg/mL) and propidium iodide (0.05 mg/mL) for 40 min at 310 K and then analyzed by flow cytometry using a FACS Calibur flow cytometer (Becton Dickinson, San Jose, CA, USA). For each sample, 10,000 events were recorded. The resulting DNA histograms were drawn and quantified using the FlowJo software (version FlowJo X 10.0.7r2, Tree Star, Ashland OR, USA). (c) Detection of the loss of the Mitochondrial Membrane Permeabilization: MCF-7 cells were treated with **1** and **2** at IC$_{50}$ values. After 48 h of incubation period of **1** and **2**, the cell medium was removed and added the Dye Loading Solution. The cells were incubated in 5% CO$_2$ at 37 °C for 30 min. Afterwards, 50 μL of Assay Buffer B was added of each well and are incubated for 30 min. The fluorescence intensity is measured at λ_{ex} = 540 and λ_{em} = 590 nm. The MMP assay kit used was purchased from sigma Aldrich "Mitochondria Membrane Potential Kit for Microplate Readers, MAK147". (d) Fluorescent studies: The fluorescence spectroscopy method using ethidium bromide (EB) was employed to determine the relative DNA binding properties of complexes **1** and **2** into CT-DNA. The emission data at 610 nm of the spectra of EB (2.3 μM) solutions which contain CT-DNA (26 μM) in the absence or presence of various concentrations of complexes **1** and **2** (0–250 μM) were recorded upon their excitation at 532 nm (Figure 5). The apparent binding constant (K_{app}) was calculated using the equation:

$$K_{EB}[EB] = K_{app}[drug] \qquad ()$$

where [drug] is the concentration of the complex at a 50% reduction of the fluorescence, $K_{EB} = 10^7$ M^{-1}, and [EB] = 2.3 μM. The concentration of the drug at a 50% reduction of the fluorescence is calculated from the diagram I_0/I vs. the concentration of the complex [Q] (Figure 5), where, I_0 and I are the fluorescence intensities of the CT-DNA in the absence and presence of complexes **1** and **2**, respectively.

The fitting of the experimental I_0/I_x ratio as a function of concentration (C) was performed utilizing the linear least-squares fitting algorithm. However, the fitting procedure depends strongly on the weighting of the data points, i.e., on the knowledge of the standard deviation (SD) of every point in the experimental spectrum (<1%) since all spectra represent the average of 5000 individually recorded spectra. As a measure for the goodness of the fit, we used the R^2 statistics.

The slope of the I_0/I_x ratio as a function of C was estimated equal to a_1 = 2100.9 ± 204.8 and a_2 = 3142.4 ± 561.4 for **1** and **2** case, respectively.

The SD of the K_{app} is calculated using the standard error propagation variance formula [50]:

$$s_f = \sqrt{\left(\frac{\partial f}{\partial x}\right)^2 (s_x)^2 + \left(\frac{\partial f}{\partial y}\right)^2 (s_y)^2 + \left(\frac{\partial f}{\partial z}\right)^2 (s_z)^2 + \cdots}$$

In general, s_f represents the standard deviation of the function f, s_x represents the standard deviation of x, s_y represents the standard deviation of y, and so forth. This formula is based on the linear characteristics of the gradient of f and therefore it is a good estimation for the standard deviation of f. Using the above methodology, the parameter K_{app} is calculated as 48,950.97 ± 4771.84 and 73,217.92 ± 13,080.62 M^{-1} for **1** and **2**, respectively. More details concerning the calculation of R^2, the error propagation issue and other statistical parameters, see e.g., reference [50]

3.5. Viscosity Measurements

The kinematic viscosity of **1** and **2** was measured with the use of an Ubbelohde-type glass capillary viscosity-meter. The value of kinematic viscosity was measured as an average from triplicate measurements of viscosity with an accuracy of 2% at constant temperature. Temperature was set at 298 K by means of an ultra-thermostat.

3.6. Computational Studies—Molecular Docking

Molecular docking studies were carried out with AutodockVina [51] along with the graphical interface AutodockTools (ADT 1.5.6) [52] using the Lamarckian genetic algorithm (LGA). The DNA structure (PDB file 1DSC) was considered as a rigid target while no torsional restraints were applied to the complexes. Docking was performed to the optimized organotin derivatives [R$_3$Sn(CH$_3$COO)],

(R = *n*-Bu (**1**), Ph– (**2**)) at the B3LYP/3-21G*/LANL2DZ(Sn) level using the Gaussian03Wsoftware package [53]. Prior to docking, water molecules and the co-crystallized ligand were removed. The search space was set as a grid box of 40 × 30 × 40 Å and spacing of 1 Å centered at the CG rich area. Default parameters were used as described in the Vina manual except from "exhaustiveness" which was set to 24.

4. Conclusions

The discovery of the antitumor activity of organotin compounds goes back to the 1980s when Gielen first reports on the subject [8,9,11,12]. Later on, the antitumor properties of orgnaotin compounds were reviewed by Saxena and Hubert [10] and Hadjiliadis et al. [13]. Continuing our studies on the antitumor activity of organotin compounds, two known compounds of formula $\{[R_3Sn(CH_3COO)]_n\}$ where R = *n*-Bu– (**1**) and R = Ph (**2**) with trigonal bipyramidal geometry around each metal center, were used to investigate their antitumor activity against MCF-7 (positive to ERs) and MDA-MB-231 (negative to ERs). The mechanism of their action against human breast tumor cells was also investigated. Compounds **1** and **2** inhibit strongly both cell lines, while their IC_{50} values are lei in nanomolar range (Table 1). The estrogen receptors play no significant role in their action. It is also concluded that (i) the tri-organotin derivatives exhibit stronger activity than the corresponding di-organotin ones (Table 1); (ii) the organotin derivatives of carboxylic acids are generally more active than those of other type of ligands (Table 1); (iii) organotin compounds are also highly toxic against normal MRC-5 cells (Table 1); (iv) the selectivity index, however, shows that **1** and **2** are more potent therapeutics (TPI values of 0.88(**1**) and 0.52(**2**)) than cisplatin (TPI = 0.20) (Table 1) and (v) among all organotins tested from our group the strongest activity against MCF-7 cells has been obtained from the triphenyltin derivative of 2-mercapto-nicotinic acid $\{[Ph_3Sn]_2(mna)\cdot[(CH_3)_2CO]\}$ (IC_{50} = 30 nM) and the tri-*n*-butyl tin compound of thiobarbituric acid $(n\text{-Bu})_3Sn(o\text{-HTBA})(H_2O)$ (IC_{50} = 68 nM and TPI = 1.6) (Table 1). The mitochondrial membrane permeability loss is not supported by MMP assay. Since **1** and **2** are acting through apoptosis on tumor cells (cell cycle studies) this should therefore be activated through a different pathway than that of the mitochondrion, such as the extrinsic apoptotic pathway or direct interaction with DNA. The relative viscosity measurements and fluorescence spectroscopic studies of DNA solutions suggest an intercalation mode of interaction between **1** and **2** and CT-DNA. Docking studies verify the intercalative mode, while the bulk and agile *n*-butyl groups significantly increase the steric hindrance prohibiting the intercalation, resulting in higher antitumor activity of the phenyl derivative instead of the corresponding *n*-Bu– (Table 1).

Supplementary Materials: Supplementary materials can be found at http://www.mdpi.com/1422-0067/19/7/2055/s1.

Author Contributions: Conceptualization, S.K.H.; Methodology, C.N.B., N.K., and S.K.H.; Validation, S.K.H., A.T., and T.B.; Investigation, G.K.L., C.N.B., N.K., C.P., N.P. A.D., and A.G.K.; Writing-Original Draft Preparation, C.N.B., N.K., and S.K.H.; Writing-Review & Editing, S.K.H.; Supervision, S.K.H.

Funding: This research received no external funding.

Acknowledgments: (i) This work was carried out in partial fulfilment of the requirements for the graduate studies of G.L. under the supervision of S.K.H. The International Graduate Program in "Biological Inorganic Chemistry", which operates at the University of Ioannina within the collaboration of the Departments of Chemistry of the Universities of Ioannina, Athens, Thessaloniki, Patras, Crete, and the Department of Chemistry of the University of Cyprus, is acknowledged for the stimulating discussions forum. (ii) C.N.B. and S.K.H. would like to thank the Unit of Bioactivity Testing of Xenobiotics, of the University of Ioannina, for providing access to the facilities. (iii) C.N.B. and S.K.H. would like to thank the Atherothrombosis Research Centre of the University of Ioannina for providing access to the fluorescence microscope and flow cytometry. (iv) A.G.K acknowledges the Laser Center of Ioannina University for the access to its infrastructure. S. Kaziannis (Lecturer) is also acknowledged by A.G.K. for his useful assistance during the luminescence experiments.

Conflicts of Interest: The authors declare no conflicts of interest.

References

1. Rosenberg, B. Platinum coordination complexes in cancer chemotherapy. *Die Naturwissenschaften* **1973**, *60*, 399–406. [CrossRef] [PubMed]
2. Lippert, B. Impact of Cisplatin on the recent development of Pt coordination chemistry: A case study. *Coord. Chem. Rev.* **1999**, *182*, 263–295. [CrossRef]
3. Arnesano, F.; Natile, G. Mechanistic insight into the cellular uptake and processing of cisplatin 30 years after its approval by FDA. *Coord. Chem. Rev.* **2009**, *253*, 2070–2081. [CrossRef]
4. Arnesano, F.; Pannunzio, A.; Natile, G. Effect of chirality in platinum drugs. *Coord. Chem. Rev.* **2015**, *284*, 286–297. [CrossRef]
5. Astolfi, L.; Ghiselli, S.; Guaran, V.; Chicca, M.; Simoni, E.; Olivetto, E.; Lelli, G.; Martini, A. Correlation of adverse effects of cisplatin administration in patients affected by solid tumours: A retrospective evaluation. *Oncol. Rep.* **2013**, *29*, 1285–1292. [CrossRef] [PubMed]
6. Galluzzi, L.; Senovilla, L.; Vitale, I.; Michels, J.; Martins, I.; Kepp, O.; Castedo, M.; Kroemer, G. Molecular mechanisms of cisplatin resistance. *Oncogene* **2012**, *31*, 1869–1883. [CrossRef] [PubMed]
7. Huang, K.-B.; Wang, F.-Y.; Tang, X.-M.; Feng, H.-W.; Chen, Z.-F.; Liu, Y.-C.; Liu, Y.-N.; Liang, H. Organometallic Gold(III) Complexes Similar to Tetrahydroisoquinoline Induce ER-Stress-Mediated Apoptosis and Pro-Death Autophagy in A549 Cancer Cells. *J. Med. Chem.* **2018**, *61*, 3478–3490. [CrossRef] [PubMed]
8. Haiduc, I.; Silvestru, C.; Gielen, M. Organotin compounds: New organometallic derivatives exhibiting anti-tuequr activity. *Bull. Soc. Chim. Belg.* **1983**, *92*, 187–189. [CrossRef]
9. Gielen, M.; Melotte, M.; Atassi, G.; Wrllem, R. Synthesis, characterization and antitumour activity of 7,7-di-n-butyi-5,9-dioxo-6,6-dioxa-7-stanna-spiro[3,5]nonane, a di-n-butyltin(IV) analog of "paraplatin", and of a series of di-*n*-butyltin(IV) derivatives of mono- and disubstituted malonic acids. *Tetrahedron* **1989**, *45*, 1219–1229. [CrossRef]
10. Saxena, A.K.; Huber, F. Organotin compounds and cancer chemotherapy. *Coord. Chem. Rev.* **1989**, *95*, 109–112. [CrossRef]
11. Gielen, M. Tin-based antitumour drugs. *Coord. Chem. Rev.* **1996**, *151*, 41–51. [CrossRef]
12. Gielen, M. Organotin compounds and their therapeutic potential: A report from the Organometallic Chemistry Department of the Free University of Brussels. *Appl. Organomet. Chem.* **2002**, *16*, 481–494. [CrossRef]
13. Hadjikakou, S.K.; Hadjiliadis, N. Antiproliferative and anti-tumor activity of organotin compounds. *Coord. Chem. Rev.* **2009**, *253*, 235–249. [CrossRef]
14. Shpakovsky, D.B.; Banti, C.N.; Beaulieu-Houle, G.; Kourkoumelis, N.; Hadjikakou, S.K.; Manoli, M.; Manos, M.J.; Tasiopoulos, A.J.; Milaeva, E.R.; Charalabopoulos, K.; et al. Synthesis, structural characterization and in vitro inhibitory studies against humanagainst human breast cancer of the bis-(2,6-di-tert-butylphenol)tin(IV) dichloride andits complexes. *Dalton Trans.* **2012**, *41*, 14568–14582. [CrossRef] [PubMed]
15. Shpakovsky, D.B.; Banti, C.N.; Mukhatova, E.M.; Gracheva, Y.A.; Osipova, V.P.; Berberova, N.T.; Albov, D.V.; Antonenko, T.A.; Aslanov, L.A.; Milaeva, E.R.; et al. Synthesis, antiradical activity and in vitro cytotoxicity of novel organotin complexes based on 2,6-di-tert-butyl-4-mercaptophenol. *Dalton Trans.* **2014**, *43*, 6880–6890. [CrossRef] [PubMed]
16. Balas, V.I.; Banti, C.N.; Kourkoumelis, N.; Hadjikakou, S.K.; Geromichalos, G.D.; Sahpazidou, D.; Male, L.; Hursthouse, M.B.; Bednarz, B.; Kubicki, M.; et al. Structural and in vitro biological studies of organotin(IV) precursors; selective inhibitory activity against human breast cancer cells, positive to estrogen receptors. *Austr. J. Chem.* **2012**, *65*, 1625–1637. [CrossRef]
17. Balas, V.I.; Verginadis, I.I.; Geromichalos, G.D.; Kourkoumelis, N.; Male, L.; Hursthouse, M.B.; Repana, K.H.; Yiannaki, E.; Charalabopoulos, K.; Bakas, T.; et al. Synthesis, structural characterization and biological studies of the triphenyltin(IV)complex with 2-thiobarbituric acid. *Eur. J. Med. Chem.* **2011**, *46*, 2835–2844. [CrossRef] [PubMed]

18. Verginadis, I.I.; Karkabounas, S.; Simos, Y.; Kontargiris, E.; Hadjikakou, S.K.; Batistatou, A.; Evangelou, A.; Charalabopoulos, K. Anticancer and cytotoxic effects of a triorganotin compound with 2-mercapto-nicotinic acid in malignant cell lines and tumor bearing Wistar rats. *Eur. J. Pharm. Sci.* **2011**, *42*, 253–261. [CrossRef] [PubMed]

19. Agiorgiti, M.S.; Evangelou, A.; Vezyraki, P.; Hadjikakou, S.K.; Kalfakakou, V.; Tsanaktsidis, I.; Batistatou, A.; Zelovitis, J.; Simos, Y.V.; Ragos, V.; et al. Cytotoxic effect, antitumour activity and toxicity of organotin derivatives with ortho- or para-hydroxy-benzoic acids. *Med. Chem. Res.* **2018**, *27*, 1122–1130. [CrossRef]

20. Abdellah, M.A.; Hadjikakou, S.K.; Hadjiliadis, N.; Kubicki, M.; Bakas, T.; Kourkoumelis, N.; Simos, Y.V.; Karkabounas, S.; Barsan, M.M.; Butler, I.S. Synthesis, characterization, and biological studies of organotin(IV) derivatives with o- or p-hydroxybenzoic acids. *Bioinorg. Chem. Appl.* **2009**, 12. [CrossRef] [PubMed]

21. Xanthopoulou, M.N.; Hadjikakou, S.K.; Hadjiliadis, N.; Milaeva, E.R.; Gracheva, J.A.; Tyurin, V.-Y.; Kourkoumelis, N.; Christoforidis, K.C.; Metsios, A.K.; Karkabounas, S.; et al. Biological studies of new organotin(IV) complexes ofthioamide ligands. *Eur. J. Med. Chem.* **2008**, *43*, 327–335. [CrossRef] [PubMed]

22. Balas, V.I.; Hadjikakou, S.K.; Hadjiliadis, N.; Kourkoumelis, N.; Light, M.E.; Hursthouse, M.; Metsios, A.K.; Karkabounas, S. Crystal structure and antitumor activity of the novel zwitterionic complex of tri-*n*-butyltin(IV) with 2-thiobarbituric acid. *Bioinorg. Chem. Appl.* **2008**, 654137. [CrossRef] [PubMed]

23. Xanthopoulou, M.N.; Hadjikakou, M.N.; Hadjiliadis, N.; Kourkoumelis, N.; Milaeva, E.R.; Gracheva, J.A.; Tyurin, V.-Y.; Verginadis, I.I.; Karkabounas, S.; Baril, M.; et al. Biological studies of organotin(IV) complexes with 2-mercaptopyrimidine. *Russ. Chem. B* **2007**, *56*, 767–773. [CrossRef]

24. Xanthopoulou, M.N.; Hadjikakou, S.K.; Hadjiliadis, N.; Kubicki, M.; Karkabounas, S.; Charalabopoulos, K.; Kourkoumelis, N.; Bakas, T. Synthesis and characterization of a new chloro-di-phenyltin(IV) complex with 2-mercapto-nicotinic acid: Study of its influence upon the catalytic oxidation of linoleic acid to hydroperoxylinoleic acid by the enzyme lipoxygenase. *J. Organomet. Chem.* **2006**, *691*, 1780–1789. [CrossRef]

25. Xanthopoulou, M.N.; Hadjikakou, S.K.; Hadjiliadis, N.; Schurmann, M.; Jurkschat, K.; Michaelides, A.; Skoulika, S.; Bakas, T.; Binolis, J.; Karkabounas, S.; et al. Synthesis, structural characterization and in vitro cytotoxity of organotin(IV) derivatives of heterocyclic thioamides, 2-mercaptobenzothiazole, 5-chloro-2-mercaptobenzothiazole, 3-methyl-2-mercaptobenzothiazole and 2-mercaptonicotinic acid. *J. Inorg. Biochem.* **2003**, *96*, 425–434. [CrossRef]

26. Banti, C.N.; Gkaniatsou, E.I.; Kourkoumelis, N.; Manos, M.J.; Tasiopoulos, A.J.; Bakas, T.; Hadjikakou, S.K. Assessment of organotins against the linoleic acid, glutathione and CT-DNA. *Inorg. Chim. Acta* **2014**, *423*, 98–106. [CrossRef]

27. Xanthopoulou, M.N.; Kourkoumelis, N.; Hadjikakou, S.K.; Hadjiliadis, N.; Kubicki, M.; Karkabounas, S.; Bakas, T. Structural and biological studies of organotin(IV) derivatives with 2-mercapto-benzoic acid and -mercapto-4-methyl-pyrimidine. *Polyhedron* **2008**, *27*, 3318–3324. [CrossRef]

28. Hadjikakou, S.K.; Jurkschat, K.; Schurmann, M. Novel organotin(IV) compounds derived from bis(organostannyl)methanes: Synthesis and crystal structures of bis[diphenyl(pyridin-2-onato)stannyl]methane and bis[bromophenyl(pyrimidine-2-thionato) stannyl]methane C$_7$H$_8$. *J. Organomet. Chem.* **2006**, *691*, 1637–1642. [CrossRef]

29. Xanthopoulou, M.N.; Hadjikakou, S.K.; Hadjiliadis, N.; Kubicki, M.; Skoulika, S.; Bakas, T.; Baril, M.; Butler, I.S. Synthesis, structural characterization, and biological studies of six and five-coordinate organotin(IV) Complexes with the Thioamides 2-Mercaptobenzothiazole, 5-Chloro-2-mercaptobenzothiazole, and 2-Mercaptobenzoxazole. *Inorg. Chem.* **2007**, *46*, 1187–1195. [CrossRef] [PubMed]

30. Carraher, C.E., Jr.; Roner, M.R. Organotin polymers as anticancer and antiviral agents. *J. Org. Chem.* **2014**, *751*, 67–82. [CrossRef]

31. Jones, L.H.; McLaren, E. Infrared Spectra of CH$_3$COONa and CD$_3$COONa and Assignments of Vibrational Frequencies. *J. Chem. Phys.* **1954**, *22*, 1796–1800. [CrossRef]

32. Greenwood, N.N.; Gibb, T.C. *Mossbauer Spectroscopy*; Chapman and Hall: London, UK, 1971.

33. Saeed, M.A.; Badshah, A.; Rauf, M.K.; Craig, D.C.; Ali, S. catena-Poly[tributyltin(IV)-l-acetato]. *Acta Cryst.* **2006**, *E62*, m469–m471. [CrossRef]

34. Molloy, K.C.; Purcell, T.G.; Quill, K. Organotin biocides 1. The structure of triphenyltin acetate. *J. Organomet. Chem.* **1984**, *267*, 237–241. [CrossRef]

35. Lee, A.V.; Oesterreich, S.; Davidson, N.E. MCF-7 Cells—Changing the Course of Breast Cancer Research and Care for 45 Years. *JNCI J. Nat. Cancer Inst.* **2015**, *107*, Djv073. [CrossRef] [PubMed]

36. Rochefort, H.; Glondu, M.; Sahla, M.E.; Platet, N.; Garcia, M. How to target estrogen receptor-negative breast cancer? *Endocr. Relat. Cancer* **2003**, *10*, 261–266. [CrossRef] [PubMed]

37. Banti, C.N.; Papatriantafyllopoulou, C.; Manoli, M.; Tasiopoulos, A.J.; Hadjikakou, S.K. Nimesulide silver metallodrugs, containing the mitochondriotropic, triaryl derivatives of pnictogen; Anticancer activity against human breast cancer cells. *Inorg. Chem.* **2016**, *55*, 8681–8696. [CrossRef] [PubMed]

38. HCilião, L.; Ribeiro, D.L.; Camargo-Godoy, R.B.O.; Specian, A.F.L.; Serpeloni, J.M.; Cólus, I.M.S. Cytotoxic and genotoxic effects of high concentrations of the immunosuppressive drugs cyclosporine and tacrolimus in MRC-5 cells. *Exp. Toxicol. Pathol.* **2015**, *67*, 179–187.

39. Gao, E.; Lin, L.; Liu, L.; Zhu, M.; Wang, B.; Gao, X. Two new palladium(II) complexes: Synthesis, characterization and their interaction with HeLa cells. *Dalton Trans.* **2012**, *41*, 11187–11194. [CrossRef] [PubMed]

40. Boer, D.R.; Canals, A.; Coll, M. DNA-binding drugs caught in action: The latest 3D pictures of drug-DNA complexes. *Dalton Trans.* **2009**, 399–414. [CrossRef] [PubMed]

41. Gilad, Y.; Hanoch, S. Docking studies on DNA intercalators. *J. Chem. Inf. Model.* **2013**, *54*, 96–107. [CrossRef] [PubMed]

42. Lian, C.; Robinson, H.; Wang, A.-H. Structure of actinomycin D bound with (GAAGCTTC)$_2$ and (GATGCTTC)$_2$ and its binding to the (CAG)$_n$:(CTG)$_n$ triplet sequence as determined by NMR analysis. *J. Am. Chem. Soc.* **1996**, *118*, 8791–8801. [CrossRef]

43. Icsel, C.; Yilmaz, V.T.; Kaya, Y.; Samli, H.; Harrison, W.T.A.; Buyukgungor, O. New palladium (II) and platinum (II) 5, 5-diethylbarbiturate complexes with 2-phenylpyridine, 2, 2′-bipyridine and 2, 2′-dipyridylamine: Synthesis, structures, DNA binding, molecular docking, cellular uptake, antioxidant activity and cytotoxicity. *Dalton Trans.* **2015**, *44*, 6880–6895. [CrossRef] [PubMed]

44. Lauria, A.; Patella, C.; Dattolo, G.; Almerico, A.M. Design and Synthesis of 4-Substituted Indolo [3, 2-e][1, 2, 3] triazolo [1, 5-a] pyrimidine Derivatives with Antitumor Activity. *J. Med. Chem.* **2008**, *51*, 2037–2046. [CrossRef] [PubMed]

45. Pages, B.J.; Ang, D.L.; Wright, E.P.; Aldrich-Wright, J.R. Metal complex interactions with DNA. *Dalton Trans.* **2015**, *44*, 3505–3526. [CrossRef] [PubMed]

46. Oxford Diffraction. *CrysAlis CCD and CrysAlis RED*; Oxford Diffraction Ltd.: Abingdon, UK, 2008.

47. Burla, M.C.; Caliandro, R.; Camalli, M.; Carrozzini, B.; Cascarano, G.L.; de Caro, L.; Giacovazzo, C.; Polidori, G.; Spagna, R. SIR2004: An improved tool for crystal structure determination and refinement. *J. Appl. Cryst.* **2005**, *38*, 381–388. [CrossRef]

48. Sheldrick, G.M. *SHELXL-2014/7, Program for Refinement of Crystal Structures*; University of Göttingen: Göttingen, Germany, 2014.

49. Farrugia, L.J. WinGX suite for small-molecule single-crystal crystallography. *J. Appl. Cryst.* **1999**, *32*, 837–838. [CrossRef]

50. Bevington, P.R.; Robinson, D.K. *Data Reduction and Error Analysis for the Physical Sciences*; McGraw Hill: New York, NY, USA, 1992; pp. 194–197.

51. OTrott, O.; Olson, A.J. AutoDockVina: Improving the speed and accuracy of docking with a new scoring function, efficient optimization and multithreading. *J. Comput. Chem.* **2010**, *31*, 455–461.

52. Morris, G.M.; Huey, R.; Lindstrom, W.; Sanner, M.F.; Belew, R.K.; Goodsell, D.S.; Olson, A.J. Autodock4 and AutoDockTools4: Automated docking with selective receptor flexibility. *J. Comput. Chem.* **2009**, *16*, 2785–2791. [CrossRef] [PubMed]

53. Frisch, M.J.; Trucks, G.W.; Schlegel, H.B.; Scuseria, G.E.; Robb, M.A.; Cheeseman, J.R.; Montgomery, J.A., Jr.; Vreven, T.; Kudin, K.N.; Burant, J.C.; et al. *GAUSSIAN 03 (Revision C.02)*; Gaussian, Inc.: Wallingford, CT, USA, 2004.

International Journal of
Molecular Sciences

MDPI

Article

Multi-Acting Mitochondria-Targeted Platinum(IV) Prodrugs of Kiteplatin with α-Lipoic Acid in the Axial Positions

Salvatore Savino [1], Cristina Marzano [2], Valentina Gandin [2], James D. Hoeschele [3], Giovanni Natile [1,*] and Nicola Margiotta [1,*]

[1] Department of Chemistry, University of Bari Aldo Moro, Via E. Orabona 4, 70125 Bari, Italy;
    salvatoresavino.s@libero.it
[2] Department of Pharmaceutical and Pharmacological Sciences, University of Padova, Via Marzolo 5,
    35131 Padova, Italy; cristina.marzano@unipd.it (C.M.); valentina.gandin@unipd.it (V.G.)
[3] Department of Chemistry, Eastern Michigan University, Ypsilanti, MI 48197, USA;
    hoeschel@chemistry.msu.edu
*   Correspondence: giovanni.natile@uniba.it (G.N.); nicola.margiotta@uniba.it (N.M.);
    Tel.: +39-080-544-2774 (G.N.); +39-080-544-2759 (N.M.)

Received: 20 June 2018; Accepted: 12 July 2018; Published: 14 July 2018

Abstract: Platinum(II) drugs are activated intracellularly by aquation of the leaving groups and then bind to DNA, forming DNA adducts capable to activate various signal-transduction pathways. Mostly explored in recent years are Pt(IV) complexes which allow the presence of two additional ligands in the axial positions suitable for the attachment of other cancer-targeting ligands. Here we have extended this strategy by coordinating in the axial positions of kiteplatin ([PtCl$_2$(*cis*-1,4-DACH)], DACH = Diaminocyclohexane) and its CBDCA (1,1-cyclobutanedicarboxylate) analogue the antioxidant α-Lipoic acid (ALA), an inhibitor of the mitochondrial pyruvate dehydrogenase kinase (PDK). The new compounds (*cis,trans,cis*-[Pt(CBDCA)(ALA)$_2$(*cis*-1,4-DACH)], **2**, and *cis,trans,cis*-[PtCl$_2$(ALA)$_2$(*cis*-1,4-DACH)], **3**), after intracellular reduction, release the precursor Pt(II) species and two molecules of ALA. The Pt residue is able to target DNA, while ALA could act on mitochondria as activator of the pyruvate dehydrogenase complex, thus suppressing anaerobic glycolysis. Compounds **2** and **3** were tested in vitro on a panel of five human cancer cell lines and compared to cisplatin, oxaliplatin, and kiteplatin. They proved to be much more effective than the reference compounds, with complex **3** most effective in 3D spheroid tumor cultures. Notably, treatment of human A431 carcinoma cells with **2** and **3** did not determine increase of cellular ROS (usually correlated to inhibition of mitochondrial PDK) and did not induce a significant depolarization of the mitochondrial membrane or alteration of other morphological mitochondrial parameters.

Keywords: platinum(IV) complexes; cisplatin; kiteplatin; α-lipoic acid; DNA; mitochondria; tumor spheroids

1. Introduction

It is well known that numerous cancer cells preferentially convert glucose to lactate, even in the presence of oxygen (aerobic glycolysis), for the generation of ATP. This phenomenon was discovered by Warburg already in the 1920s and is hence reported as the Warburg effect [1,2]. Yet, aerobic glycolysis is not very efficient from an energetical point of view since it leads to the production of only two molecules of ATP per molecule of glucose while the complete oxidation of glucose to carbon dioxide and water produces 32 molecules of ATP. For this reason, cancer cells need huge amounts of glucose to satisfy their high demand of energy, but, at the same time, they are able to adapt to hypoxic conditions. Cancer therapy can take advantage of the key feature of glucose metabolism of cancer cells. The pyruvate dehydrogenase complex is located in the mitochondrial matrix of eukaryotes; the complex acts to convert pyruvate (a product of glycolysis in the cytosol) to acetyl-coA, which is then oxidized in the mitochondria to produce energy in the citric acid cycle. Pyruvate dehydrogenase kinase (PDK) is a kinase enzyme which acts to inactivate the pyruvate dehydrogenase enzyme by phosphorylating it using ATP; by downregulating the pyruvate dehydrogenase, PDK will decrease the oxidation of pyruvate in mitochondria and increase the conversion of pyruvate to lactate in the cytosol. An inhibitor of PDK will activate those enzymes that are able to shift the metabolism toward the complete oxidation of glucose. A compound that has been used for such purpose (inhibition of PDK) is dichloroacetate (DCA; Scheme 1), an orphan drug. DCA works to counteract the increased production of lactate exhibited by tumor cells (anaerobic respiration) by activating the pathway to pull the intermediates into the citric acid cycle and finish off with oxidative phosphorylation (aerobic respiration) [3].

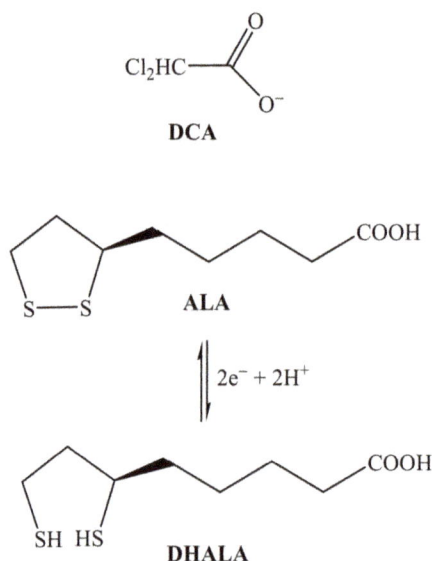

Scheme 1. Molecular structures of dichloroacetate (DCA), α-lipoic acid (ALA), and α-dihydrolipoic acid (DHALA).

Also the dithiol compound (R)-(+)-α-lipoic acid (6,8-dithio-octanoic acid; ALA, Scheme 1) is a potential activator of the pyruvate dehydrogenase complex and can be potentially used to suppress anaerobic glycolysis [4,5]. ALA is synthesized from octanoic acid in the mitochondria and it is also taken from food. Apart from being capable to chelate metal ions, ALA has a unique reductive power and its disulfide group can be easily reduced to form α-dihydrolipoic acid (DHALA; Scheme 1) and both constitute a low redox potential pair ($E'_0 = -0.29$ V) [6] capable to scavenge a variety of reactive oxygen species (ROS) [7–10].

ROS and, in general, oxidative stress also play a crucial role in cancer cells being correlated with cell growth and apoptosis [11]. In particular, ALA was shown to activate apoptosis in human cancer cell lines by inducing a reversible cell-cycle arrest but failed to trigger apoptosis in non-transformed cell lines [12]. Apoptosis was potentiated by ALA also in human leukemia cells and, because of its antioxidant properties, it was suggested that this compound can promote a reducing intracellular environment that is necessary for the activation of caspases [13]. On the other hand, in neurons and in hepatocytes [14,15], ALA was demonstrated to exert a protective activity against apoptosis, thus showing an opposite activity in tumor and healthy cells.

In another study, exposure of HT-29 human colon cancer cells to ALA caused a dose-dependent increase of caspase-3-like activity associated with DNA fragmentation. Moreover, in this tumor cell line ALA was not able to scavenge cytosolic $O_2^{\cdot-}$ whereas it was capable to increase the generation of the superoxide anion radical in the mitochondria [16] preceded by an increased influx of lactate or pyruvate into the organelles. Oppositely to HT-29 colon cancer cells, no apoptosis was observed in nontransformed human colonocytes, thus providing evidences for a selective induction of apoptosis in the cancer cell line by a prooxidant mechanism initiated by an increased uptake of oxidizable metabolites in the mitochondria.

In vitro cell proliferation inhibition by ALA was also reported for neuroblastoma cell lines Kelly, SK-N-SH, and Neuro-2a, for breast cancer cell lines SkBr3 [16] and MDA-MB-231, as well as for leukemia cells Jurkat and CCRF-CEM [17,18]. Moreover, in a xenograft mouse model with subcutaneous SkBr3 cells, daily treatment with ALA significantly retarded tumor growth, further supporting the potential anticancer activity of this compound.

Platinum(II) complexes are well-known antiproliferative agents [19] and, among them, cisplatin, carboplatin, and oxaliplatin (Scheme 2) have received Food and Drug Administration approval and are used in the clinic worldwide [19–21]. However, the appearance of resistance and induction of severe side effects limit the use of these drugs [20]. In order to overcome at least some of these drawbacks, new platinum drugs based on the platinum(IV) core have been developed. In this context, we have recently focused on Pt(IV) complexes [22,23] derived from kiteplatin, [PtCl$_2$(*cis*-1,4-DACH)] (DACH = diaminocyclohexane; Scheme 2) including a derivative having two DCA ligands in axial positions [24]; the aim was to obtain a dual acting complex endowed with the ability to target both nuclear DNA and mitochondria. Kiteplatin contains an isomeric form of the diamine ligand 1*R*,2*R*-DACH present in oxaliplatin and is effective against both cisplatin-resistant (ovarian C13*) and oxaliplatin-resistant (colon LoVo-OXP) cancer cell lines, suggesting that the spectrum of activity of this compound could be different from that of cisplatin and oxaliplatin [25–28]. The newly synthesized Pt(IV) complexes with DCA, *cis,trans,cis*-[PtCl$_2$(DCA)$_2$(*cis*-1,4-DACH)] and *cis,trans,cis*-[Pt(CBDCA)(DCA)$_2$(*cis*-1,4-DACH)] (CBDCA = 1,1-cyclobutanedicarboxylate), were tested in vitro against a series of different tumor cell lines, some of which were selected for their resistance to cisplatin and oxaliplatin, and the antitumor activity of the lead compound, *cis,trans,cis*-[PtCl$_2$(DCA)$_2$(*cis*-1,4-DACH)], was also assessed in vivo in a syngeneic murine model of solid tumor (the Lewis Lung Carcinoma).

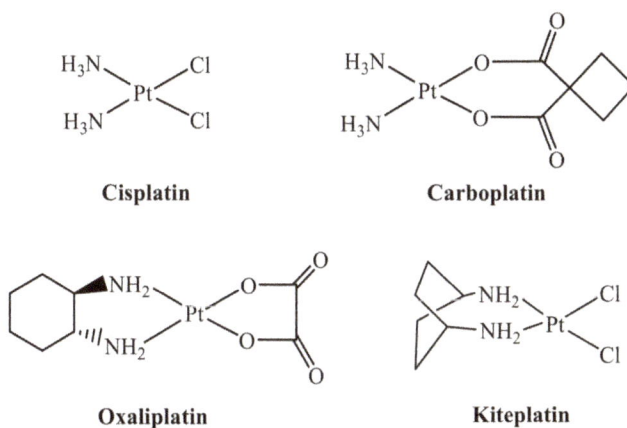

Scheme 2. Sketches of clinically used platinum(II) drugs (Cisplatin, Carboplatin, and Oxaliplatin) and of Kiteplatin.

Tested compounds induced a substantial increase of ROS production, blockage of oxidative phosphorylation, hypopolarization of the mitochondrial membrane, and caspase-3/7-mediated apoptotic cell death. These effects could be a consequence of the DCA released after intracellular reduction of the Pt(IV) complexes.

On this basis we decided to explore the conjugation of ALA (another activator of the pyruvate dehydrogenase complex like DCA) in the axial positions of a Pt(IV) derivative of kiteplatin and in this paper we report the synthesis, characterization, and biological activity of two kiteplatin derivatives, *cis,trans,cis*-[Pt(CBDCA)(ALA)$_2$(*cis*-1,4-DACH)] (**2**; Scheme 3) and *cis,trans,cis*-[PtCl$_2$(ALA)$_2$(*cis*-1,4-DACH)] (**3**; Scheme 3).

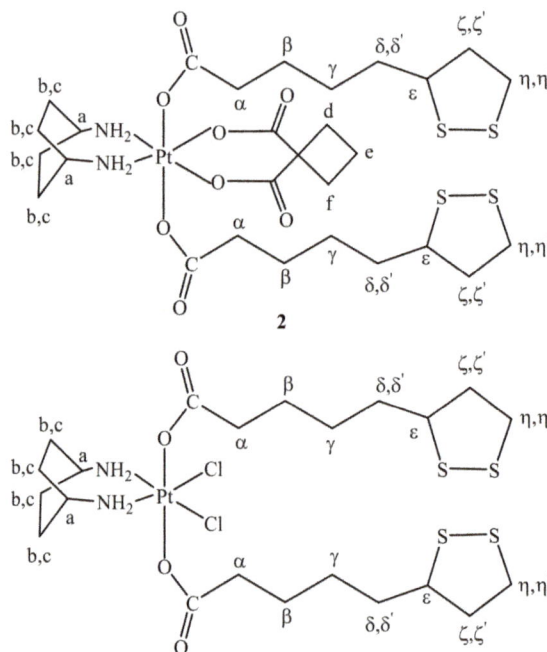

Scheme 3. Sketches of the two novel kiteplatin Pt(IV) derivatives investigated in this work: *cis,trans,cis*-[Pt(CBDCA)(ALA)$_2$(*cis*-1,4-DACH)] (**2**) and *cis,trans,cis*-[PtCl$_2$(ALA)$_2$(*cis*-1,4-DACH)] (**3**).

It could be expected that the new compounds could mitigate some side effects of platinum-drug therapy such as ototoxicity and nephrotoxicity. In fact, it has been shown that pretreatment with ALA significantly reduces apoptotic cell death of the inner and outer hair cells in cisplatin-treated organ of Corti explants and attenuates ototoxicity via marked lowering of the expression levels of proinflammatory cytokines and other cisplatin-induced biomarkers of ototoxicity in cisplatin-treated HEI-OC1 cells [29]. Moreover, it has been shown that lipoic acid also prevents cisplatin-induced nephrotoxicity in rats [30].

2. Results and Discussion

2.1. Synthesis and Characterization

It was demonstrated that compounds that support pyruvate dehydrogenase reaction (such as DCA and ALA) are promising agents in cancer therapy since they are able to target the glucose metabolism of cancer cells [31]. In particular, the antioxidant ALA is able to induce apoptosis in HT-29 human colon cancer cells via an increased ROS production in mitochondria by enhancing the uptake of pyruvate and lactate from glycolysis into the mitochondria [16]. Oxidation of the monocarboxylates pyruvate and lactate in the citric acid cycle increases the delivery of reductive equivalents to the respiratory chain with the final result of drastically increasing the mitochondrial production of $O_2^{\cdot -}$ which, in turn, triggers apoptosis in the cancer cells.

We have already reported on kiteplatin-Pt(IV) complexes having DCA in the axial positions with the lead compound, *cis,trans,cis*-[PtCl$_2$(DCA)$_2$(*cis*-1,4-DACH)] showing encouraging antitumor activity also in vivo in the solid syngeneic murine model Lewis Lung Carcinoma [24]. In this work we have extended the investigation to Pt(IV) kiteplatin derivatives containing, in the axial positions, the antioxidant ALA: namely *cis,trans,cis*-[Pt(CBDCA)(ALA)$_2$(*cis*-1,4-DACH)] (**2**) and *cis,trans,cis*-[PtCl$_2$(ALA)$_2$(*cis*-1,4-DACH)] (**3**).

Lipoic acid has been already used as a ligand in platinum complexes. In particular, an amide formed by lipoic acid and aniline was used as a linker between a platinum complex and gold nanoparticles [32], while polynuclear platinum complexes with lipoic acid were patented for the prophylaxis or treatment of cancer [33]. To the best of our knowledge, this is the first time that mononuclear platinum complexes with lipoic acid have been prepared and tested in vitro for their potential application as antitumor drugs.

Naturally occurring α-Lipoic acid is in the (*R*)-(+) configuration, however, in this work we have used the commercially available racemic (±)-α-Lipoic acid that was activated into its anhydride by DCC conjugation and then reacted with the Pt(IV)-dihydroxido derivative of kiteplatin *cis,trans,cis*-[Pt(CBDCA)(OH)$_2$(*cis*-1,4-DACH)] (**1**). The NMR characterization of compound **1** is reported in the Experimental Section and in the Supplementary Information (Figures S1–S3).

Cis,trans,cis-[Pt(CBDCA)(ALA)$_2$(*cis*-1,4-DACH)] (**2**) has been characterized by elemental analysis, ESI-MS, and NMR. The ESI-MS (+) spectrum showed the presence of a peak at $m/z = 885.15$ corresponding to [**2** + Na]$^+$ and the experimental isotopic pattern was in good agreement with the theoretical one (data not shown). The NMR characterization of compound **2** was obtained with the help of a 2D COSY experiment (Figure 1). Assignment of protons started from the methylenic protons of coordinated ALA. The triplet integrating for 4 protons falling at 2.22 ppm and the multiplet at 1.48 ppm, correlated by a COSY cross peak, were assigned to CH$_2$ in α and β positions, respectively, of the lipoato ligand (see Scheme 3 for numbering of protons). The multiplet located at 1.32 ppm was assigned to the CH$_2$γ protons since it shows COSY cross peaks with the CH$_2$ in β position and the multiplets at 1.62 and 1.51 ppm; these latter were assigned to the protons in δ' and δ positions, respectively. The 1,2-dithiolane ring showed 5 different signals. The multiplet resonating at 3.58 ppm was assigned to the ε CH based on a COSY a cross peak with the proton δ and two additional cross peaks with the multiplets resonating at 2.39 and 1.85 ppm assigned, respectively, to the methylenic protons ζ' and ζ. The characterization of the lipoato ligand ends with the attributions of the multiplets at 3.17 and 3.11 ppm to the methylenic protons η' and η, these latter signals showing cross peaks with the protons ζ and ζ'.

With reference to the *cis*-1,4-DACH ligand, the singlet with platinum satellites ($^2J_{H-Pt}$ = 66 Hz; see Figure 2) located at 7.76 ppm was assigned to the aminic protons, at lower field with respect to the corresponding signal in compound **1** (6.47 ppm). The deshielding is probably due to the presence of the carboxylic groups of coordinated ALA in axial positions (hydrogen bonds between axial C=O and coordinated NH$_2$ were found in similar systems) [34]. The methynic (H$_a$) and methylenic (H$_{b,c}$) protons of coordinated DACH give, respectively, a singlet at 2.84 and a broad signal (integrating for eight protons) at 1.63 ppm. Finally, the multiplet at 2.40 ppm (partially overlapping with ζ') and the quintet at 1.78 ppm were assigned to the cyclobutane protons of CBDCA.

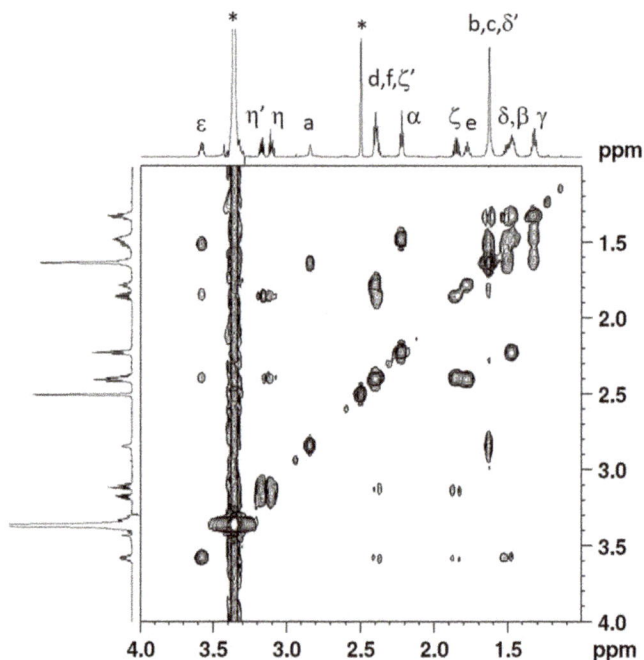

Figure 1. Selected region of the 2D COSY (700 MHz) spectrum obtained for *cis,trans,cis*-[Pt(CBDCA)(ALA)₂(*cis*-1,4-DACH)] (**2**) in DMSO-d₆. Numbering of protons is reported in Scheme 3. The asterisks indicate residual solvent peaks.

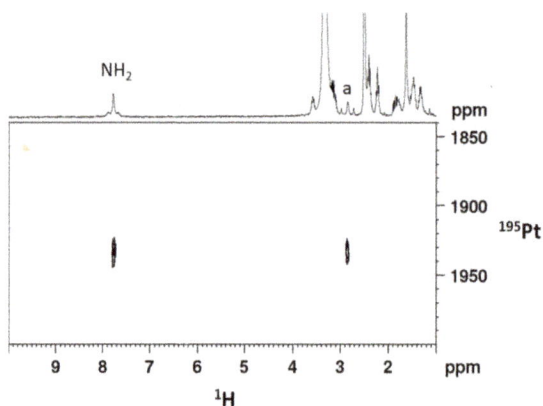

Figure 2. [¹H, ¹⁹⁵Pt]-HSQC 2D spectrum (¹H 300 MHz) obtained for *cis,trans,cis*-[Pt(CBDCA)(ALA)₂(*cis*-1,4-DACH)] (**2**) in DMSO-d₆.

The [¹H-¹⁹⁵Pt]-HSQC 2D NMR spectrum of compound **2** in DMSO-d₆ (Figure 2) exhibits two cross peaks, located at 7.76/1932.7 and 2.84/1932.7 ppm (¹H/¹⁹⁵Pt), correlating the aminic and methynic protons of *cis*-1,4-DACH with the platinum atom. The ¹⁹⁵Pt chemical shift is found at lower field with respect to that of the precursor compound **1** (1617.6 ppm in DMSO-d₆) but is in good agreement with that reported for similar Pt(IV) dicarboxylato derivatives with the platinum atom in a N₂O₄ coordination environment [35].

The assignment of ^{13}C signals has been accomplished by a [^{1}H, ^{13}C]-HSQC 2D NMR spectrum (Figure 3) and data are reported in the Experimental Section.

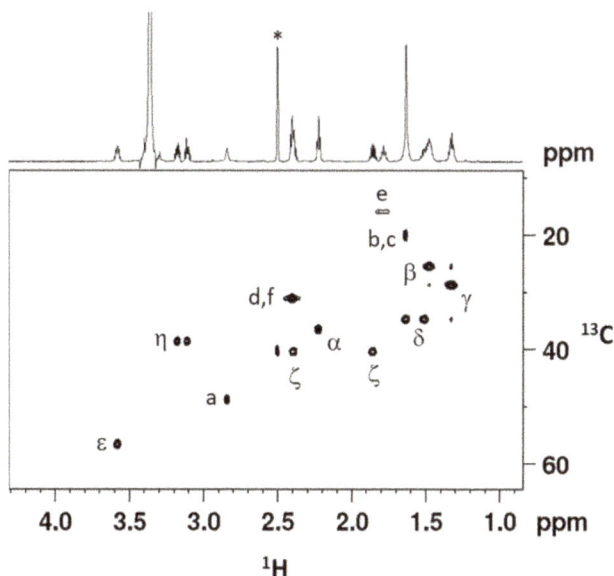

Figure 3. [^{1}H, ^{13}C]-HSQC (^{1}H 700 MHz) of *cis,trans,cis*-[Pt(CBDCA)(ALA)$_2$(*cis*-1,4-DACH)] (**2**) in DMSO-d$_6$. Numbering of protons is reported in Scheme 3. The asterisk indicates the residual solvent peak.

Compound **3**, *cis,trans,cis*-[PtCl$_2$(ALA)$_2$(*cis*-1,4-DACH)], was prepared with a procedure similar to that used for compound **2**, the only differences being the starting platinum(IV) derivative of kiteplatin, *cis,trans,cis*-[PtCl$_2$(OH)$_2$(*cis*-1,4-DACH)], and the purification step that required the use of a less polar solvent such as *n*-pentane. Compound **3** has been characterized by elemental analysis, ESI-MS, and NMR. The ESI-MS (+) spectrum showed the presence of a peak at *m/z* = 813.07 corresponding to [**3** + Na]$^+$ and the experimental isotopic pattern of the peak resulted to be in agreement with the theoretical one (data not shown). The NMR characterization (in DMSO-d$_6$) of compound **3** was obtained with the help of a 2D COSY experiment (Figure S4) and is similar to that reported for complex **2** with the exclusion of the signals belonging to the CBDCA ligand. The assignment of ^{13}C signals has been accomplished by a [^{1}H, ^{13}C]-HSQC 2D NMR spectrum (Figure S5). The [^{1}H-^{195}Pt]-HSQC 2D NMR spectrum of compound **3** in DMSO-d6 (Figure S6) exhibits two cross peaks, located at 8.19/1217.6 and 2.98/1217.6 ppm (^{1}H/^{195}Pt), correlating the aminic and methynic protons of *cis*-1,4-DACH with the platinum atom. The ^{195}Pt chemical shift is found at lower field with respect to that of the precursor compound *cis,trans,cis*-[PtCl$_2$(OH)$_2$(*cis*-1,4-DACH)] (964.6 ppm in DMSO-d$_6$) [36] and is in good agreement with that reported for similar Pt(IV) derivatives with the platinum atom in a Cl$_2$N$_2$O$_2$ coordination environment [37,38]. All the NMR data are reported in the Experimental Section while the NMR spectra are reported in the Supplementary Information (Figures S4–S6).

2.2. Biological Assays

The in vitro antitumor potency of the ALA Pt(IV) kiteplatin-derivatives **2** and **3** was evaluated on a panel of human cancer cells and compared to that of cisplatin (CDDP), oxaliplatin (OXP), and kiteplatin as well as to that of [Pt(CBDCA)(*cis*-1,4-DACH)]. Cell lines representative of lung (H157), colon (HCT-15), breast (MCF-7), cervical (A431), and ovarian (2008) carcinoma have been included. The cytotoxicity was evaluated by means of the MTT test for 72 h incubation with different concentrations of the tested compounds. IC$_{50}$ values, calculated from dose-survival curves, are reported in Table 1.

Table 1. In vitro antitumor activity.

Compound	IC$_{50}$ (μM) $\pm$ S.D.					
	2008	MCF-7	A431	HCT-15	H157	Average
[Pt(CBDCA)(*cis*-1,4-DACH)]	25.47 $\pm$ 4.11	20.58 $\pm$ 3.15	13.56 $\pm$ 2.18	30.58 $\pm$ 5.25	19.21 $\pm$ 4.15	21.9
2	3.15 $\pm$ 0.85	3.05 $\pm$ 0.89	2.25 $\pm$ 0.52	2.31 $\pm$ 0.56	3.44 $\pm$ 0.97	2.8
3	0.34 $\pm$ 0.09	1.12 $\pm$ 0.57	0.10 $\pm$ 0.04	0.61 $\pm$ 0.15	0.99 $\pm$ 0.31	0.6
CDDP	2.22 $\pm$ 1.02	10.58 $\pm$ 0.82	2.10 $\pm$ 0.87	15.28 $\pm$ 2.63	2.12 $\pm$ 0.89	6.5
OXP	1.65 $\pm$ 0.46	4.52 $\pm$ 0.95	3.71 $\pm$ 0.76	1.15 $\pm$ 0.43	5.99 $\pm$ 1.85	3.4
Kiteplatin	1.89 $\pm$ 1.04	3.10 $\pm$ 1.42	3.95 $\pm$ 1.11	2.66 $\pm$ 0.95	2.08 $\pm$ 0.66	2.7

CDDP: cisplatin, OXP: oxaliplatin. S.D. = standard deviation. Cells (3–5 $\times$ 10^4 mL^{-1}) were treated for 72 h with different concentrations of tested compounds. Cytotoxicity was assessed by the MTT test. IC$_{50}$ values were calculated by a four parameter logistic model ($p < 0.05$).

The two ALA derivatives **2** and **3** proved to be much more effective than the reference compounds kiteplatin and [Pt(CBDCA)(*cis*-1,4-DACH)]. The higher potency of complex **3**, compared to that of complex **2**, reflects the one order magnitude higher potency of kiteplatin compared to that of [Pt(CBDCA)(*cis*-1,4-DACH)]. However, an effect due to the release of ALA after reduction of the Pt(IV) complexes is also evident. Notably, over the five tested cell lines the IC$_{50}$ values of **2** and **3** were, in the order, about 7 and 4 times lower than those of the reference Pt(II) complexes. In addition, both Pt(IV) complexes showed, on average, an in vitro antitumor potency superior to those of the reference metallodrugs CDDP and OXP. Among Pt(IV) ALA derivatives, complex **3** was the most effective with an in vitro antitumor potential roughly an order of magnitude higher than that of CDDP and about 5.5 times greater than that of OXP.

The marked cell-killing effect observed against human A431 squamous cervical carcinoma cells prompted us to evaluate the in vitro antitumor activity of the Pt(IV) ALA derivatives on 3D cell cultures. As opposed to 2D monolayer cultures, cells growing in 3D culture systems form spheroids that are comprised of cells in various stages. The outer layers of the spheroid, being highly exposed to the medium, are mainly comprised of viable, proliferating cells whereas the core cells receiving less oxygen, growth factors, and nutrients, tend to be in a quiescent or hypoxic state [39]. Such cellular heterogeneity resembles that of in vivo tumors, making 3D cell cultures more predictive than conventional 2D monolayer cultures in screening antitumor drugs. Table 2 summarizes the IC$_{50}$ values obtained after treatment of 3D cell spheroids of human A431 cervical cancer cells with the Pt(IV) ALA complexes as well as kiteplatin, [Pt(CBDCA)(*cis*-1,4-DACH)], CDDP, and OXP used as references.

Table 2. Cytotoxicity towards human A431 cancer cell spheroids.

	IC$_{50}$ (μM) $\pm$ S.D.					
	[Pt(CBDCA)(*cis*-1,4-DACH)]	2	3	Kiteplatin	CDDP	OXP
A431	98.9 $\pm$ 6.8	58.4 $\pm$ 3.8	30.2 $\pm$ 5.1	65.2 $\pm$ 5.8	71.1 $\pm$ 3.9	65.3 $\pm$ 5.2

Spheroids (3 $\times$ 10^3 cells/well) were treated for 72 h with increasing concentrations of tested compounds. The growth inhibitory effect was evaluated by means of MTT test. IC$_{50}$ values were calculated from the dose-survival curves by the four parameter logistic model ($p < 0.05$). S.D. = standard deviation.

Consistently with 2D studies, complex **3** proved to be the most effective compound, showing an efficacy (in decreasing cancer spheroid viability) about 2 times higher than those of CDDP, OXP, and kiteplatin. Conversely, complex **2** was slightly more effective than CDDP, OXP, and kiteplatin but markedly less cytotoxic than **3**.

Based on previous findings highlighting the ability of ALA derivatives to affect PDK (thus leading to ROS production and mitochondria hampering) [16–18], we investigated the effects of Pt(IV)

Int. J. Mol. Sci. **2018**, *19*, 2050

ALA kiteplatin derivatives on mitochondria. In particular, we investigated the ROS production and the alteration of the mitochondrial membrane potential and of the mitochondrial morphological parameters. A preliminary NMR investigation revealed that the ALA ligands conjugated in the axial positions of the Pt(IV) complexes maintained their oxidized state also in the presence of glutathione. This finding was in line with literature data reporting the potentials of redox systems such as NAD+/NADH, GSSG/GSH, dehydroascorbate/ascorbate, etc. [6]. Hence, we are confident that the ALA ligand is released after the complexes have entered the tumor cells and can be reduced only after reduction of the Pt(IV) complexes to their Pt(II) counterparts.

Treatment of A431 cells with derivatives **2** and **3** did not determine any substantial increase in cellular ROS basal production (Figure 4A), whereas 2 h treatment with antimycin, a classical inhibitor of the mitochondrial respiratory chain at the level of complex III, caused a remarkable increase of the hydrogen peroxide content (about 6 times greater than that of control cells). Consistently, treatment with **2** and **3** did not induce a significant increase of A431 human cancer cells with depolarized mitochondria (Figure 4B) or any morphological alteration of mitochondria parameters (Figure 4C). Indeed, mitochondria of A431 cancer cells treated with **3** were conserved in shape and ultrastructure (cristae).

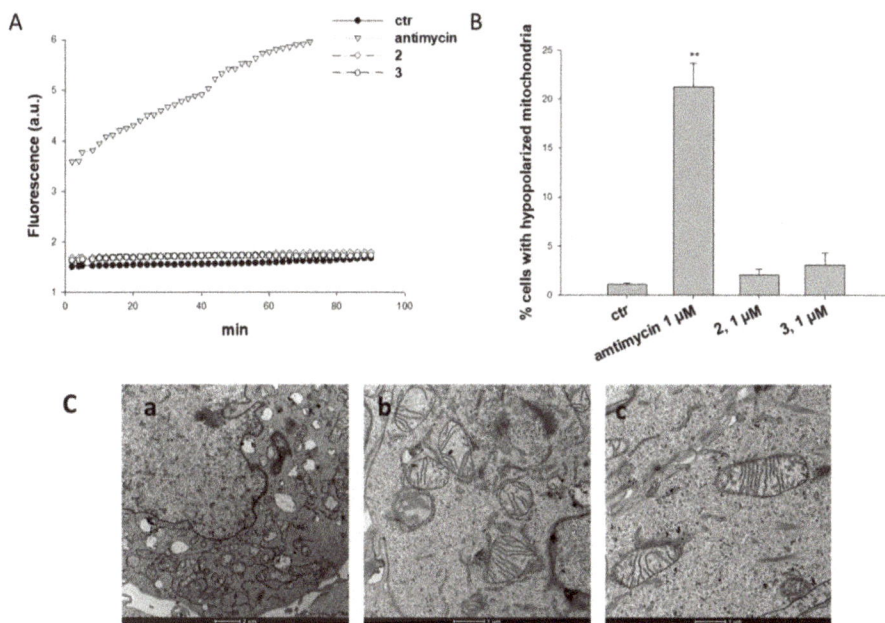

Figure 4. Effects induced by Pt(IV) ALA kiteplatin derivatives at mitochondrial level. (**A**) ROS production in A431 cells. Cells were pre-incubated in PBS/10 mM glucose medium for 20 min at 37 °C in the presence of 10 mM CM-H_2DCFDA and then treated with **2**, **3**, or antimycin (1 μM). Fluorescence of DCF was measured at 485 nm (excitation) and 527 nm (emission). (**B**) Effects on mitochondrial membrane potential. A431 cells were treated for 24 h with **2**, **3**, or antimycin (1 μM) and stained with TMRM (10 nM). Fluorescence was estimated at 490 nm (excitation) and 590 nm (emission). Data are the means of five independent experiments. Error bars indicate S.D. ** $p < 0.01$. (**C**) TEM analysis of A431 cells treated for 24 h with 1 μM of **2** and **3**. Cells were processed through standard procedures as reported in the Experimental section. (**a**) Control; (**b**) **2**; (**c**) **3**.

Overall, these results suggest that **2** and **3** do not activate a macroscopic antimitochondrial mechanism.

3. Materials and Methods

3.1. Chemicals, Instrumentation, and Platinum Complexes Precursors

Commercial reagent grade chemicals ((±)-α-Lipoic acid, dicyclohexylcarbodiimmide (DCC), 4-(dimethylamino)pyridine; Sigma-Aldrich, Milan, Italy) and solvents were used as received without further purification. ^{1}H-NMR, COSY, and [^{1}H, ^{13}C]-HSQC 2D NMR spectra were recorded on Bruker Avance DPX 300 MHz and Bruker Avance III 700 MHz instruments. ^{1}H and ^{13}C chemical shifts were referenced using the internal residual peak of the solvent (DMSO-d$_6$: 2.50 ppm for ^{1}H and 39.51 ppm for ^{13}C). [^{1}H, ^{195}Pt]-HSQC spectra were recorded on Bruker Avance DPX 300 MHz instrument (Bruker Italia S.r.L., Milano, Italy). ^{195}Pt NMR spectra were referenced to K$_2$PtCl$_4$ (external standard placed at −1620 ppm with respect to Na$_2$[PtCl$_6$]) [37]. Electrospray Mass Spectrometry: ESI-MS was performed with a dual electrospray interface and a quadrupole time-of-flight mass spectrometer (Agilent 6530 Series Accurate-Mass Quadrupole Time-of- Flight (Q-TOF) LC/MS; Agilent Technologies Italia S.p.A., Cernusco sul Naviglio, Italy). Elemental analyses were carried out with an Eurovector EA 3000 CHN Instrument (Eurovector S.p.A., Milano, Italy).

Kiteplatin, [Pt(CBDCA)(cis-1,4-DACH)] (DACH = diaminocyclohexane; CBDCA = 1,1-cyclobutanedicarboxylate) [40] and cis,trans,cis-[PtCl$_2$(OH)$_2$(cis-1,4-DACH)] [36] were prepared according to already reported procedures and all analytical data were in good agreement with the given formulation.

3.2. Synthesis of cis,trans,cis-[Pt(CBDCA)(OH)$_2$(cis-1,4-DACH)] (1)

This compound was prepared according to a procedure reported in the literature with slight modifications [41]. Briefly, a solution of [Pt(CBDCA)(cis-1,4-DACH)] (100 mg; 0.22 mmol) in 40 mL of H$_2$O was treated with a solution of H$_2$O$_2$ in H$_2$O (30% *w/w*, 450 μL). The mixture was stirred at room temperature for 24 h in the dark. The resulting suspension was filtered and the filtrate was concentrated under reduced pressure to a minimum volume. Addition of acetone induced the formation of a white precipitate that was isolated by filtration of the mother liquor, washed with acetone, and dried under vacuum. Yield 77% (82 mg, 0.17 mmol). Anal.: calculated for C$_{12}$H$_{22}$N$_2$O$_6$Pt (1): C, 29.69; H, 4.57; N, 5.77%. Found: C, 29.53; H, 4.71; N, 5.63%. ESI-MS: calculated for C$_{12}$H$_{22}$N$_2$O$_6$PtNa [1 + Na]$^+$: 508.10. Found: 508.10 *m/z*. ^{1}H-NMR (DMSO-d$_6$): 6.47 (4H, NH$_2$), 2.80 (2H, CH*a*), 2.52 (4H, CH*d,f*), 2.02 (4H, CH*b*), 1.77 (2H, CH*e*), 1.48 (4H, CH*c*), 0.35 (2H, OH) ppm. ^{195}Pt NMR (DMSO-d$_6$): 1617.60 ppm. ^{13}C-NMR (DMSO-d$_6$): 47.57 (C*a*), 31.67 (C*d,f*), 19.23 (C*b,c*), 15.69 (C*e*) ppm.

3.3. Synthesis of cis,trans,cis-[Pt(CBDCA)(ALA)$_2$(cis-1,4-DACH)] (ALA = (±)-α-Lipoic Acid) (2)

α-Lipoic anhydride was prepared according to a procedure reported in the literature with slight modifications [42]. A mixture of (±)-α-lipoic acid (727 mg, 3.52 mmol) and DCC (494 mg, 2.39 mmol) was stirred in 26 mL of methylene chloride for about 15 h at room temperature. The solution was filtered, to remove the byproduct dicyclohexylurea (DCU), treated with 4-(dimethylamino)pyridine (1.4 mg, 0.0117 mmol) and 1 (57 mg, 0.117 mmol) and left under stirring at room temperature for 24 h. The resulting suspension was filtered and the solution treated with 120 mL of diethyl ether which induced the formation of a white precipitate that was isolated by filtration of the mother liquor, washed several times with diethyl ether, and dried under vacuum. Yield 35% (35 mg, 0.041 mmol). Anal.: calculated for C$_{28}$H$_{46}$N$_2$O$_8$PtS$_4$·1/3diethyl ether (2·1/3 C$_4$H$_{10}$O): C, 39.73; H, 5.61; N, 3.25%. Found: C, 39.73; H, 5.45; N, 3.43%. ESI-MS: calculated for C$_{28}$H$_{46}$N$_2$O$_8$PtS$_4$Na [2 + Na]$^+$: 885.16. Found: *m/z* 885.15. ^{1}H-NMR (DMSO-d$_6$): 7.76 (4H, NH$_2$), 3.58 (2H, CHε; see Scheme 3 for numbering of protons), 3.17 (2H, CH*η'*), 3.11 (2H, CH*η*), 2.84 (2H, CH*a*), 2.40 (4H, CH*d,f*), 2.39 (2H, CH*ζ'*), 2.22 (2H, CH*α*), 1.85 (2H, CH*ζ*), 1.78 (2H, CH*e*), 1.63 (8H, CH*b,c*), 1.62 (2H, CH*δ'*), 1.51 (2H, CH*δ*), 1.48 (4H, CH*β*), 1.32 (4H, CH*γ*) ppm. ^{195}Pt NMR (DMSO-d$_6$): 1932.70 ppm. ^{13}C-NMR (DMSO-d$_6$): 55.93 (C*ε*), 48.02 (C*a*), 39.69 (C*ζ*), 37.98 (C*η*), 35.69 (C*α*), 33.92 (C*δ*), 30.47 (C*d,f*), 28.09 (C*γ*), 24.73 (C*β*), 19.35 (C*b,c*), 15.21 (C*e*) ppm.

3.4. Synthesis of cis,trans,cis-[PtCl$_2$(ALA)$_2$(cis-1,4-DACH)] (**3**)

This compound was prepared as described for complex **2** with some differences in the purification step. Briefly, a solution of ALA (448 mg, 2.17 mmol) and DCC (299 mg, 1.45 mmol) was stirred in 16 mL of methylene chloride for about 15 h at room temperature. The byproduct DCU was removed by filtration and the solution treated with 4-(dimethylamino)pyridine (0.88 mg, 0.0072 mmol) and *cis,trans,cis*-[PtCl$_2$(OH)$_2$(*cis*-1,4-DACH)] (30 mg, 0.072 mmol) and left under stirring at room temperature for 24 h. The suspension was filtered and the resulting yellow solution was concentrated to 10 mL under reduced pressure. Addition of 40 mL of diethyl ether and 10 mL of *n*-pentane induced the formation of a yellow precipitate that was isolated by filtration of the mother liquor, washed several times with diethyl ether, and dried under vacuum. Yield 53% (30 mg, 0.038 mmol). Anal.: calculated for C$_{22}$H$_{40}$Cl$_2$N$_2$O$_4$PtS$_4$·1/3diethyl ether (**3**·1/3 C$_4$H$_{10}$O): C, 34.36; H, 5.36; N, 3.44%. *Found:* C, 34.56; H, 5.19; N, 3.90%. ESI-MS: calculated for C$_{22}$H$_{40}$Cl$_2$N$_2$O$_4$PtS$_4$Na [**3** + Na]$^+$: 813.07. Found: *m*/*z* 813.07. ^{1}H-NMR (DMSO-d$_6$): 8.19 (4H, NH$_2$), 3.59 (2H, CHε; see Scheme 3 for numbering of protons), 3.23–3.08 (4H, CHη',η), 2.98 (2H, CHa), 2.43 (2H, CHζ'), 2.26 (4H, CHα), 1.88 (2H, CHζ), 1.67 (2H, CHδ'), 1.61 (8H, CHb,c), 1.54 (2H, CHδ), 1.52 (4H, CHβ), 1.38 (4H, CHγ) ppm. ^{195}Pt NMR (DMSO-d$_6$): 1217.60 ppm. ^{13}C-NMR (DMSO-d$_6$): 55.79 (Cε), 49.47 (Ca), 39.62 (Cζ), 37.72 (Cη), 35.93 (Cα), 33.92 (Cδ), 27.93 (Cγ), 24.84 (Cβ), 19.76 (Cb,c) ppm.

3.5. Experiments with Cultured Human Cells

Pt(IV) compounds **2** and **3** were dissolved in DMSO just before running the experiment and a calculated amount of drug solution was added to the cell growth medium to a final DMSO concentration of 0.5%, which had no detectable effect on cell viability. Cisplatin, kiteplatin, and [Pt(CBDCA)(*cis*-1,4-DACH)] were dissolved in 0.9% NaCl solution.

3.6. Cell Culture Studies

Human lung (H157), colon (HCT-15), and breast (MCF-7) carcinoma cell lines were obtained by American Type Culture Collection (ATCC, Rockville, MD, USA). The human ovarian 2008 adenocarcinoma cells were kindly provided by Prof. G. Marverti (Dept. of Biomedical Science of Modena University, Modena, Italy). Human cervical (A431) carcinoma cells were kindly provided by Prof. F. Zunino (Istituto Nazionale per lo Studio e la Cura dei Tumori, Milan, Italy). Cell lines were maintained using the following culture media containing 10% fetal calf serum (Euroclone, Milan, Italy), antibiotics (50 units/mL penicillin and 50 µg/mL streptomycin), and L-glutamine (2 mM): (i) RPMI for HCT-15, A431, MCF-7 and 2008 cells; (ii) DMEM medium for A375 cells.

3.7. Cytotoxicity Assays

The growth inhibitory effect toward tumor cell lines was evaluated by means of the MTT as previously described [24]. Cancer cells were seeded in 96-well microplates in growth medium (100 µL, $3–8 \times 10^3$ cells/well, depending upon the growth characteristics of the cell line) and then incubated in a 5% carbon dioxide atmosphere at 37 °C. Following 24 h, the medium was replaced with a fresh one containing the compound to be tested. Triplicate cultures were established for each treatment. After 72 h, 10 µL of a 5 mg/mL MTT saline solution were added to each well and microplates were incubated for five additional hours. Subsequently 100 µL of a sodium dodecyl sulfate (SDS) solution in 0.01 M HCl were added to each well. After an overnight incubation, the inhibition of cell growth induced by the tested compound was evaluated by measuring the absorbance at 570 nm using a BioRad 680 microplate reader (BioRad Laboratories S.r.L.; Segrate, Italy). The average absorbance for each drug dose was expressed as a percentage of the control and plotted versus drug concentration. IC$_{50}$ values were obtained from the dose-response curves by means of the 4-PL model ($p < 0.05$). IC$_{50}$ values are the drug concentrations that reduce the mean absorbance at 570 nm to 50% of those of the untreated control wells.

3.8. Spheroid Cultures

Spheroids were initiated in liquid overlay by seeding 3×10^3 A431 cells/well in phenol red free RPMI-1640 medium (Sigma Chemical Co.; Sigma-Aldrich, Milan, Italy), containing 10% FCS and supplemented with 20% methyl cellulose stock solution. A total of 150 μL of this cell suspension was transferred to each well of a round bottom, non-tissue culture 96 well-plate (Greiner Bio-one, Kremsmünster, Austria) to allow spheroid formation within 72 h.

3.9. ROS Production

The production of ROS was measured in A431 cells (10^4 cells per well) grown for 24 h in 96-well plates in RPMI medium without phenol red (Sigma Chemical Co.). Cells were then washed with PBS and loaded with 10 μM 5-(and-6)-chloromethyl-2′,7′-dichlorodihydrofluorescein diacetate, acetyl ester (CM–H_2DCFDA; Molecular Probes-Invitrogen) for 25 min, in the dark. Afterwards, cells were washed with PBS and incubated with increasing concentrations of tested complexes. Fluorescence increase was estimated with a plate reader (Fluoroskan Ascent FL, Labsystem, Finland) at 485 (excitation) and 527 nm (emission). Antimycin (1 μM, Sigma Chemical Co.), a potent inhibitor of Complex III in the electron transport chain, was used as positive control.

3.10. Mitochondrial Membrane Potential (ΔΨ)

The ΔΨ was assayed using the Mito-ID® Membrane Potential Kit according to the manufacturer's instructions (Enzo Life Sciences, Farmingdale, NY, USA). Briefly, A431 cells (5×10^3 cells per well) were seeded in 96-well plates; after 24 h, cells were washed with PBS and loaded with Mito-ID Detection Reagent for 30 min at 37 °C in the dark. Afterwards, cells were washed with PBS and incubated with increasing concentrations of tested complexes. Fluorescence was estimated using a plate reader (Fluoroskan Ascent FL, ThermoScientific, Vantaa, Finland) at 490 nm (excitation) and 590 nm (emission).

3.11. Transmission Electron Microscopy (TEM) Analyses

About 10^6 A431 cells were seeded in 24-well plates and, after 24 h incubation, treated with the tested compounds and incubated for additional 24 h. Cells were then washed with cold PBS, harvested, and directly fixed with 1.5% glutaraldehyde buffer with 0.2 M sodium cacodylate, pH 7.4. After washing with buffer and post-fixation with 1% OsO_4 in 0.2 M cacodylate buffer, specimens were dehydrated and embedded in epoxy resin (Epon Araldite; Fisher Scientific Italia, Rodano (MI), Italy). Sagittal serial sections (1 μm) were counterstained with toluidine blue; thin sections (90 nm) were given contrast by staining with uranyl acetate and lead citrate. Micrographs were taken with a Hitachi H-600 electron microscope (Hitachi, Tokyo, Japan) operating at 75 kV. All photos were typeset in Corel Draw 11.

3.12. Statistical Analysis

All values are the means ± S.D. of no less than three measurements starting from three different cell cultures. Multiple comparisons were made by ANOVA followed by the Tukey–Kramer multiple comparison test (** $p < 0.01$) using GraphPad Prism 5.03 for Windows software (GraphPad Software, La Jolla, CA, USA).

4. Conclusions

A considerable amount of evidence has demonstrated that platinum drugs are activated intracellularly by aquation of the leaving groups and subsequent covalent binding to DNA, forming DNA adducts capable to activate various signal-transduction pathways such as those involved in DNA-damage recognition and repair, cell-cycle arrest, and programmed cell death or apoptosis. The Pt-DNA adducts cause unwinding and bending of double helix DNA that are recognized by many cellular proteins. Additional intracellular targets have been used in recent years to obtain more potent Pt complexes, mostly explored have been Pt(IV) complexes which allow the presence of two additional ligands in the axial positions suitable for the attachment of other cancer-targeting ligands [43,44]. In a previous work, this result was obtained by coordinating in the axial positions the ligand DCA, an orphan drug capable to inhibit the mitochondrial PDK [24]. Here we have extended this strategy by coordinating in the axial positions the antioxidant ALA. The new compounds (*cis,trans,cis*-[Pt(CBDCA)(ALA)$_2$(*cis*-1,4-DACH)] (**2**) and *cis,trans,cis*-[PtCl$_2$(ALA)$_2$(*cis*-1,4-DACH)] (**3**)), after intracellular reduction, are capable to release kiteplatin (or its CBDCA analogue) and two molecules of α-Lipoic acid. The Pt(II) residue reaches its target DNA, while ALA could act on mitochondria as activator of the pyruvate dehydrogenase complex and could suppress anaerobic glycolysis.

Compounds **2** and **3** were prepared and thoroughly characterized by means of spectroscopic and spectrometric techniques and their in vitro cytotoxicity was tested on a panel of five human cancer cell lines and compared to that of cisplatin, oxaliplatin, and kiteplatin. Compounds **2** and **3** were much more effective than the reference compounds, with complex **3** proving to be the most effective also in 3D spheroid tumor cells. However, treatment of A431 cells with **2** and **3** did not determine an increase in cellular ROS basal production, usually correlated with the inhibition of mitochondrial PDK. In addition, treatment of A431 cells with **2** and **3** did not induce a significant depolarization of the mitochondrial membrane or any morphological alteration of mitochondria.

The overall results hence suggest that the potentiated activity of the Pt(IV) conjugates **2** and **3**, with respect to their Pt(II) precursors, can be due to other types of interactions promoted by the release of the ALA ligands (at micromolar concentration reached into the treated tumor cells), which, however, do not appear to significantly affect the macroscopic mitochondrial membrane potential and the mitochondrial morphological parameters. This aspect deserves further investigation. Moreover, since it has been shown that pretreatment with ALA significantly attenuates the effect of cisplatin on HEI-OC1 cells and prevents cisplatin-induced nephrotoxicity in rats, we are planning to evaluate the toxicity profiles by in vivo investigations.

Supplementary Materials: Supplementary materials can be found at http://www.mdpi.com/xxx/s1.

Author Contributions: Data curation: S.S., C.M., V.G., J.D.H., G.N. and N.M. Investigation: S.S., V.G. and N.M. Methodology: V.G. Supervision: N.M. Writing—original draft: S.S., V.G., G.N. and N.M.

Funding: This research received no external funding.

Acknowledgments: The Universities of Bari and Padova, the Italian Ministero dell'Università e della Ricerca (MIUR), and the Inter-University Consortium for Research on the Chemistry of Metal Ions in Biological Systems (C.I.R.C.M.S.B.) are gratefully acknowledged.

Conflicts of Interest: The authors declare no conflicts of interest.

Abbreviations

ALA	(±)-α-Lipoic acid
CBDCA	1,1-cyclobutanedicarboxylate
CCCP	Carbonylcyanide m-chlorophenyl hydrazone
CDDP	cisplatin
CM–H2DCFDA	5-(and-6)-chloromethyl-2′,7′-dichlorodihydrofluorescein diacetate, acetyl ester
COSY	correlation spectroscopy
DCA	dichloroacetate
DCC	dicyclohexylcarbodiimmide
DACH	diaminocyclohexane
DHALA	dihydro α-lipoic acid
DCU	dicyclohexylurea
DMEM	Dulbecco's Modified Eagle Medium
DMSO	dimethylsulfoxide
ESI-MS	Electrospray Ionisation Mass Spectrometry

FCS	Fetal calf serum
HSQC	Heteronuclear single quantum coherence spectroscopy
MTT	3-(4,5-dimethylthiazol-2-yl)-2,5-diphenyltetrazolium bromide
OXP	oxaliplatin
PBS	phosphate buffer saline
ROS	reactive oxygen species
RPMI	Roswell Park Memorial Institute
SDS	sodium dodecyl sulfate
TEM	Transmission electron microscopy

References

1. Warburg, O.; Wind, F.; Negelein, E. The metabolism of tumors in the body. *J. Gen. Physiol.* **1927**, *8*, 519–530. [CrossRef] [PubMed]

2. Warburg, O.; Posener, K.; Negelein, E. Ueber den Stoffwechsel der Carcinomzelle. *Biochem. Z.* **1924**, *152*, 309–344.

3. Michelakis, E.D.; Webster, L.; Mackey, J.R. Dichloroacetate (DCA) as a potential metabolic-targeting therapy for cancer. *Br. J. Cancer* **2008**, *99*, 989–994. [CrossRef] [PubMed]

4. Korotchkina, L.G.; Sidhu, S.; Patel, M.S. R-lipoic acid inhibits mammalian pyruvate dehydrogenase kinase. *Free Radic. Res.* **2004**, *38*, 1083–1092. [CrossRef] [PubMed]

5. Novotny, L.; Rauko, P.; Cojocel, C. Alpha-Lipoic acid: The potential for use in cancer therapy. *Neoplasma* **2008**, *55*, 81–86. [PubMed]

6. Bilska, A.; Włodek, L. Lipoic acid—The drug of the future? *Pharmacol. Rep.* **2005**, *57*, 570–577. [PubMed]

7. Biewenga, G.P.; Haenen, G.R.; Bast, A. The pharmacology of the antioxidant lipoic acid. *Gen. Pharmacol.* **1997**, *29*, 315–331. [CrossRef]

8. Shay, K.P.; Moreau, R.F.; Smith, E.J.; Smith, A.R.; Hagen, T.M. Alpha-lipoic acid as a dietary supplement: Molecular mechanisms and therapeutic potential. *Biochim. Biophys. Acta* **2009**, *1790*, 1149–1160. [CrossRef] [PubMed]

9. Handelman, G.J.; Han, D.; Tritschler, H.; Packer, L. Alpha-lipoic acid reduction by mammalian cells to the dithiol form, and release into the culture medium. *Biochem. Pharmacol.* **1994**, *47*, 1725–1730. [CrossRef]

10. Packer, L.; Witt, E.H.; Tritschler, H.J. Alpha-Lipoic acid as a biological antioxidant. *Free Radic. Biol. Med.* **1995**, *19*, 227–250. [CrossRef]

11. Gackowski, D.; Banaszkiewicz, Z.; Rozalski, R.; Jawien, A.; Olinski, R. Persistent oxidative stress in colorectal carcinoma patients. *Int. J. Cancer* **2002**, *101*, 395–397. [CrossRef] [PubMed]

12. Van de Mark, K.; Chen, J.S.; Steliou, K.; Perrine, S.P.; Faller, D.V. α-Lipoic acid induces p27Kip-dependent cell cycle arrest in non-transformed cell lines and apoptosis in tumor cell lines. *J. Cell. Physiol.* **2003**, *194*, 325–340. [CrossRef] [PubMed]

13. Sen, C.K.; Sashwati, R.; Packer, L. Fas mediated apoptosis of human Jurkat T-cells: Intracellular events and potentiation by redox-active α-lipoic acid. *Cell Death Differ.* **1999**, *6*, 481–491. [CrossRef] [PubMed]

14. Pierce, R.H.; Campbell, J.S.; Stephenson, A.B.; Franklin, C.C.; Chaisson, M.; Poot, M.; Kavanagh, T.J.; Rabinovitch, P.S.; Fausto, N. Disruption of redox homeostasis in tumor necrosis factor-induced apoptosis in a murine hepatocyte cell line. *Am. J. Pathol.* **2000**, *157*, 221–236. [CrossRef]

15. Piotrowski, P.; Wierzbicka, K.; Smiałek, M. Neuronal death in the rat hippocampus in experimental diabetes and cerebral ischaemia treated with antioxidants. *Folia Neuropathol.* **2001**, *39*, 147–154. [PubMed]

16. Wenzel, U.; Nickel, A.; Daniel, H. α-Lipoic acid induces apoptosis in human colon cancer cells by increasing mitochondrial respiration with a concomitant $O_2^{\cdot-}$ generation. *Apoptosis* **2005**, *10*, 359–368. [CrossRef] [PubMed]

17. Pack, R.A.; Hardy, K.; Madigan, M.C.; Hunt, N.H. Differential effects of the antioxidant alpha-lipoic acid on the proliferation of mitogen-stimulated peripheral blood lymphocytes and leukaemic T cells. *Mol. Immunol.* **2002**, *38*, 733–745. [CrossRef]

18. Na, M.H.; Seo, E.Y.; Kim, W.K. Effects of alpha-lipoic acid on cell proliferation and apoptosis in MDA-MB-231 human breast cells. *Nutr. Res. Pract.* **2009**, *3*, 265–271. [CrossRef] [PubMed]

19. Lippert, B. *Cisplatin: Chemistry and Biochemistry of a Leading Anticancer Drug*; Verlag Helvetica Chimica Acta: Zürich, Switzerland, 1999; pp. 29–69. ISBN 9783906390420.

20. Wang, D.; Lippard, S.J. Cellular processing of platinum anticancer drugs. *Nat. Rev. Drug Discov.* **2005**, *4*, 307–320. [CrossRef] [PubMed]

21. Arnesano, F.; Natile, G. Mechanistic insight into the cellular uptake and processing of cisplatin 30 years after its approval by FDA. *Coord. Chem. Rev.* **2009**, *253*, 2070–2081. [CrossRef]

22. Margiotta, N.; Savino, S.; Marzano, C.; Pacifico, C.; Hoeschele, J.D.; Gandin, V.; Natile, G. Cytotoxicity-boosting of kiteplatin by Pt(IV) prodrugs with axial benzoate ligands. *J. Inorg. Biochem.* **2016**, *160*, 85–93. [CrossRef] [PubMed]

23. Margiotta, N.; Savino, S.; Denora, N.; Marzano, C.; Laquintana, V.; Cutrignelli, A.; Hoeschele, J.D.; Gandin, V.; Natile, G. Encapsulation of lipophilic kiteplatin Pt(IV) prodrugs in PLGA-PEG micelles. *Dalton Trans.* **2016**, *45*, 13070–13081. [CrossRef] [PubMed]

24. Savino, S.; Gandin, V.; Hoeschele, J.D.; Marzano, C.; Natile, G.; Margiotta, N. Dual-acting antitumor Pt(IV) prodrugs of kiteplatin with dichloroacetate axial ligands. *Dalton Trans.* **2018**, *47*, 7144–7158. [CrossRef] [PubMed]

25. Ranaldo, R.; Margiotta, N.; Intini, F.P.; Pacifico, C.; Natile, G. Conformer distribution in (*cis*-1,4-DACH)bis(guanosine-5′-phosphate) platinum(II) adducts: A reliable model for DNA adducts of antitumoral cisplatin. *Inorg. Chem.* **2008**, *47*, 2820–2830. [CrossRef] [PubMed]

26. Kasparkova, J.; Suchankova, T.; Halamikova, A.; Zerzankova, L.; Vrana, O.; Margiotta, N.; Natile, G.; Brabec, V. Cytotoxicity, cellular uptake, glutathione and DNA interactions of an antitumor large-ring PtII chelate complex incorporating the cis-1,4-diaminocyclohexane carrier ligand. *Biochem. Pharmacol.* **2010**, *79*, 552–564. [CrossRef] [PubMed]

27. Margiotta, N.; Marzano, C.; Gandin, V.; Osella, D.; Ravera, M.; Gabano, E.; Platts, J.A.; Petruzzella, E.; Hoeschele, J.D.; Natile, G. Revisiting [PtCl₂(*cis*-1,4-DACH)]: An underestimated antitumor drug with potential application to the treatment of oxaliplatin-refractory colorectal cancer. *J. Med. Chem.* **2012**, *55*, 7182–7192. [CrossRef] [PubMed]

28. Brabec, V.; Malina, J.; Margiotta, N.; Natile, G.; Kasparkova, J. Thermodynamic and mechanistic insights into translesion DNA synthesis catalyzed by Y-family DNA polymerase across a bulky double-base lesion of an antitumor platinum drug. *Chem. Eur. J.* **2012**, *18*, 15439–15448. [CrossRef] [PubMed]

29. Kim, J.; Cho, H.-J.; Sagong, B.; Kim, S.-J.; Lee, J.-T.; So, H.-S.; Lee, I.-K.; Kim, U.-K.; Lee, K.-Y.; Choo, Y.-S. Alpha-lipoic acid protects against cisplatin-induced ototoxicity via the regulation of MAPKs and proinflammatory cytokines. *Biochem. Biophys. Res. Commun.* **2014**, *449*, 183–189. [CrossRef] [PubMed]

30. Somani, S.M.; Husain, K.; Whitworth, C.; Trammell, G.L.; Malafa, M.; Rybak, L.P. Dose-dependent protection by lipoic acid against cisplatin-induced nephrotoxicity in rats: Antioxidant defense system. *Pharmacol. Toxicol.* **2008**, *86*, 234–241. [CrossRef]

31. Feuerecker, B.; Pirsig, S.; Seidl, C.; Aichler, M.; Feuchtinger, A.; Bruchelt, G.; Senekowitsch-Schmidtke, R. Lipoic acid inhibits cell proliferation of tumor cells in vitro and in vivo. *Cancer Biol. Ther.* **2012**, *13*, 1425–1435. [CrossRef] [PubMed]

32. Siemeling, U.; Bretthauer, F.; Bruhn, C.; Fellinger, T.-P.; Tong, W.-L.; Chan, M.C.W. Gold Nanoparticles Bearing an α-Lipoic Acid-based Ligand Shell: Synthesis, Model Complexes and Studies Concerning Phosphorescent Platinum(II)-Functionalisation. *Z. Naturforsch. B* **2010**, *65*. [CrossRef]

33. Lal, M.; Palepu, N. Platinum Compound. WO2004/006859, 22 January 2004.

34. Ang, W.H.; Pilet, S.; Scopelliti, R.; Bussy, F.; Juillerat-Jeanneret, L.; Dyson, P.J. Synthesis and characterization of Platinum(IV) anticancer drugs with functionalized aromatic carboxylate ligands: Influence of the ligands on drug efficacies and uptake. *J. Med. Chem.* **2005**, *48*, 8060–8069. [CrossRef] [PubMed]

35. Gramatica, P.; Papa, E.; Luini, M.; Monti, E.; Gariboldi, M.B.; Ravera, M.; Gabano, E.; Gaviglio, L.; Osella, D. Antiproliferative Pt(IV) complexes: Synthesis, biological activity, and quantitative structure–activity relationship modeling. *J. Biol. Inorg. Chem.* **2010**, *15*, 1157–1169. [CrossRef] [PubMed]

36. Petruzzella, E.; Margiotta, N.; Ravera, M.; Natile, G. NMR investigation of the spontaneous thermal- and/or photoinduced reduction of trans dihydroxido Pt(IV) derivatives. *Inorg. Chem.* **2013**, *52*, 2393–2403. [CrossRef] [PubMed]

37. Pregosin, P.S. Platinum-195 nuclear magnetic resonance. *Coord. Chem. Rev.* **1982**, *44*, 247–291. [CrossRef]

38. Gabano, E.; Marengo, E.; Bobba, M.; Robotti, E.; Cassino, C.; Botta, M.; Osella, D. 195Pt NMR spectroscopy: A chemometric approach. *Coord. Chem. Rev.* **2006**, *250*, 2158–2174. [CrossRef]

39. Kim, J. Bin Three-dimensional tissue culture models in cancer biology. *Semin. Cancer Biol.* **2005**, *15*, 365–377. [CrossRef] [PubMed]

40. Shamsuddin, S.; Santillan, C.C.; Stark, J.L.; Whitmire, K.H.; Siddik, Z.H.; Khokhar, A.R. Synthesis, characterization, and antitumor activity of new platinum(IV) trans-carboxylate complexes: Crystal structure of [Pt(*cis*-1,4-DACH)trans-(acetate)$_2$Cl$_2$]. *J. Inorg. Biochem.* **1998**, *71*, 29–35. [CrossRef]

41. Shamsuddin, S.; Takahashi, I.; Siddik, Z.H.; Khokhar, A.R. Synthesis, characterization, and antitumor activity of a series of novel cisplatin analogs with cis-1,4-diaminocyclohexane as nonleaving amine group. *J. Inorg. Biochem.* **1996**, *61*, 291–301. [CrossRef]

42. Liu, F.; Wang, M.; Wang, Z.; Zhang, X. Polymerized surface micelles formed under mild conditions. *Chem. Commun.* **2006**, 1610. [CrossRef] [PubMed]

43. Curci, A.; Denora, N.; Iacobazzi, R.M.; Ditaranto, N.; Hoeschele, J.D.; Margiotta, N.; Natile, G. Synthesis, characterization, and in vitro cytotoxicity of a Kiteplatin-Ibuprofen Pt(IV) prodrug. *Inorganica Chim. Acta* **2018**, *472*, 221–228. [CrossRef]

44. Savino, S.; Denora, N.; Iacobazzi, R.M.; Porcelli, L.; Azzariti, A.; Natile, G.; Margiotta, N. Synthesis, characterization, and cytotoxicity of the first oxaliplatin Pt(IV) derivative having a TSPO ligand in the axial position. *Int. J. Mol. Sci.* **2016**, *17*, 1010. [CrossRef] [PubMed]

International Journal of
Molecular Sciences

MDPI

Article

Syntheses, Crystal Structures, and Antitumor Activities of Copper(II) and Nickel(II) Complexes with 2-((2-(Pyridin-2-yl)hydrazono)methyl) quinolin-8-ol

Qi-Yuan Yang [†], Qian-Qian Cao [†], Qi-Pin Qin, Cai-Xing Deng, Hong Liang and Zhen-Feng Chen *

State Key Laboratory for Chemistry and Molecular Engineering of Medicinal Resources, School of Chemistry and Pharmacy, Guangxi Normal University, 15 Yucai Road, Guilin 541004, China; 2016110029@stu.gxnu.edu.cn (Q.-Y.Y.); 2014011049@stu.gxnu.edu.cn (Q.-Q.C.); 2014110004@stu.gxnu.edu.cn (Q.-P.Q.); 2016010981@stu.gxnu.edu.cn (C.-X.D.); hliang@gxnu.edu.cn (H.L.)
* Correspondence: chenzf@gxnu.edu.cn; Fax: +86-773-2120958
† These authors contributed equally to this work.

Received: 12 May 2018; Accepted: 5 June 2018; Published: 26 June 2018

Abstract: Two transition metal complexes with 2-((2-(pyridin-2-yl)hydrazono)methyl)quinolin-8-ol (L), $[Cu(L)Cl_2]_2$ (**1**) and $[Ni(L)Cl_2] \cdot CH_2Cl_2$ (**2**), were synthesized and fully characterized. Complex **1** exhibited high in vitro antitumor activity against SK-OV-3, MGC80-3 and HeLa cells with IC_{50} values of 3.69 ± 0.16, 2.60 ± 0.17, and 3.62 ± 0.12 μM, respectively. In addition, complex **1** caused cell arrest in the S phase, which led to the down-regulation of Cdc25 A, Cyclin B, Cyclin A, and CDK2, and the up-regulation of p27, p21, and p53 proteins in MGC80-3 cells. Complex **1** induced MGC80-3 cell apoptosis via a mitochondrial dysfunction pathway, as shown by the significantly decreased level of bcl-2 protein and the loss of $\Delta\psi$, as well as increased levels of reactive oxygen species (ROS), intracellular Ca^{2+}, cytochrome C, apaf-1, caspase-3, and caspase-9 proteins in MGC80-3 cells.

Keywords: quinolinyl hydrazine; copper(II) complex; cytotoxicity; apoptosis

1. Introduction

Numerous platinum(II) complexes have been successfully used for the treatment of different types of cancers [1]. Platinum complexes stand out among chemotherapeutic agents for its high efficacy in combination therapy. However, they also show drawbacks like toxicity and drug resistance [2]. Especially, the clinical use of cisplatin is severely limited by its unwanted side effects, including ototoxicity and nephrotoxicity, which reduce patient tolerance during treatment and interfere with the long-term quality of life [3]. Therefore, it is necessary to explore other nonplatinum complexes that could offer high efficacy with fewer side effects.

Many studies show that the copper and nickel complexes play an important role in the endogenous oxidative DNA damage associated with aging and cancer [4–8]. For example, complexes with Cu(II) ion show high DNA binding and DNA cleavage activities [9], and copper complexes induced reversible condensation of DNA and apoptosis in osteosarcoma MG-63 cell lines [10]. Many nickel complexes bearing biological activity have been reported including Ni(II) complexes with antitumor activity [11,12]. Nickel complexation with lidocaine enhances the DNA binding affinity, DNA cleavage activity, and cytotoxic properties of lidocaine [13]; Nickel complexes also show considerable cytotoxic activity against the human hepatocarcinoma cells (Hep-G2), human leukemic cells (HL-60), and human prostatic carcinoma cells (PC-3) [14]. Therefore, the synthesis and biological testing of copper and nickel complexes have become an important area of current bioinorganic chemistry research [15–17].

The compound 8-Hydroxyquinoline (HQ) has attracted considerable interest as a privileged structure (Scheme 1), and 8-hydroxyquinoline derivatives (HQs) have been explored for a broad range of biological applications [18], such as metal-chelators for neuroprotection, chelators of metalloproteins, inhibitors of 2OG-dependent enzymes, *Mycobacterium tuberculosis* inhibitors, botulinum neurotoxin inhibitors and anticancer, anti-HIV, antifungal, antileishmanial, and antischistosomal agents [19–21]. The HQs with anticancer or anti-Alzheimer activities include mainly halogenated derivatives [22,23], diperazino and alkyno derivatives [24,25], nitro derivatives [26–28], carboxylic and carboxamido derivatives [29–31], amino and imino derivatives [32,33], sulfoxine and sulfonamide derivatives [34–36], Bis- and poly-HQs [37,38], HQ bioconjugates [39–41], and other HQ derivatives [42]. In addition, it is well known that quinolinylhydrazones show various important biological activities and the quinoline ring plays an important role in the development of new anticancer agents [43–47]. For example, the quinolinylhydrazones exhibit significant cytotoxicity in comparison with similar reported systems and the apoptosis induction in MCF-7 cancer cells increased when it was coordinated with the gold nanoparticle surface [48]. Recently, the synthesis of 2-((2-(pyridin-2-yl)hydrazono)methyl)quinolin-8-ol (L) was reported [49]. The metal complexes of HQs show enhanced tumor cytotoxicity [50–56], including ruthenium [50,51], gold [52], platinum [53], copper [43,48,49], and vanadium [44] complexes. However, there are few reports on the synthesis and antitumor activity of Cu(II) and Ni(II) complexes. Chan et al. found that 8-hydroxy-2-quinolinecarbaldehyde (Scheme 1) showed the highest in vitro cytotoxicity against the human cancer cell lines, including MDA231, T-47D, Hs578t, SaoS2, K562, SKHep1, and Hep3B [42].

Therefore, as part of our continuing work on the synthesis, characterization and medicinal application of metal complexes with HQ [45–47], we report the synthesis and characterization of Cu(II) and Ni(II) complexes with 2-((2-(pyridin-2-yl)hydrazono)methyl)quinolin-8-ol (L) and the in vitro cytotoxicities against seven tumor cells and their antitumor mechanism.

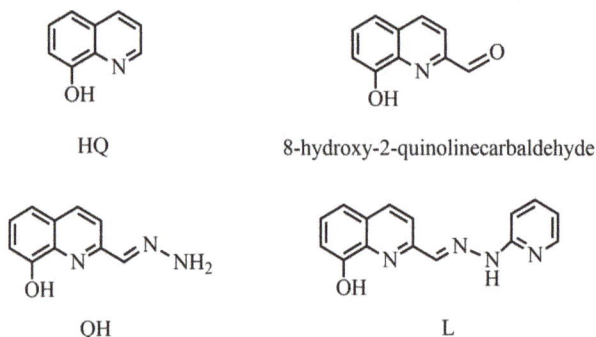

HQ

8-hydroxy-2-quinolinecarbaldehyde

QH

L

Scheme 1. The structures of 8-hydroxyquinoline (HQ), quinolinyl hydrazine (QH) and L.

2. Results

2.1. Synthesis

As outlined in Scheme 2, complexes **1**, **2** were synthesized by the reaction of L with $CuCl_2 \cdot 2H_2O$ and $NiCl_2 \cdot 6H_2O$ in hot methanol, respectively. They were satisfactorily characterized by mass spectrometry (MS), elemental analysis (EA), infrared spectroscopy (IR), and single-crystal X-ray diffraction analysis. The absorptions around 1550–1650 cm^{-1} of the IR (Figures S3–S5) were assigned to the imine bond stretching vibrations of L. The imine bonds of complexes **1** and **2** underwent a left-shift of 10–60 cm^{-1} upon coordination, indicating the participation of this group in coordination. The single-crystal structure analysis suggested that the Cu(II) complex was $[Cu(L)Cl_2]_2$ (**1**) and the Ni(II) complex was $[Ni(L)Cl_2] \cdot CH_2Cl_2$ (**2**).

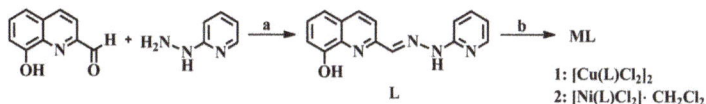

Scheme 2. The synthetic routes for ligand (L) and its metal complexes **1** and **2**. Reagents are as follows: (a) EtOH, r.t, 16 h; (b) CuCl$_2$ or NiCl$_2$, MeOH/CH$_2$Cl$_2$.

2.2. Crystal Structures of Complexes 1 and 2

The crystal data and refinement details of complexes **1** and **2** are summarized in Table S1 (Supporting Information), and the selected bond lengths and angles are listed in Tables S2 and S3. The crystal structures of complexes **1** and **2** are shown in Figures 1 and 2. Complexes **1** and **2** have different coordination pattern. Complex **1** was a dinuclear L-Cu-Cl-(μ-Cl)$_2$-Cu-Cl-L complex, and the Cu(II) ions were coordinated by three Cl and two N atoms from L in a distorted square pyramidal geometry.

In complex **2**, the central Ni^{II} adopted an approximately five-coordinated tetragonal pyramidal geometry.

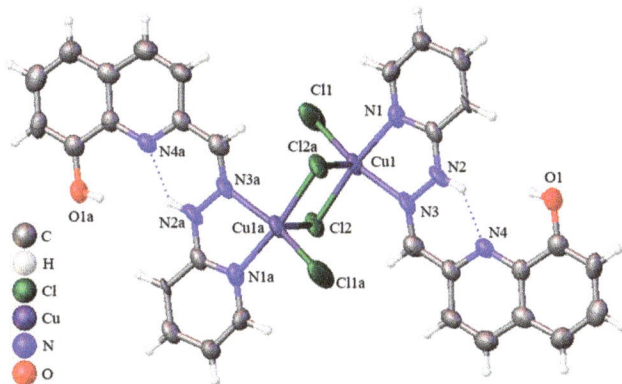

Figure 1. The crystal structure of Cu(II) complex **1**.

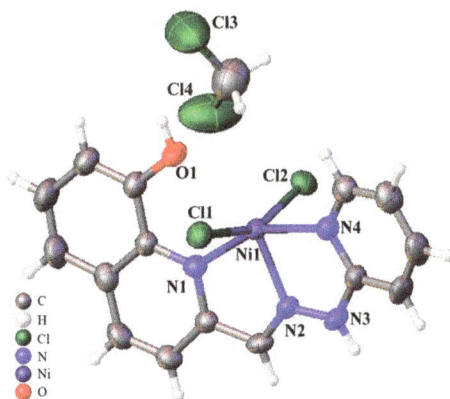

Figure 2. The crystal structure of complex **2**.

2.3. Stability in Solution

Ligand L, complexes **1** and **2** were tested for their stabilities in both dimethyl sulfoxide (DMSO) and Tris-HCl buffer solution (TBS) (TBS solution with pH at 7.35, containing 1% DMSO) by means of UV-Vis spectroscopy. The time-dependent (in the time course of 0, 12, 24, 36 and 48 h) UV-Vis spectra of each complex dissolved in TBS solution are shown in Figure S1. There were no obvious changes in the spectral characteristics and the peak absorptions for ligand L, complexes **1** and **2** over the time course. In addition, the stabilities of L, complexes **1** and **2** were monitored by high performance liquid chromatography (HPLC) detected at 245 nm, and no significant change was observed for these three compounds in TBS at 0, 24, and 48 h (Figure S2). Combining the ESI-MS data, the results suggested that complex **2** was stable in TBS solution, and complexes **1** was stable in TBS solution as mononuclear species because it was dissociated in water solution and Tris-HCl buffer (see the results of Figure S9).

2.4. In Vitro Cytotoxicity

The in vitro cytotoxicities of L, complexes **1** and **2** were evaluated by MTT assay in seven human tumor cell lines Hep-G2, SK-OV-3, MGC80-3, HeLa, T-24, BEL-7402, and NCI-H460 and one normal liver cell line HL-7702. Each compound was prepared as 2.0 mM DMSO stock solution before it was diluted in PBS buffer to 20 μM aqueous solutions (containing 2.5% DMSO). These 20 μM aqueous solutions were stable and no precipitate was formed.

The in vitro antitumor activities of complex **1** were further evaluated by determining the corresponding IC_{50} values. As shown in Table 1, the IC_{50} values of complex **1** against SK-OV-3, MGC80-3 and HeLa were 3.69 ± 0.16, 2.60 ± 0.17, and 3.62 ± 0.12 μM, respectively, which were approximately 11.6, 15.6, and 16.2 fold increases compared with that of the free L. In addition, complex **1** exhibited stronger cytotoxicities than cisplatin towards the SK-OV-3, MGC80-3, and HeLa tumor cells. In summary, complex **1** exhibited a lower IC_{50} value for MGC80-3 cells than other cells and higher cytotoxicity than complex **2**. Thus, complex **1** was chosen to study the underlying cellular and molecular mechanisms of its cytotoxicity. (As a support material, Inhibitory rates (%) of compounds were shown in Table S4)

Table 1. The IC_{50} [a] (μM) values of L, complexes **1** and **2** to the selected tumor cells for 48 h.

Compounds	Hep-G2	SK-OV-3	MGC80-3	HeLa	T-24	BEL-7402	NCI-H460	HL-7702
L	58.40 ± 0.69	42.94 ± 2.64	40.93 ± 0.94	58.73 ± 1.29	85.93 ± 15.11	47.85 ± 0.37	36.93 ± 3.93	48.63 ± 0.34
1	4.51 ± 0.38	3.69 ± 0.16	2.60 ± 0.17	3.62 ± 0.12	4.41 ± 0.06	5.92 ± 0.01	5.01 ± 0.16	12.78 ± 0.55
2	>100	39.77 ± 2.15	38.99 ± 2.42	27.13 ± 6.51	18.97 ± 3.47	51.68 ± 0.66	36.31 ± 3.75	31.73 ± 2.11
Cisplatin [b]	9.55 ± 0.46	16.32 ± 1.37	12.37 ± 1.53	9.45 ± 2.05	28.15 ± 1.67	19.4 ± 0.58	9.59 ± 0.48	15.87 ± 0.36

[a] IC_{50} values are presented as the mean ± SD (standard error of the mean) from five separate experiments. [b] Cisplatin was dissolved at a concentration of 1 mM in 0.154 M NaCl.

2.5. Cell Cycle Analysis and Expressions of the Related Proteins

The IC_{50} value of complex **1** towards the MGC80-3 cells was in the low micromolar range. To determine the cell cycle phase of growth arrest by complex **1**, the DNA content of cells was estimated by flow cytometry after the cells were stained with propidiumiodide (PI). As shown in Figure 3, complex **1** caused a dose-dependent accumulation of MGC80-3 cells in the S phase, whereas most of the control cells were in the G1 and G2/M phase of the cell cycle. Additionally, the cell population of S phase increased from 20.77% in the control to 60.18% in the MGC80-3 cells treated with 8 μM of complex **1** for 24 h. After incubating the cells with complex **1** (8 μM) for 24 h, the cell population of the G2/M phase was decreased to 0.00%. These results indicated that the MGC80-3 cells were mainly blocked in the S phase.

The protein expression levels of ATR, ATM, Cdc25 A, Cyclin B, Cyclin A, CDK2, p27, p21, and p53 protein in MGC80-3 cells after treated with complex **1** (2.0, 2.6, 5.2, and 8.0 μM) for 24 h were determined by Western blot and the results are shown in Figure 4, which demonstrated that complex **1** caused a dose-dependent inhibition on the protein expression levels of Cdc25 A, Cyclin B, Cyclin A, and CDK2, and decreased levels of p27, p21, and p53 proteins.

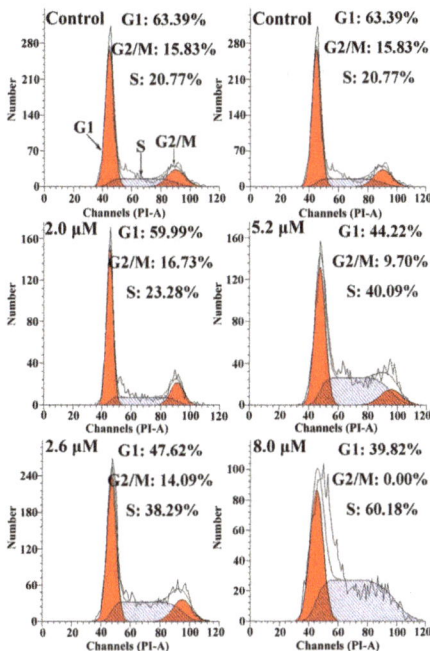

Figure 3. The cell cycle analysis by flow cytometry of MGC80-3 cells treated with complex **1** (2.0, 2.6, 5.2, and 8.0 μM) for 24 h.

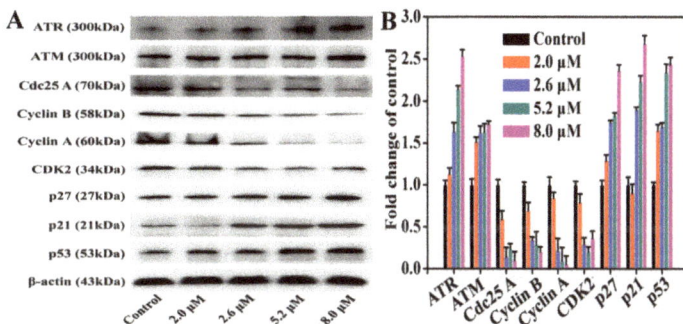

Figure 4. The expressions of ATR, ATM, Cdc25 A, Cyclin B, Cyclin A, CDK2, p27, p21, and p53 protein in MGC80-3 cells after treated with complex **1** (2.0, 2.6, 5.2, and 8.0 μM) for 24 h were analyzed by Western blot. (**A**) The same blots were stripped and re-probed with a β-actin antibody to show equal protein loading; (**B**) The whole-cell extracts were prepared and analyzed by Western blot analysis using antibodies against cell cycle protein regulator proteins. The same blots were stripped and re-probed with the β-actin antibody to show equal protein loading. Western blotting bands from three independent measurements were quantified with Image J in (**A**).

2.6. Apoptosis Assay

Apoptosis assay can provide important information for the preliminary investigation of the mode of action [55–57]. To determine whether the death of MGC80-3 cells induced by complex **1** resulted from apoptosis or necrosis, common biochemical markers of apoptosis were monitored, including mitochondrial membrane depolarization, chromatin condensation, and phosphatidylserine exposure. The cells subjected to annexin V-FITC and PI staining were classified as necrotic cells (Q1; annexin V−/PI+), early apoptotic cells (Q2; annexin V+/PI−), late apoptotic cells (Q3; annexin V+/PI+), and intact cells (Q4; annexin V−/PI−). The assay showed (Figure 5) that complex **1** (1.5, 2.0, 2.6, and 3.6 μM) induced the apoptotic death of MGC80-3 cells as measured by annexin V staining and flow cytometry. After treatment with complex **1** for 24 h, the populations of apoptotic cells (Q2+Q3) changed from 7.08% to 27.39% with the increase of complex **1** concentration, but the population of apoptotic cells (Q2+Q3) of control was only 1.70%. The significantly increased percentages of apoptotic cells confirmed that complex **1** effectively induced MGC80-3 cell apoptosis in a dose-dependent manner, which was consistent with the results of the MTT assay.

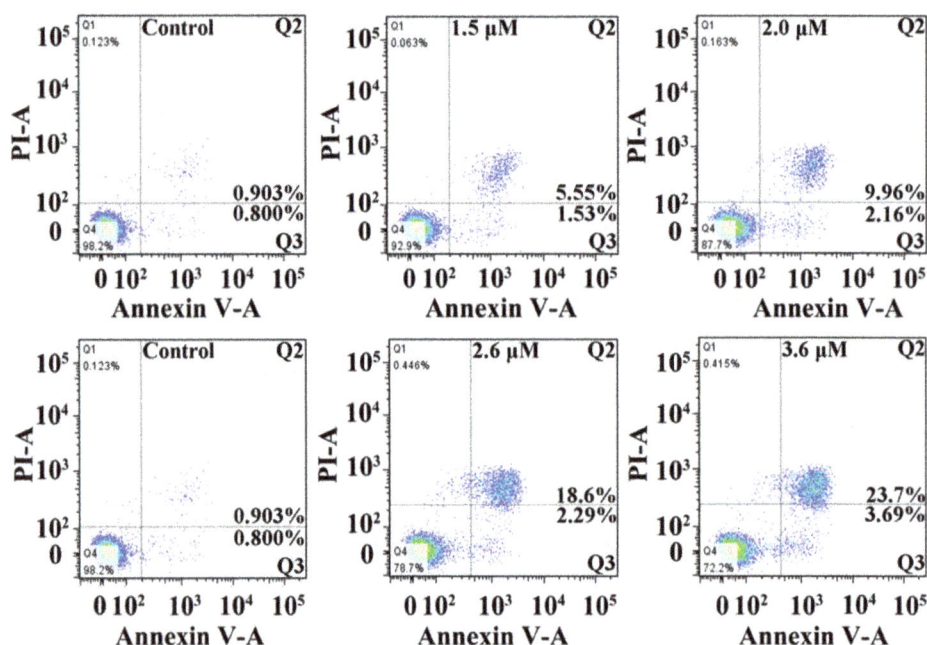

Figure 5. The Annexin V/propidium iodide assay and flow cytometry assay of MGC80-3 cells treated with Cu(II) complex **1** (1.5, 2.0, 2.6 and 3.6 μM).

2.7. Loss of Mitochondrial Membrane Potential in MGC80-3 Cells

Growing evidence has shown that mitochondria play a key role in the progression of apoptosis, and the loss of mitochondrial membrane potential ($\Delta\psi$) is involved in apoptotic cell death due to the cytotoxicity of the antitumor compounds [58–60]. The changes in $\Delta\psi$ induced by complex **1** are shown in Figures 6 and 7. JC-1 staining was used as a fluorescent probe [58]. After the MGC80-3 cells were treated with complex **1** for 24 h, the $\Delta\psi$ decreased significantly with the increase of dose (from 2.0 to 8.0 μM) of complex **1**, suggesting that the induction of apoptosis by complex **1** was associated with the intrinsic (mitochondrial) pathway.

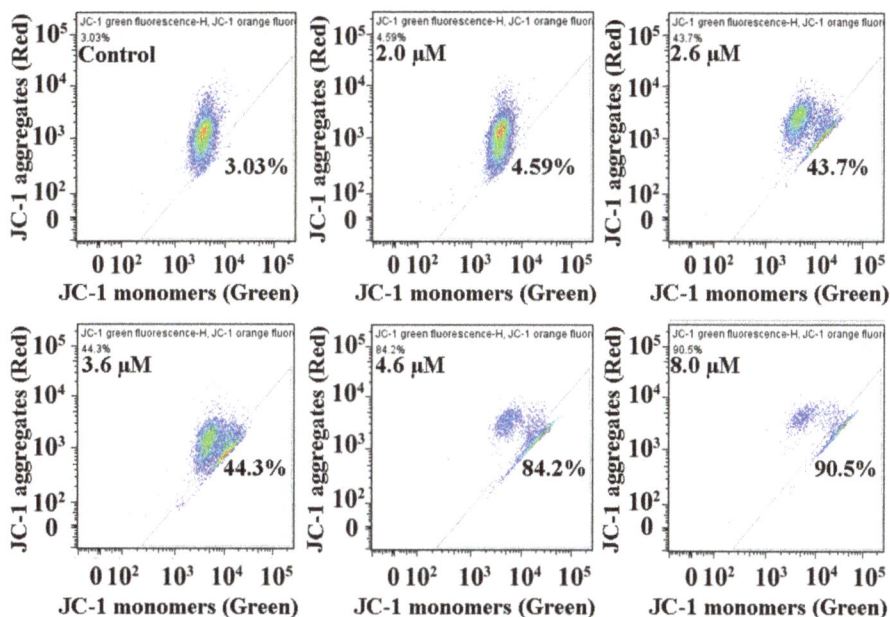

Figure 6. The collapse of mitochondrial membrane potential in MGC80-3 cells treated with Cu(II) complex **1** for 24 h, as determined by JC-1 staining.

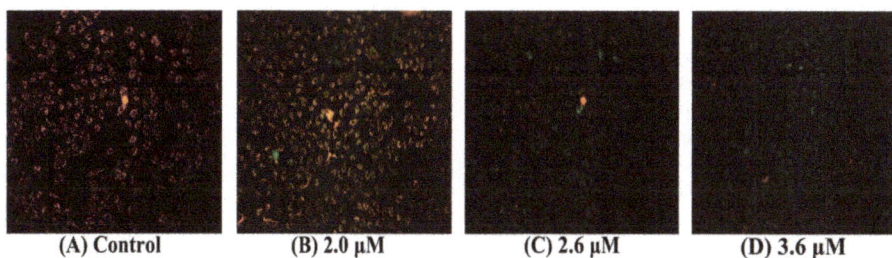

Figure 7. The loss of $\Delta\psi$ in MGC80-3 cells treated with complex **1** (2.0, 2.6 and 3.6 μM) for 24 h, and the cells were examined under a fluorescence microscope (Nikon Te2000, 200×) after being stained with JC-1.

2.8. Intracellular Ca²⁺

The mitochondrial membrane potential $\Delta\psi$ can alter the intracellular Ca^{2+} level, which has been recognized as a factor in cell death, apoptosis, and injury mediated by various pathways [61,62]. We examined the effects of complex **1** on the mobilization of intracellular Ca^{2+} in MGC80-3 cells. As shown in Figure 8, the level of intracellular free Ca^{2+} in MGC80-3 cells was lower than that of the control group, but it increased steadily in a dose-dependent manner (2.0, 2.3, and 3.6 μM of complex **1**). Therefore, the changes of the intracellular Ca^{2+} level could be involved in the induction of apoptosis by complex **1** in MGC80-3 cells.

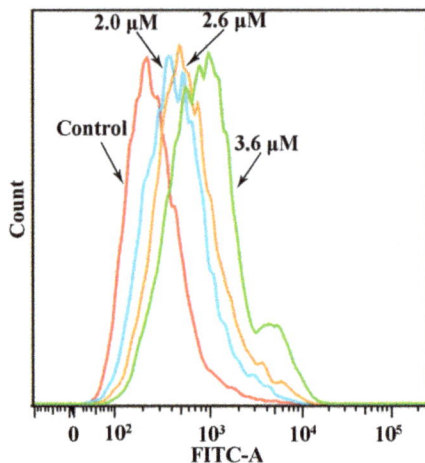

Figure 8. The effect of complex **1** (2.0, 2.6, and 3.6 μM) on the intracellular free Ca^{2+} level in MGC80-3 cells for 24 h.

2.9. Reactive Oxygen Species (ROS) Level

The dysregulation of ROS generation could dramatically affect cancer cell structure and result in cell damage, and consequently cell death and apoptosis [63,64]. To determine whether ROS generation is involved in the apoptosis or death of MGC80-3 cells induced by complex **1**, the ROS level was measured by a fluorescent marker after the MGC80-3 cells were treated with complex **1** (2.0, 2.6, and 3.6 μM) for 24 h. As shown in Figures 9 and 10, the levels of ROS in MGC80-3 cells were higher than that in the control after treatment, and the levels of ROS increased in a dose-dependent manner (from 2.0 to 3.6 μM of complex **1**). The results confirmed that complex **1** stimulated ROS-induced apoptosis in MGC80-3 cells.

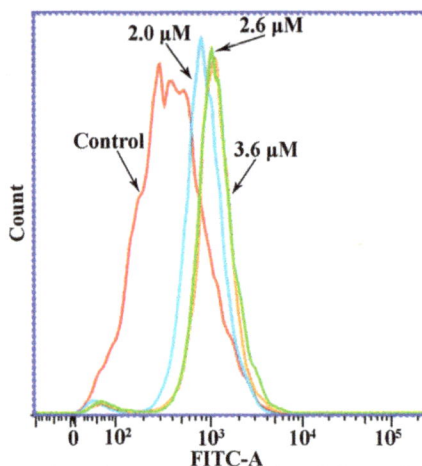

Figure 9. The ROS generation assay by flow cytometry analysis of MGC80-3 cells treated with complex **1** (2.0, 2.6, and 3.6μM). Results are expressed as relative fluorescent intensities (from left to right).

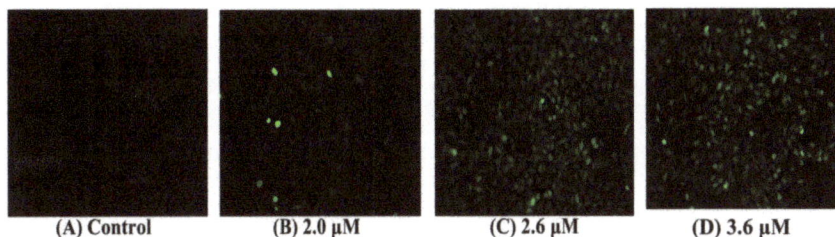

Figure 10. The ROS generation assay of MGC80-3 cells treated with complex **1** (2.0, 2.6, and 3.6 μM), and the cells were examined under a fluorescence microscope (Nikon Te2000, 200×).

2.10. Western Blot Assay

To further investigate the mechanism of action of complex **1**, the cytochrome C (Cyt C), bcl-2, bax, and apaf-1 proteins in the mitochondria-related apoptotic pathway were assayed by Western blot [65]. Figure 11 shows that the levels of bax, Cyt C, and apaf-1 proteins increased significantly and the level of bcl-2 protein decreased significantly in the MGC80-3 cells after treatment with complex **1** (1.5, 2.0, 2.6, and 3.6 μM) for 24 h. Additionally, the levels of bax, Cyt C, and apaf-1 proteins increased in a dose-dependent manner. These results further demonstrated that complex **1** may be involved in mitochondria-related apoptosis [65].

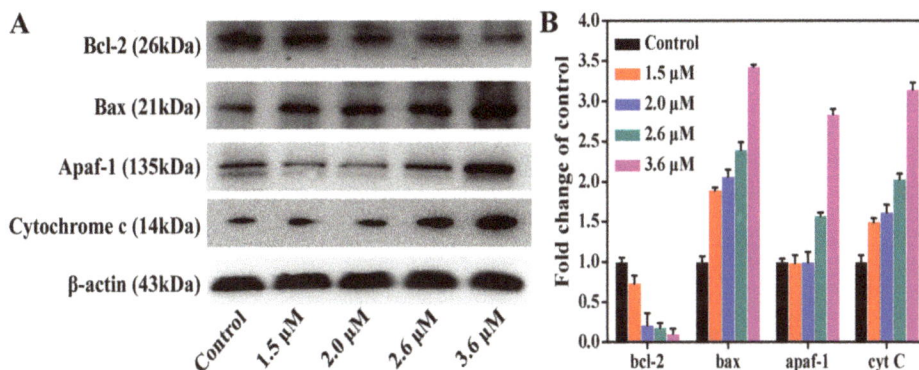

Figure 11. The western blot assay of apoptosis-related protein levels in Hep-G2 cells treated with complex **1** (1.5, 2.0, 2.6, and 3.6 μM) for 24 h. (**A**) Western blot was used to determine the expression levels of bax, cytochrome c, apaf-1, and bcl-2 in MGC80-3 cells treated with complex **1** (1.5, 2.0, 2.6, and 3.6 μM) for 24 h; (**B**) Densitometric analysis of apoptotic-related proteins normalized to β-actin. The relative expression of each protein is represented by the density of the protein band/density of β-actin band.

2.11. Assessment of Caspase-3/9/8 Activation

To determine whether caspase-3/9 were involved in the induced apoptotic cell death, MGC80-3 cells were analyzed by flow cytometry after treatment with complex **1** (1.5, 2.0, and 2.6 μM) for 24 h. The results showing peaks of activated caspase-3 (FITC-DEVD-FMK probes), activated caspase-8 (FITC-IETD-FMK probes), and activated caspase-9 (FITC-LEHD-FMK probes) for the treated cells are shown in Figure 12. It is notable that the proportion of cells with activated caspase-3, caspase-9, and caspase-8 increased from 5.04% to 18.70%, 2.59% to 23.9%, and 6.45% to 21.60%, respectively. Therefore, complex **1** could induce cell apoptosis by triggering the caspase-3/9/8 activity in MGC80-3 cells [66–69].

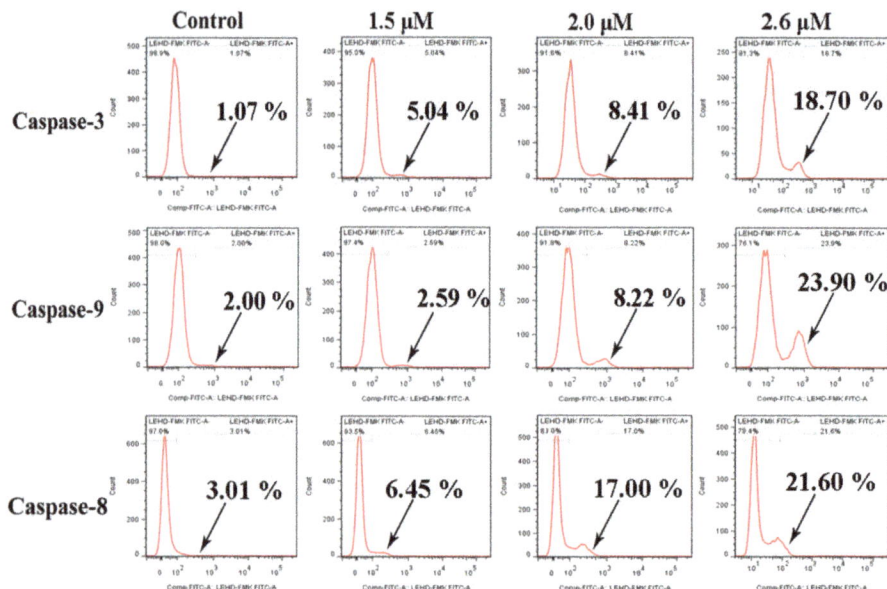

Figure 12. The activation of caspase-3, caspase-8, and caspase-9 in MGC80-3 cells treated with complex **1** (1.5, 2.0, and 2.6 µM) for 24 h.

Taken together, complex **1** induced apoptosis in MGC80-3 cells likely by disrupting mitochondrial function, which led to a significantly decreased level of bcl-2 protein and loss of $\Delta\psi$, as well as a significant increase in the levels of ROS, intracellular Ca^{2+}, Cyt C protein, apaf-1 protein, activated caspase-3, and activated caspase-9 in MGC80-3 cells.

3. Materials and Methods

3.1. Materials

All chemical reagents, including chloride salts and solvents, were of analytical grade. All materials were used as received without further purification unless specifically noted. All the synthetic complexes were dissolved in dimethyl sulfoxide (DMSO) for the preparation of stock solution at a concentration of 2.0 mM.

3.2. Instrumentation

Elemental analyses (C, H, N) were carried out on a Perkin Elmer Series II CHNS/O 2400 elemental analyzer. NMR spectra were recorded on a Bruker AV-500 NMR spectrometer. Fluorescence measurements were performed on a Shimadzu RF-5301/PC spectrofluorophotometer. The region between 200 and 400 nm was scanned for each sample. UV-Vis spectra were recorded on a TU-1901 ultraviolet spectrophotometer.

3.3. Synthesis

3.3.1. Synthesis of L

The 2-((2-(pyridin-2-yl)hydrazono)methyl)quinolin-8-ol (L) was obtained from the condensation reaction of 8-hydroxyquinoline-2-carbaldehyde with 2-hydrazinylpyridine in good yield (89.0%) [44], as shown in Scheme 2. m.p. 238 °C; ^{1}H-NMR (500 MHz, DMSO-d_6) δ 11.46 (s, 1H, OH), 9.72 (s, 1H, NH), 8.27 (d, *J* = 9.0 Hz, 2H), 8.18 (d, *J* = 4.0 Hz, 1H), 8.14 (d, *J* = 8.6 Hz, 1H), 7.74–7.68 (m, 1H),

7.42–7.36 (m, 3H), 7.09 (dd, J = 7.2, 1.5 Hz, 1H), 6.85 (dd, J = 6.9, 5.1 Hz, 1H); ^{13}C-NMR (125 MHz, DMSO-d_6) δ 157.10, 153.61, 153.16, 148.32, 139.65, 138.60, 138.52, 136.57, 128.71, 127.87, 118.22, 117.91, 116.27, 112.41, 107.19; HRMS(EI): Calcd for $C_{15}H_{13}N_4O$ [L + H]$^+$, m/z 265.1089, found m/z 265.1086. IR (cm^{-1}): ν_{NH} = 3049 cm^{-1}; $\nu_{C=N}$ = 1580 cm^{-1}. (The ^{1}H-NMR, ^{13}C-NMR and MS were shown in Figures S6–S8)

3.3.2. Synthesis of Complex **1**

The mixture of ligand L (0.26 g, 1.0 mmol) and CuCl$_2$·2H$_2$O (0.17 g, 1.0 mmol) in 20 mL methanol was maintained at reflux (70 °C) for 6 h to afford complex **1** as black crystals in 70% yield. The black crystals of complex **1** suitable for X-ray diffraction analysis were subsequently harvested. ESI-MS m/z: 427.0361[Cu(L)Cl + H + 2MeOH]$^+$. Anal. Calcd for $C_{30}H_{24}Cl_4Cu_2N_8O_2$: C, 45.18; H, 3.03; N, 14.05; O, 4.01; Found: C, 45.16; H, 3.02; N, 15.07. IR (cm^{-1}): ν_{NH} = 3101 cm^{-1}; $\nu_{C=N}$ = 1636 cm^{-1}. (The MS was shown in Figure S9)

3.3.3. Synthesis of Complex **2**

By means of the similar procedure, complex **2** was obtained from NiCl$_2$·6H$_2$O as black crystals in 75% yield. The black crystals of complex **2** suitable for X-ray diffraction analysis were subsequently harvested. ESI-MS m/z: 321.0281 [Ni(L) + H]$^+$. Anal. calcd for $C_{16}H_{14}Cl_4N_4NiO$: C, 40.13; H, 2.95; N, 11.70; Found: C, 40.14; H, 2.94; N, 11.69. IR (cm^{-1}): ν_{NH} = 3060 cm^{-1}; $\nu_{C=N}$ = 1615 cm^{-1}. (The MS was shown in Figure S10)

3.4. X-ray Crystallography

Complexes **1** (0.31 × 0.22 × 0.10 mm) and **2** (0.34 × 0.18 × 0.17 mm) were measured on an Agilent SuperNova CCD area detector (Rigaku Corporation, Tokyo, Japan) equipped with a graphite-monochromatic Mo-Kα radiation source (λ = 0.71073 Å) at room temperature 293(2) K. All non-hydrogen atoms' positions and anisotropic thermal parameters were refined on F2 by full-matrix least-squares techniques with the SHELX-97 program package [70]. The hydrogen atoms were added theoretically, riding on the concerned atoms. The semi-empirical methods from equivalents were used to correct absorption. The crystallographic data and refinement details of the structures are summarized in Tables S1–S3 (Supporting Information).

3.5. In Vitro Cytotoxicity

Seven tumor cells Hep-G2, SK-OV-3, MGC80-3, HeLa, T-24, BEL7402, and NCI-H460 and one normal liver cell HL-7702 were obtained from the Shanghai Cell Bank in the Chinese Academy of Sciences. Cells were grown in triplicate in 96-well plates (Gibco, Carlsbad, CA, USA) and incubated at 37 °C for 48 h in a humidified atmosphere containing 5% CO$_2$ and 95% air. To investigate the potential activity of L and complexes **1** and **2**, cisplatin was employed as a reference metallodrug. Cytotoxicity assays were carried out in 96-well flat-bottomed microtite plates that were supplemented with culture medium and cells. Ligand L, complexes **1** and **2**, and cisplatin were dissolved in the culture medium at various concentrations (1.25, 2.5, 5.0, 10.0, and 20.0 μM) with 1% DMSO and the resulting solutions were subsequently added to a set of wells. The control wells contained supplemented medium with 1% DMSO. The microtitre plates were then incubated at 37 °C under a humidified atmosphere containing 5% CO$_2$ and 95% air for 2 days. Cytotoxicity screening was conducted through 3-(4,5-dimethylthiazol-2-yl)-2,5-diphenyltetrazolium bromide (MTT) assay. After each incubation period, the MTT solution (10 mL, 5 mg·mL^{-1}) was added into each well and the cultures were incubated at 37 °C in a humidified atmosphere containing 5% CO$_2$ and 95% air for a further 48 h. After the removal of the supernatant, DMSO (150 mL) was added to dissolve the formazan crystals.

The absorbance at 490 and 630 nm was read on a plate reader. Relative to the negative control, cytotoxicity was estimated based on the percentage cell survival in a dose-dependent manner. The final

IC_{50} values were calculated by the Bliss method ($n = 5$). All tests were repeated in at least three independent experiments.

3.6. Cell Cycle Arrest

The MGC80-3 cells were maintained in Dulbecco's modified Eagle's medium with 10.0% fetal calf serum under 5% CO_2 at 37 °C. The cells were harvested by trypsinization, rinsed with PBS, and centrifuged at $3000\times g$ for 10 min. The pellet (105–106 cells) was suspended in PBS (1.0 mL) and kept on ice for 5 min. The cell suspension was then fixed by the dropwise addition of 9 mL precooled (4 °C) 100% ethanol with vigorous shaking, and the fixed samples were kept at 4 °C until use. For staining, the cells were centrifuged, resuspended in PBS, digested with 150 mL RNase A (250 μg·mL^{-1}), treated with 150 mL PI (0.15 mM), and then incubated for 30 min at 4 °C. PI-positive cells were counted with a fluorescence-activated cell sorter (FACS). The population of cells in each cell cycle was determined by the Cell Modi FIT software (Becton Dickinson, version 1.0, San Jose, CA, USA).

3.7. Other Experimental Methods

The supporting information provides the detailed procedures of other experimental methods, including the measurement of mitochondrial membrane potential (by JC-1 staining), ROS generation, intracellular free Ca^{2+}, Western blot, and caspase-3/9 activity. The procedures were similar to those given in the previous work of Chen et al. [71].

3.8. Statistics

Data processing included the Student's *t*-test with $p \leq 0.05$ taken as significance level, using SPSS 13.0 (IBM, Armonk, NY, USA).

4. Conclusions

Two transition metal complexes with 2-((2-(pyridin-2-yl) hydrazono) methyl)quinolin-8-ol (L), [Cu(L)Cl$_2$]$_2$ (**1**) and [Ni(L)Cl$_2$]·CH$_2$Cl$_2$ (**2**), were synthesized and fully characterized. In vitro antitumor screening revealed that complex **1** exhibited higher inhibitory activities than cisplatin against SK-OV-3, MGC80-3, and HeLa cells. In addition, complex **1** caused MGC80-3 cell arrest in the S phase, which led to the significant down-regulation of the related proteins. Complex **1** can down-regulate the expression of the bcl-2 protein and upregulate the levels of the bax, Cyt C, and apaf-1 proteins in MGC80-3 cells. We found that complex **1** induced MGC80-3 cell apoptosis via a mitochondrial dysfunction pathway, which was mediated by $\Delta\psi$, ROS, and intracellular Ca^{2+}. Moreover, complex **1** could induce cell apoptosis by triggering the caspase-3/9/8 activity in MGC80-3 cells. Therefore, complex **1** is a potent anticancer drug candidate.

Supplementary Materials: Can be found at http://www.mdpi.com/1422-0067/19/7/1874/s1, Full cif depositions (excluding structure factors) lodged with the Cambridge Crystallographic Data Centre (CCDC 1848527 (for complexes 1), 1848516 (for complexes 2)) contain the supplementary crystallographic data for this paper. These data can be obtained free of charge from The Cambridge Crystallographic Data Centre via www.ccdc.cam.ac.uk/data_request/cif.

Author Contributions: Q.-Y.Y., Q.-Q.C., Q.-P.Q., C.-X.D., H.L., Z.-F.C. conceived, designed the experiments, performed the experiments, analyzed the data, contributed reagents/materials/analysis tools, wrote and approved the final manuscript.

Acknowledgments: This work was supported by the National Natural Science Foundation of China (Grants 81473102, 21431001), IRT_16R15, CMEMR2012-A22, Natural Science Foundation of Guangxi Province (Grant No. 2012GXNSFDA053005) and Innovation Project of Guangxi Graduate Education (Grant No. YCBZ2018033) as well as "BAGUI Scholar" program of Guangxi Province of China.

Conflicts of Interest: The authors declare no conflict of interest.

References

1. Chaudhuri, A.R.; Callen, E.; Ding, X.; Gogola, E.; Duarte, A.A.; Lee, J.E.; Wong, N.; Lafarga, V.; Calvo, J.A.; Panzarino, N.J.; et al. Replication fork stability confers chemoresistance in BRCA-deficient cells. *Nature* **2016**, *535*, 382–387. [CrossRef] [PubMed]

2. Cheff, D.M.; Hall, M.D. A Drug of Such Damned Nature.1 Challenges and Opportunities in Translational Platinum Drug Research. *J. Med. Chem.* **2017**, *60*, 4517–4532. [CrossRef] [PubMed]

3. Li, Y.; Li, A.; Wu, J.; He, Y.; Yu, H.; Chai, R.; Li, H. MiR-182-5p protects inner ear hair cells from cisplatin-induced apoptosis by inhibiting FOXO3a. *Cell Death Dis.* **2016**, *7*, e2362. [CrossRef] [PubMed]

4. Li, X.; Fang, C.; Zong, Z.; Cui, L.; Bi, C.; Fan, Y. Synthesis, characterization and anticancer activity of two ternary copper(II) Schiff base complexes. *Inorg. Chim. Acta* **2015**, *432*, 198–207. [CrossRef]

5. Saleem, K.; Wani, W.A.; Haque, A.; Lone, M.N.; Hsieh, M.F.; Jairajpuri, M.A.; Ali, I. Synthesis, DNA binding, hemolysis assays and anticancer studies of copper(II), nickel(II) and iron(III) complexes of a pyrazoline-based ligand. *Future Med. Chem.* **2013**, *5*, 135–146. [CrossRef] [PubMed]

6. Muralisankar, M.; Haribabu, J.; Bhuvanesh, N.S.P.; Karvembu, R.; Sreekanth, A. Synthesis, X-ray crystal structure, DNA/protein binding, DNA cleavage and cytotoxicity studies of N(4) substituted thiosemicarbazone based copper(II)/nickel(II) complexes. *Inorg. Chim. Acta* **2016**, *449*, 82–95. [CrossRef]

7. Wani, W.A.; Al-Othman, Z.; Ali, I.; Saleem, K.; Hsieh, M.F. Copper(II), nickel(II), and ruthenium(III) complexes of an oxopyrrolidine-based heterocyclic ligand as anticancer agents. *J. Coord. Chem.* **2014**, *67*, 2110–2130. [CrossRef]

8. Haleel, A.; Arthi, P.; Reddy, N.D.; Veenac, V.; Sakthivelc, N.; Arund, Y.; Perumald, P.T.; Rahiman, K. DNA binding, molecular docking and apoptotic inducing activity of nickel(II), copper(II) and zinc(II) complexes of pyridine-based tetrazolo[1,5-a] pyrimidine ligands. *RSC Adv.* **2014**, *4*, 60816–60830. [CrossRef]

9. Jin, Q.M.; Lu, Y.; Jin, J.L.; Guo, H.; Lin, G.W.; Wang, Y.; Lu, T. Synthesis, characterization, DNA binding ability and cytotoxicity of the novel platinum(II); copper(II), cobalt(II) and nickel(II) complexes with 3-(1H-benzo[d]imidazol-2-yl)-β-carboline. *Inorg. Chim. Acta* **2014**, *421*, 91–99. [CrossRef]

10. Rajalakshmi, S.; Kiran, M.S.; Nair, B.U. DNA condensation by copper(II) complexes and their anti-proliferative effect on cancerous and normal fibroblast cells. *Eur. J. Med. Chem.* **2014**, *80*, 393–406. [CrossRef] [PubMed]

11. Totta, X.; Papadopoulou, A.A.; Hatzidimitriou, A.G.; Papadopoulosb, A.; Psomas, G. Synthesis, structure and biological activity of nickel(II) complexes with mefenamato and nitrogen-donor ligands. *J. Inorg. Biochem.* **2015**, *145*, 79–93. [CrossRef] [PubMed]

12. Hsu, C.W.; Kuo, C.F.; Chuang, S.M.; Hou, M.H. Elucidation of the DNA-interacting properties and anticancer activity of a Ni(II)-coordinated mithramycin dimer complex. *Biometals* **2013**, *26*, 1–12. [CrossRef] [PubMed]

13. Tabrizi, L.; McArdle, P.; Erxleben, A.; Chiniforoshan, H. Nickel(II) and cobalt(II) complexes of lidocaine: Synthesis, structure and comparative invitro evaluations of biological perspectives. *Eur. J. Med. Chem.* **2015**, *103*, 516–529. [CrossRef] [PubMed]

14. Zhu, T.F.; Wang, Y.; Ding, W.J.; Xu, J.; Chen, R.H.; Xie, J.; Zhu, W.J.; Jia, L.; Ma, T.L. Anticancer Activity and DNA-Binding Investigations of the Cu(II) and Ni(II) Complexes with Coumarin Derivative. *Chem. Biol. Drug Des.* **2015**, *85*, 385–393. [CrossRef] [PubMed]

15. Alomar, K.; Landreau, A.; Allain, M.; Boueta, G.; Larcher, G. Synthesis, structure and antifungal activity of thiophene-2,3-dicarboxaldehyde bis(thiosemicarbazone) and nickel(II), copper(II) and cadmium(II) complexes: Unsymmetrical coordination mode of nickel complex. *J. Inorg. Biochem.* **2013**, *126*, 76–83. [CrossRef] [PubMed]

16. Ramírez-Macías, I.; Maldonado, C.R.; Marín, C.; Olmoa, F.; Gutiérrezsánchezc, R.; Rosalesa, M.J.; Quirósb, M.; Salasb, J.M. In vitro anti-leishmania evaluation of nickel complexes with a triazolopyrimidine derivative against *Leishmania infantum* and *Leishmania braziliensis*. *J. Inorg. Biochem.* **2012**, *112*, 1–9. [CrossRef] [PubMed]

17. Betanzos-Lara, S.; Gómez-Ruiz, C.; Barrón-Sosa, L.R.; Gracia-Mora, I.; Flores-Álamo, M.; Barba-Behrens, N. Cytotoxic copper(II), cobalt(II), zinc(II), and nickel(II) coordination compounds of clotrimazole. *J. Inorg. Biochem.* **2012**, *114*, 82–93. [CrossRef] [PubMed]

18. Turnaturi, R.; Oliveri, V.; Vecchio, G. Biotin-8-hydroxyquinoline conjugates and their metal complexes: Exploring the chemical properties and the antioxidant activity. *Polyhedron* **2016**, *110*, 254–260. [CrossRef]

19. Xu, H.; Chen, W.; Zhan, P.; Liu, X. 8-Hydroxyquinoline: A privileged structure with a broad-ranging pharmacological potential. *MedChemComm* **2015**, *6*, 61–74. [CrossRef]

20. Prachayasittikul, V.; Prachayasittikul, S.; Ruchirawat, S.; Prachayasittikul, V. 8-Hydroxyquinolines: A review of their metal chelating properties and medicinal applications. *Drug Des. Dev. Ther.* **2013**, *7*, 1157–1178. [CrossRef] [PubMed]

21. Solomon, R.; Lee, H. Quinoline as a Privileged Scaffold in Cancer Drug Discovery. *Curr. Med. Chem.* **2011**, *18*, 1488–1508. [CrossRef] [PubMed]

22. Schimmer, A.D. Clioquinol—A Novel Copper-Dependent and Independent Proteasome Inhibitor. *Cancer Drug Targets* **2011**, *11*, 325–331. [CrossRef]

23. Matlack, K.E.; Tardiff, D.F.; Narayan, P.; Hamamichi, S.; Caldwell, K.A.; Caldwell, G.A.; Lindquist, S. Clioquinol promotes the degradation of metal-dependent amyloid-β (Aβ) oligomers to restore endocytosis and ameliorate Aβ toxicity. *Proc. Natl. Acad. Sci. USA* **2014**, *111*, 4013–4018. [CrossRef] [PubMed]

24. Wang, L.; Esteban, G.; Ojima, M.; Bautista-Aguilera, O.M.; Inokuchi, T.; Moraleda, I.; Iriepa, I.; Samadi, A.; Youdim, M.B.; Romero, A.; et al. Donepezil + propargylamine + 8-hydroxyquinoline hybrids as new multifunctional metal-chelators; ChE and MAO inhibitors for the potential treatment of Alzheimer's disease. *Eur. J. Med. Chem.* **2014**, *80*, 543–561. [CrossRef] [PubMed]

25. Wu, M.Y.; Esteban, G.; Brogi, S.; Shionoya, M.; Wang, L.; Campiani, G.; Unzeta, M.; Inokuchi, T.; Butini, S.; Marco-Contelles, J. Donepezil-like multifunctional agents: Design, synthesis, molecular modeling and biological evaluation. *Eur. J. Med. Chem.* **2016**, *121*, 864–879. [CrossRef] [PubMed]

26. Jiang, H.; Taggart, J.E.; Zhang, X.; Benbrook, D.M.; Lind, S.E.; Ding, W.Q. Nitroxoline (8-hydroxy-5-nitroquinoline) is more a potent anti-cancer agent than clioquinol (5-chloro-7-iodo-8-quinoline). *Cancer Lett.* **2011**, *312*, 11–17. [CrossRef] [PubMed]

27. Sošić, I.; Mirković, B.; Arenz, K.; Štefane, B.; Kos, J.; Gobec, S. Development of New Cathepsin B Inhibitors: Combining Bioisosteric Replacements and Structure-Based Design To Explore the Structure–Activity Relationships of Nitroxoline Derivatives. *J. Med. Chem.* **2013**, *56*, 521–533. [CrossRef] [PubMed]

28. Knez, D.; Brus, B.; Coquelle, N.; Sošič, I.; Šink, R.; Brazzolotto, X.; Mravljak, J.; Colletier, J.P.; Gobec, S. Structure-based development of nitroxoline derivatives as potential multifunctional anti-Alzheimer agents. *Bioorg. Med. Chem.* **2015**, *23*, 4442–4452. [CrossRef] [PubMed]

29. Rotili, D.; Tomassi, S.; Conte, M.; Benedetti, R.; Tortorici, M.; Ciossani, G.; Valente, S.; Marrocco, B.; Labella, D.; Novellino, E.; et al. Pan-Histone Demethylase Inhibitors Simultaneously Targeting Jumonji C and Lysine-Specific Demethylases Display High Anticancer Activities. *J. Med. Chem.* **2013**, *57*, 42–55. [CrossRef] [PubMed]

30. Schiller, R.; Scozzafava, G.; Tumber, A.; Wickens, J.R.; Bush, J.T.; Rai, G.; Lejeune, C.; Choi, H.; Yeh, T.L.; Chan, M.C.; et al. A Cell-Permeable Ester Derivative of the JmjC Histone Demethylase Inhibitor IOX1. *Chem. Med. Chem.* **2014**, *9*, 566–571. [CrossRef] [PubMed]

31. Sliman, F.; Blairvacq, M.; Durieu, E.; Meijer, L.; Rodrigo, J.; Desmaële, D. Identification and structure-activity relationship of 8-hydroxy-quinoline-7-carboxylic acid derivatives as inhibitors of Pim-1 kinase. *Bioorg. Med. Chem. Lett.* **2010**, *20*, 2801–2805. [CrossRef] [PubMed]

32. Li, X.M.; Wood, T.E.; Sprangers, R.; Jansen, G.; Franke, N.E.; Mao, X.L.; Wang, X.M.; Zhang, Y.; Verbrugge, S.E.; Adomat, H.; et al. Effect of Noncompetitive Proteasome Inhibition on Bortezomib Resistance. *J. Natl. Cancer Inst.* **2010**, *102*, 1069–1082. [CrossRef] [PubMed]

33. Liu, Y.C.; Yang, Z.Y. Synthesis, crystal structure; antioxidation and DNA binding properties of binuclear Ho(III) complexes of Schiff-base ligands derived from 8-hydroxyquinoline-2-carboxaldehyde and four aroylhydrazines. *J. Inorg. Biochem.* **2009**, *103*, 1014–1022. [CrossRef] [PubMed]

34. Jacobsen, J.A.; Fullagar, J.L.; Miller, M.T.; Cohen, S.M. Identifying Chelators for Metalloprotein Inhibitors Using a Fragment-Based Approach. *J. Med. Chem.* **2011**, *54*, 591–602. [CrossRef] [PubMed]

35. Shaw, A.Y.; Chang, C.Y.; Hsu, M.Y.; Lu, P.J.; Yang, C.N.; Chen, H.L.; Lo, C.W.; Shiau, C.W.; Chern, M.K. Synthesis and structure-activity relationship study of 8-hydroxyquinoline-derived Mannich bases as anticancer agents. *Eur. J. Med. Chem.* **2010**, *45*, 2860–2867. [CrossRef] [PubMed]

36. Ariyasu, S.; Sawa, A.; Morita, A.; Hanaya, K.; Hoshi, M.; Takahashi, I.; Wang, B.; Aoki, S. Design and synthesis of 8-hydroxyquinoline-based radioprotective agents. *Bioorg. Med. Chem.* **2014**, *22*, 3891–3905. [CrossRef] [PubMed]

37. Cacciatore, I.; Fornasari, E.; Baldassarre, L.; Cornacchia, C.; Fulle, S.; Di Filippo, E.S.; Pietrangelo, T.; Pinnen, F. A Potent (*R*)-alpha-bis-lipoyl Derivative Containing 8-Hydroxyquinoline Scaffold: Synthesis and Biological Evaluation of Its Neuroprotective Capabilities in SH-SY5Y Human Neuroblastoma Cells. *Pharmaceuticals* **2013**, *6*, 54–69. [CrossRef] [PubMed]

38. Du Moulinet D'Hardemare, A.; Gellon, G.; Philouze, C.; Serratrice, G. Oxinobactin and Sulfoxinobactin; Abiotic Siderophore Analogues to Enterobactin Involving 8-Hydroxyquinoline Subunits: Thermodynamic and Structural Studies. *Inorg. Chem.* **2012**, *51*, 12142–12151. [CrossRef] [PubMed]

39. Fernández-Bachiller, M.I.; Pérez, C.; González-Munoz, G.C.; Conde, S.; López, M.G.; Villarroya, M.; García, A.G.; Rodriguez-Franco, M.I. Novel Tacrine−8-Hydroxyquinoline Hybrids as Multifunctional Agents for the Treatment of Alzheimer's Disease; with Neuroprotective; Cholinergic; Antioxidant; and Copper-Complexing Properties. *J. Med. Chem.* **2010**, *53*, 4927–4937. [CrossRef] [PubMed]

40. Calvaresi, E.C.; Hergenrother, P.J. Glucose conjugation for the specific targeting and treatment of cancer. *Chem. Sci.* **2013**, *4*, 2319–2333. [CrossRef] [PubMed]

41. Oliveri, V. New Glycoconjugates for the Treatment of Diseases Related to Metal Dyshomeostasis. *ChemistryOpen* **2015**, *4*, 792–795. [CrossRef] [PubMed]

42. Chan, S.H.; Chui, C.H.; Chan, S.W.; Kok, S.H.L.; Chan, D.; Tsoi, M.Y.T.; Leung, P.H.M.; Lam, A.K.Y.; Chan, A.S.C.; Lam, K.H.; et al. Synthesis of 8-Hydroxyquinoline Derivatives as Novel Antitumor Agents. *ACS Med. Chem. Lett.* **2012**, *4*, 170–174. [CrossRef] [PubMed]

43. Núñez, C.; Oliveira, E.; García-Prdo, J.; Diniza, M.; Lorenzof, J.; Capeloa, J.L.; Lodeiroa, C. A novel quinoline molecular probe and the derived functionalized gold nanoparticles: Sensing properties and cytotoxicity studies in MCF-7 human breast cancer cells. *J. Inorg. Biochem.* **2014**, *137*, 115–122. [CrossRef] [PubMed]

44. Kao, M.H.; Chen, T.Y.; Cai, Y.R.; Hu, C.H.; Liu, Y.W.; Jhong, Y.; Wu, A.T. A turn-on Schiff-base fluorescence sensor for Mg^{2+} ion and its practical application. *J. Lumin.* **2016**, *169*, 156–160. [CrossRef]

45. Zhang, H.R.; Meng, T.; Liu, Y.C.; Qin, Q.P.; Chen, Z.F.; Liu, Y.N.; Liang, H. Synthesis, Structure Characterization and Antitumor Activity Study of a New Iron(III) Complex of 5-Nitro-8-hydroxylquinoline (HNOQ). *Chem. Pharm. Bull.* **2016**, *64*, 1208–1217. [CrossRef] [PubMed]

46. Zhang, H.R.; Meng, T.; Liu, Y.C.; Chen, Z.F.; Liu, Y.N.; Liang, H. Synthesis, characterization and biological evaluation of a cobalt(II) complex with 5-chloro-8-hydroxyquinoline as anticancer agent. *Appl. Organomet. Chem.* **2016**, *30*, 740–747. [CrossRef]

47. Zhang, H.R.; Liu, Y.C.; Chen, Z.F.; Meng, T.; Zou, B.Q.; Liu, Y.N.; Liang, H. Studies on the structures, cytotoxicity and apoptosis mechanism of 8-hydroxylquinoline rhodium(III) complexes in T-24 cells. *New J. Chem.* **2016**, *409*, 6005–6014. [CrossRef]

48. Heidary, D.K.; Howerton, B.S.; Glazer, E.C. Coordination of Hydroxyquinolines to a Ruthenium *bis*-dimethyl-phenanthroline Scaffold Radically Improves Potency for Potential as Antineoplastic Agents. *J. Med. Chem.* **2014**, *57*, 8936–8946. [CrossRef] [PubMed]

49. Dömötör, O.; Pape, V.F.; May, N.V.; Szakács, G.; Enyedy, É.A. Comparative solution equilibrium studies of antitumor ruthenium(η6-p-cymene) and rhodium(η5-$C_5$$Me_5$) complexes of 8-hydroxyquinolines. *Dalton Trans.* **2017**, *46*, 4382–4396. [CrossRef] [PubMed]

50. Martín-Santos, C.; Michelucci, E.; Marzo, T.; Messori, L.; Szumlas, P.; Bednarski, P.J.; Mas-Ballesté, R.; Navarro-Ranninger, C.; Cabrer, S. Gold(III) complexes with hydroxyquinoline, aminoquinoline and quinoline ligands: Synthesis, cytotoxicity, DNA and protein binding studies. *J. Inorg. Biochem.* **2015**, *153*, 339–345. [CrossRef] [PubMed]

51. Qin, Q.P.; Chen, Z.F.; Qin, J.L.; He, X.J.; Li, Y.L.; Liu, Y.C.; Liang, H. Studies on antitumor mechanism of two planar platinum(II) complexes with 8-hydroxyquinoline: Synthesis, characterization, cytotoxicity, cell cycle and apoptosis. *Eur. J. Med. Chem.* **2015**, *92*, 302–313. [CrossRef] [PubMed]

52. Tardito, S.; Barilli, A.; Bassanetti, I.; Tegoni, M.; Bussolati, O.; Franchi-Gazzola, R.; Marchiò, L. Copper-Dependent Cytotoxicity of 8-Hydroxyquinoline Derivatives Correlates with Their Hydrophobicity and Does Not Require Caspase Activation. *J. Med. Chem.* **2012**, *55*, 10448–10459. [CrossRef] [PubMed]

53. Rogolino, D.; Cavazzoni, A.; Gatti, A.; Tegoni, M.; Pelosi, G.; Verdolino, V.; Carcelli, M. Anti-proliferative effects of copper(II) complexes with hydroxyquinoline-thiosemicarbazone ligands. *Eur. J. Med. Chem.* **2017**, *128*, 140–153. [CrossRef] [PubMed]

54. Correia, I.; Adao, P.; Roy, S.; Wahba, M.; Matos, C.; Maurya, M.R.; Pessoa, J.C. Hydroxyquinoline derived vanadium (IV and V) and copper(II) complexes as potential anti-tuberculosis and anti-tumor agents. *J. Inorg. Biochem.* **2014**, *141*, 83–93. [CrossRef] [PubMed]

55. Laplante, S.R.; Fader, L.D.; Fandrick, K.R.; Fandrick, D.R.; Hucke, O.; Kemper, R.; Miller, S.P.F.; Edwards, P.J. Assessing Atropisomer Axial Chirality in Drug Discovery and Development. *J. Med. Chem.* **2016**, *54*, 7005–7022. [CrossRef] [PubMed]

56. Sommerwerk, S.; Heller, L.; Kuhfs, J.; Csuk, R. Urea derivates of ursolic, oleanolic and maslinic acid induce apoptosis and are selective cytotoxic for several human tumor cell lines. *Eur. J. Med. Chem.* **2016**, *119*, 1–16. [CrossRef] [PubMed]

57. Ooi, K.K.; Yeo, C.I.; Mahandaran, T.; Ang, K.P.; Akim, A.M.; Cheah, Y.K.; Seng, H.L.; Tiekink, E.R. G_2/M cell cycle arrest on HT-29 cancer cells and toxicity assessment of triphenylphosphanegold(I) carbonimidothioates, $Ph_3PAu[SC(OR) = NPh]$, R = Me, Et, and iPr; during zebrafish development. *J. Inorg. Biochem.* **2017**, *166*, 173–181. [CrossRef] [PubMed]

58. Zhang, C.; Han, B.J.; Zeng, C.C.; Lai, S.H.; Li, W.; Tang, B.; Wan, D.; Jiang, G.B.; Liu, Y. Synthesis, characterization, in vitro cytotoxicity and anticancer effects of ruthenium(II) complexes on BEL-7402 cells. *J. Inorg. Biochem.* **2016**, *157*, 62–72. [CrossRef] [PubMed]

59. Zheng, C.P.; Liu, Y.N.; Liu, Y.; Qin, X.Y.; Zhou, Y.H.; Liu, J. Dinuclear ruthenium complexes display loop isomer selectivity to c-MYC DNA G-quadriplex and exhibit anti-tumour activity. *J. Inorg. Biochem.* **2016**, *156*, 122–132. [CrossRef] [PubMed]

60. Yan, J.; Chen, J.; Zhang, S.; Hu, J.H.; Ling, H.; Li, X.S. Synthesis, Evaluation; and Mechanism Study of Novel Indole-Chalcone Derivatives Exerting Effective Antitumor Activity Through Microtubule Destabilization in Vitro and in Vivo. *J. Med. Chem.* **2016**, *59*, 5264–5283. [CrossRef] [PubMed]

61. Zhang, Y.L.; Qin, Q.P.; Cao, Q.Q.; Han, H.H.; Liu, Z.L.; Liu, Y.C.; Liang, H.; Chen, Z.F. Synthesis, crystal structure, cytotoxicity and action mechanism of a Rh(III) complex with 8-hydroxy-2-methylquinoline as a ligand. *Med. Chem. Commun.* **2017**, *8*, 184–190. [CrossRef]

62. Horvat, A.; Zorec, R.; Vardjan, N. Adrenergic stimulation of single rat astrocytes results in distinct temporal changes in intracellular Ca2+ and cAMP-dependent PKA responses. *Cell Calcium* **2016**, *59*, 156–163. [CrossRef] [PubMed]

63. Hu, X.Z.; Xu, Y.; Hu, D.C.; Hui, Y.; Yang, F.X. Apoptosis induction on human hepatoma cells Hep G2 of decabrominated diphenyl ether (PBDE-209). *Toxicol. Lett.* **2007**, *171*, 19–28. [CrossRef] [PubMed]

64. Kawiak, A.; Zawacka-Pankau, J.; Wasilewska, A.; Stasilojc, G.; Bigda, J.; Lojkowska, E. Induction of Apoptosis in HL-60 Cells through the ROS-Mediated Mitochondrial Pathway by Ramentaceone from *Drosera aliciae*. *J. Nat. Prod.* **2012**, *75*, 9–14. [CrossRef] [PubMed]

65. Spierings, D.; McStay, G.; Saleh, M.; Bender, C.; Chipuk, J.; Maurer, U.; Green, D.R. Connected to Death: The (Unexpurgated) Mitochondrial Pathway of Apoptosis. *Science* **2005**, *310*, 66–67. [CrossRef] [PubMed]

66. Qin, J.L.; Qin, Q.P.; Wei, Z.Z.; Yu, C.C.; Meng, T.; Wu, C.X.; Liang, Y.L.; Liang, H.; Chen, Z.F. Stabilization of c-myc G-Quadruplex DNA, inhibition of telomerase activity, disruption of mitochondrial functions and tumor cell apoptosis by platinum(II) complex with 9-amino-oxoisoaporphine. *Eur. J. Med. Chem.* **2016**, *124*, 417–427. [CrossRef] [PubMed]

67. Zhang, G.H.; Cai, L.J.; Wang, Y.F.; Zhou, Y.H.; An, Y.F.; Liu, Y.C.; Peng, Y.; Chen, Z.F.; Liang, H. Novel compound PS-101 exhibits selective inhibition in non-small-cell lung cancer cell by blocking the EGFR-driven antiapoptotic pathway. *Biochem. Pharmacol.* **2013**, *86*, 1721–1730. [CrossRef] [PubMed]

68. Chen, Z.F.; Qin, Q.P.; Qin, J.L.; Liu, Y.C.; Huang, K.B.; Li, Y.L.; Meng, T.; Zhang, G.H.; Peng, Y.; Luo, X.J.; et al. Stabilization of G-Quadruplex DNA, Inhibition of Telomerase Activity, and Tumor Cell Apoptosis by Organoplatinum(II) Complexes with Oxoisoaporphine. *J. Med. Chem.* **2015**, *58*, 2159–2179. [CrossRef] [PubMed]

69. Huang, K.B.; Chen, Z.F.; Liu, Y.C.; Xie, X.L.; Liang, H. Dihydroisoquinoline copper(II) complexes: Crystal structures, cytotoxicity, and action mechanism. *RSC Adv.* **2015**, *5*, 81313–81323. [CrossRef]

70. Sheldrick, G.M. *SHELXS-97, Program for Solution of Crystal Structures*; University of Göttingen: Göttingen, Germany, 1997.

71. Qin, Q.P.; Qin, J.L.; Meng, T.; Yang, G.A.; Wei, Z.Z.; Liu, Y.C.; Liang, H.; Chen, Z.F. Preparation of 6/8/11-Amino/Chloro-Oxoisoaporphine and Group-10 Metal Complexes and Evaluation of Their in Vitro and in Vivo Antitumor Activity. *Sci. Rep.* **2016**, *6*, 37644. [CrossRef] [PubMed]

International Journal of
Molecular Sciences

MDPI

Article

Palladacyclic Conjugate Group Promotes Hybridization of Short Oligonucleotides

Madhuri Hande, Sajal Maity and Tuomas Lönnberg *

Department of Chemistry, University of Turku, Vatselankatu 2, 20014 Turku, Finland;
nimamadhuri@gmail.com (M.H.); sajal.k.maity@utu.fi (S.M.)
* Correspondence: tuanlo@utu.fi; Tel.: +358-29-450-3191

Received: 18 May 2018; Accepted: 26 May 2018; Published: 28 May 2018

Abstract: Short oligonucleotides with cyclopalladated benzylamine moieties at their 5′-termini have been prepared to test the possibility of conferring palladacyclic anticancer agents sequence-selectivity by conjugation with a guiding oligonucleotide. Hybridization of these oligonucleotides with natural counterparts was studied by UV and CD (circular dichroism) melting experiments in the absence and presence of a competing ligand (2-mercaptoethanol). Cyclopalladated benzylamine proved to be strongly stabilizing relative to unmetalated benzylamine and modestly stabilizing relative to an extra A•T base pair. The stabilization was largely abolished in the presence of 2-mercaptoethanol, suggesting direct coordination of Pd(II) to a nucleobase of the complementary strand. In all cases, fidelity of Watson-Crick base pairing between the two strands was retained. Hybridization of the cyclopalladated oligonucleotides was characterized by relatively large negative enthalpy and entropy, consistent with stabilizing Pd(II) coordination partially offset by the entropic penalty of imposing conformational constraints on the flexible diethylene glycol linker between the oligonucleotide and the palladacyclic moiety.

Keywords: DNA; oligonucleotide; hybridization; organometallic; palladacycle; palladium

1. Introduction

The groundbreaking discovery of the antitumor activity of cisplatin [1,2] has been followed by efforts to develop more potent anticancer agents based on transition metal complexes [3–11]. In particular, problems associated with the presently available platinum anticancer compounds, notably acquired or intrinsic resistance, limited spectrum of activity and relatively high degree of toxicity [12–14], have prompted interest in transition metals other than platinum for chemotherapeutic use [4,6–8,15]. Palladium is an attractive candidate because its coordination chemistry is similar to that of platinum [16,17]. Pd(II) complexes are, however, kinetically approximately five orders of magnitude more labile than the respective Pt(II) complexes [18]. While the relatively rapid ligand-exchange of Pd(II) should allow formation of thermodynamic (rather than kinetic) products and thus higher selectivity than attainable with Pt(II)-based drugs, it is also likely to result in a different mode of action, at least with simple analogues [19]. No clinically approved palladium-containing drugs are presently available.

The possibility of using palladacyclic complexes as anticancer agents to circumvent the problems caused by the kinetic lability of Pd(II) has received attention over the past decade. The high stability of palladacyclic compounds in physiological media and the resultant low toxicity to normal cells make them promising candidates for future therapeutic agents [9,20]. The selectivity of these agents could be further improved by conjugation to a guiding oligonucleotide. The feasibility of this approach has already been demonstrated with a number of Pt(II)-carrying DNA and PNA oligonucleotides [21–26] but with palladacyclic modifications we are only aware of a single recent example, a short DNA oligonucleotide incorporating a single cyclopalladated phenylpyridine residue in the middle of the sequence [27]. In that case, coordination of Pd(II) to the opposite base on a complementary strand was inferred from the abnormally high UV and CD (circular dichroism) signals but the expected stabilization of the double helix by such coordination could not be demonstrated unambiguously. Possibly a stable Pd(II)-mediated base pair was formed but could not be readily accommodated within the base stack, leading to disruption of the double helix.

Herein we describe the synthesis and hybridization properties of short oligonucleotides incorporating cyclopalladated benzylamine "warheads" at their 5′-termini. At monomer level, palladacyclic benzylamine derivatives have already been found to exhibit antitumor activity [28,29]. The 5′-terminal position was chosen to avoid disruption of the double helix by suboptimal coordination geometry. For the same reason, a relatively long and flexible diethylene glycol spacer was used between the cyclopalladated benzylamine and the oligonucleotide.

2. Results

2.1. Synthesis of the Benzylamine Phosphoramidite Building Block

Synthesis of the protected phosphoramidite building block of benzylamine (**1**) is outlined in Scheme 1. First, benzyl bromide was allowed to react with an excess of 2-(2-aminoethoxy)ethanol to give 2-[2-(benzylamino)ethoxy]ethanol (**2**). The secondary amino function was then protected as a trifluoroacetamide by treatment with ethyl trifluoroacetate. Finally, the protected intermediate **3** was phosphitylated by conventional methods to afford the phosphoramidite building block **1**.

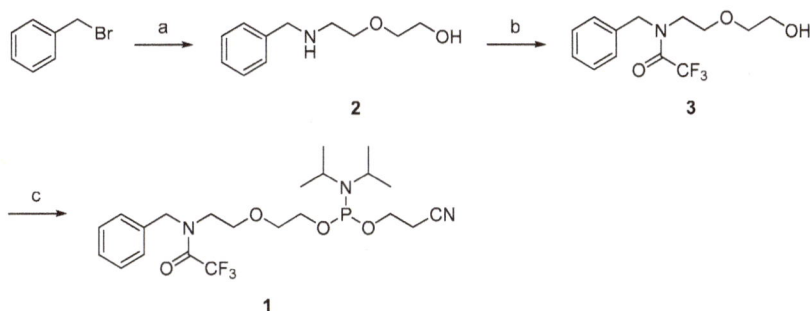

Scheme 1. Synthesis of the benzylamine phosphoramidite **1**. Reagents and conditions: (a) 2-(2-aminoethoxy)ethanol, MeCN, 25 °C, 16 h; (b) ethyl trifluoroacetate, Et$_3$N, MeOH, 25 °C, 16 h; (c) 2-cyanoethyl-*N*,*N*-diisopropylchlorophosphoramidite, Et$_3$N, CH$_2$Cl$_2$, N$_2$ atmosphere, 25 °C, 3 h.

2.2. Cyclopalladation of 2-[2-(benzylamino)ethoxy]ethanol

Cyclopalladation was first tested at monomer level with 2-[2-(benzylamino)ethoxy]ethanol (**2**) by treatment with an equimolar amount of lithium tetrachloropalladate in a mixture of water and acetonitrile. Near-quantitative conversion of the starting material was achieved overnight. ^{1}H NMR (nuclear magnetic resonance) spectrum of the product revealed loss of one ortho proton of the phenyl ring and ^{13}C NMR spectrum a downfield shift of the respective carbon signal. Both results are consistent with replacement of the ortho proton with Pd(II). The most likely structure of the product is the chlorido-bridged dimer **4** (Scheme 2), as reported previously on related compounds [30–32]. While only mononuclear Pd(II) species could be unambiguously identified in the mass spectrum, the splitting of several peaks in both ^{1}H and ^{13}C NMR is consistent with formation of a dimer, present in both cisoid and transoid forms.

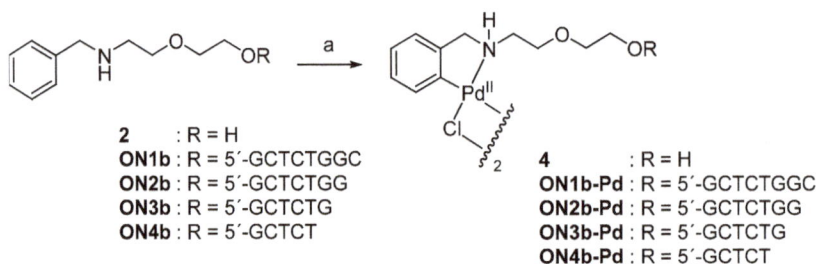

Scheme 2. Cyclopalladation of 2-[2-(benzylamino)ethoxy]ethanol (**2**) and the corresponding modified oligonucleotides **ON1b**, **ON2b**, **ON3b** and **ON4b**. Reagents and conditions: a) Li$_2$PdCl$_4$, MeCN, H$_2$O, 25 °C, 16 h.

2.3. Oligonucleotide Synthesis

The sequences of the oligonucleotides used in the present study are summarised in Table 1. Synthesis of the modified oligonucleotides **ON1b**, **ON2b**, **ON3b** and **ON4b**, having a 5′-terminal benzylamine moiety, was carried out on an automated DNA synthesizer using conventional phosphoramidite strategy. Treatment with concentrated aq. ammonia was employed for removal of the base and phosphate protections and release of the oligonucleotides from the solid support. Cyclopalladation of oligonucleotides **ON1b**, **ON2b**, **ON3b** and **ON4b** was carried out as described above for the monomer **2** (Scheme 2), except that 2.0 equivalents of lithium tetrachloropalladate was used. All modified oligonucleotides were purified by reversed-phase high performance liquid chromatography (RP-HPLC), characterized by electrospray ionization mass spectrometry (ESI-MS) and quantified by UV spectrophotometry.

Table 1. Oligonucleotides used in this study.

Oligonucleotide	Sequence [1]
ON1a	5′-AGCTCTGGC-3′
ON2a	5′-AGCTCTGG-3′
ON3a	5′-AGCTCTG-3′
ON4a	5′-AGCTCT-3′
ON1b	5′-BGCTCTGGC-3′
ON2b	5′-BGCTCTGG-3′
ON3b	5′-BGCTCTG-3′
ON4b	5′-BGCTCT-3′
ON1b-Pd	5′-B^{Pd}GCTCTGGC-3′
ON2b-Pd	5′-B^{Pd}GCTCTGG-3′
ON3b-Pd	5′-B^{Pd}GCTCTG-3′
ON4b-Pd	5′-B^{Pd}GCTCT-3′
ON5a	5′-GCCAGAGCTCG-3′
ON5c	5′-GCCAGCGCTCG-3′
ON5g	5′-GCCAGGGCTCG-3′
ON5t	5′-GCCAGTGCTCG-3′

[1] B refers to unmetalated and B^{Pd} to cyclopalladated benzylamine residue. The residues varied in the hybridization studies have been underlined.

2.4. Hybridization Studies

The impact of the 5′-terminal palladacyclic "warheads" on the hybridization properties of short oligonucleotides was assessed by recording melting temperatures of duplexes formed by oligonucleotides **ON1b-Pd**, **ON2b-Pd**, **ON3b-Pd** and **ON4b-Pd** with the natural counterparts **ON2a**, **ON2c**, **ON2g** and **ON2t**. For reference, similar experiments were also carried out on respective duplexes formed by oligonucleotides **ON1b**, **ON2b**, **ON3b**, **ON4b**, **ON1a**, **ON2a**, **ON3a** and **ON4a**, having either an unmetalated benzylamine or an adenine residue at their 5′-termini. In all assemblies, the 5′-terminal residue was placed opposite to a thymine residue of a trinucleotide overhang of the complementary oligonucleotide (Figure 1). A single base pair within the double helical region, on the other hand, was varied to test the sensitivity of hybridization to a single-nucleotide mismatch. All experiments were performed at pH 7.4 (20 mM cacodylate buffer) and ionic strength of 0.10 M (adjusted with sodium perchlorate) and each sample was first annealed by heating to 90 °C and then slowly cooling down to room temperature.

$$
\begin{array}{ll}
\text{5′-X G C T C T G G C-3′} & \textbf{ON1x} \\
\phantom{\text{5′-X G}}\bullet\ \bullet\ \bullet\ \bullet\ \bullet\ \bullet\ \bullet\ \bullet & \\
\text{3′-G C T C G Y G A C C G-5′} & \textbf{ON5y}
\end{array}
$$

$$
\begin{array}{ll}
\text{5′-X G C T C T G G-3′} & \textbf{ON2x} \\
\phantom{\text{5′-X G}}\bullet\ \bullet\ \bullet\ \bullet\ \bullet\ \bullet\ \bullet & \\
\text{3′-G C T C G Y G A C C G-5′} & \textbf{ON5y}
\end{array}
$$

$$
\begin{array}{ll}
\text{5′-X G C T C T G-3′} & \textbf{ON3x} \\
\phantom{\text{5′-X G}}\bullet\ \bullet\ \bullet\ \bullet\ \bullet\ \bullet & \\
\text{3′-G C T C G Y G A C C G-5′} & \textbf{ON5y}
\end{array}
$$

$$
\begin{array}{ll}
\text{5′-X G C T C T-3′} & \textbf{ON4x} \\
\phantom{\text{5′-X G}}\bullet\ \bullet\ \bullet\ \bullet\ \bullet & \\
\text{3′-G C T C G Y G A C C G-5′} & \textbf{ON5y}
\end{array}
$$

Figure 1. Outline of the hybridization assays used. X is either adenine or unmetalated or cyclopalladated benzylamine and Y is any canonical nucleobase. The bullets indicate Watson-Crick base pairing.

The longest matched duplexes **ON1x•ON5a** all exhibited sigmoidal melting profiles, with T_m (melting temperature) values ranging from 36 to 41 °C (Figure 2A–C). The shorter duplexes did not fully hybridize even at the lowest temperature applicable (10 °C) but their T_m values could still be determined with reasonable accuracy as inflection points of the melting curves. The melting temperatures of the mismatched duplexes, on the other hand, were high enough to be determined reliably only in the case of the longest duplexes **ON1x•ON5y**. The A•C mismatch was particularly destabilizing and precluded determination of the T_m in all cases, regardless of the length of the duplex. Melting temperatures are summarized in Figure 2D for the longest duplexes and in the Supplementary Materials for all duplexes.

Melting temperatures of the longest matched duplexes **ON1a•ON5a**, **ON1b•ON5a** and **ON1b-Pd•ON5a**, were 40.4 ± 0.7 °C, 35.8 ± 0.6 °C and 41.0 ± 0.6 °C, respectively. In other words, the cyclopalladated benzylamine residue was modestly stabilizing relative to an adenine residue and strongly stabilizing relative to the unmetalated benzylamine residue. To explore the origin of this stabilization, the UV melting experiments were repeated in the presence of 2-mercaptoethanol (100 μM). 2-Mercaptoethanol is a strong ligand for soft transition metal ions and would, hence, be expected to disrupt coordination of Pd(II) to nucleobases. If such coordination is important for duplex stability, a decrease in T_m on addition of 2-mercaptoethanol should be observed.

Melting temperatures of the longest matching duplexes **ON1x•ON5a** in the absence and presence of 2-mercaptoethanol are presented in Figure 3 (all melting temperatures are presented in the Supporting Information). As expected, melting temperatures of duplexes **ON1a•ON5a** and **ON1b•ON5a** did not change appreciably on addition of 2-mercaptoethanol. With **ON1b-Pd•ON5a**, on the other hand, a clear drop in T_m was observed, consistent with stabilizing coordination of Pd(II) in the absence of competing ligands.

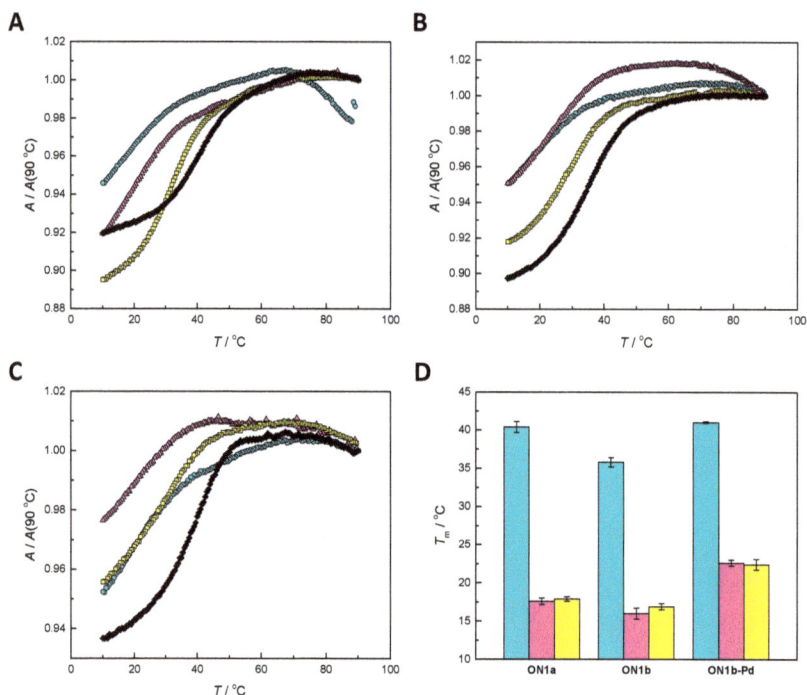

Figure 2. UV melting profiles for duplexes formed by **ON5a** with (**A**) **ON1a** (cyan circles), **ON2a** (magenta triangles), **ON3a** (yellow squares) and **ON4a** (black diamonds); (**B**) **ON1b** (cyan circles), **ON2b** (magenta triangles), **ON3b** (yellow squares) and **ON4b** (black diamonds) and (**C**) **ON1b-Pd** (cyan circles), **ON2b-Pd** (magenta triangles), **ON3b-Pd** (yellow squares) and **ON4b-Pd** (black diamonds); (**D**) melting temperatures of duplexes formed by **ON1a**, **ON1b** and **ON1b-Pd** with **ON5a** (cyan), **ON5g** (magenta) and **ON5t** (yellow); pH = 7.4 (20 mM cacodylate buffer); [oligonucleotides] = 3.0 μM; I(NaClO$_4$) = 0.10 M. The error bars represent standard deviations of three independent measurements.

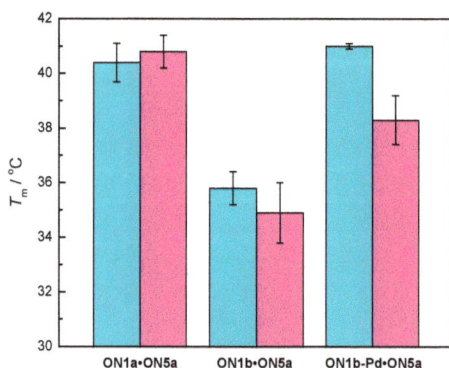

Figure 3. Melting temperatures of duplexes **ON1a•ON5a**, **ON1b•ON5a** and **ON1b-Pd•ON5a** in the absence (cyan) and presence (magenta) of 2-mercaptoethanol; pH = 7.4 (20 mM cacodylate buffer); [oligonucleotides] = 3.0 μM; [2-mercaptoethanol] = 0/100 μM; I(NaClO$_4$) = 0.10 M. The error bars represent standard deviations of three independent measurements.

To further elucidate the role of the palladacyclic "warhead", a detailed thermodynamic analysis of the hybridization of **ON1a•ON5a**, **ON1b•ON5a** and **ON1b-Pd•ON5a** was carried out as described in the literature [33]. The resultant enthalpies and entropies of hybridization are presented in Table 2. With **ON1b-Pd•ON5a**, both values were significantly more negative than with **ON1a•ON5a** or **ON1b•ON5a**.

Table 2. Thermodynamic parameters of hybridization for **ON1a•ON5a**, **ON1b•ON5a** and **ON1b-Pd•ON5a**; pH = 7.4 (20 mM cacodylate buffer); [oligonucleotides] = 3.0 µM; I(NaClO$_4$) = 0.10 M.

Duplex	$\Delta H°/\text{kJ·mol}^{-1}$	$\Delta S°/\text{J·mol}^{-1}\text{·K}^{-1}$
ON1a•ON5a	−200 ± 20	−530 ± 50
ON1b•ON5a	−200 ± 10	−550 ± 40
ON1b-Pd•ON5a	−250 ± 20	−670 ± 50

Secondary structure of duplexes formed by the cyclopalladated oligonucleotides was studied CD spectropolarimetrically over a temperature range of 10–90 °C at 10 °C intervals. All the other conditions were identical to those of the UV melting experiments. With the longest matched duplexes, the spectra obtained at 10 °C were clearly characteristic of a B-type double helix, with prominent negative and positive signals at 260 and 280 nm, respectively (Figure 4 for **ON1b-Pd•ON5a**, all spectra are presented in the Supplementary Materials). Similar, but weaker, signals were observed with the shorter and/or mismatched duplexes, reflecting their lower melting temperatures. In all cases, the signals diminished on increasing temperature, consistent with unwinding of the double helix.

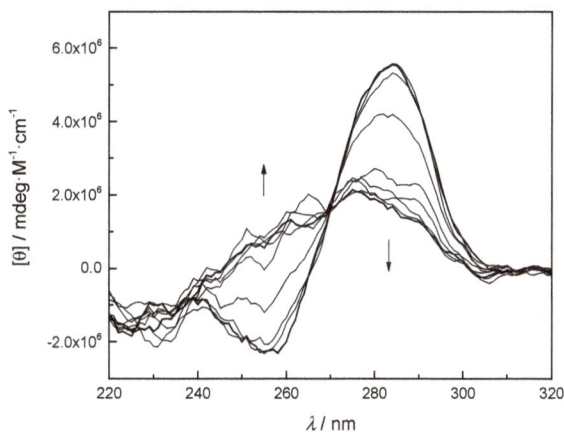

Figure 4. CD (circular dichroism) spectra of **ON1b-Pd•ON5a**, recorded at 10 °C intervals between 10 and 90 °C; pH = 7.4 (20 mM cacodylate buffer); [oligonucleotides] = 3.0 µM; I(NaClO$_4$) = 0.10 M. Spectra acquired at extreme temperatures are indicated by thicker lines and thermal shifts of the minima and maxima by arrows.

3. Discussion

3.1. Duplex Stabilization by the Palladacyclic "Warhead"

The sensitivity of the duplex stabilization by the cyclopalladated benzylamine moiety to the presence of 2-mercaptoethanol suggests direct coordination of Pd(II) to a base moiety of the complementary strand. The thymine base directly opposite to the palladacyclic residue appears as the most likely candidate but, given the length and flexibility of the diethylene glycol linker, coordination to other bases of the 3'-overhang cannot be ruled out. The most likely donor atoms within these bases are the N3 of cytosine and thymine and the N1 of guanine [34–36].

Duplex stabilization by the cyclopalladated benzylamine moiety is rather modest when compared to stabilizations achieved previously with metal mediated base pairing [37–46]. One should, however, bear in mind that within a double helix the most stable metal mediated base pairs are formed between two nucleosides or nucleoside analogues, preorganized to place the donor atoms at appropriate positions. In the present case the flexible diethylene glycol linker first has to adopt a conformation conducive to Pd(II) coordination, consistent with the observed highly negative entropy of hybridization. As a result, the stabilizing enthalpic contribution by Pd(II) coordination is almost entirely offset by the entropic penalty.

3.2. Impact of the Palladacyclic "Warhead" on Sequence Selectivity

Highly stabilizing modifications are known to be able to override sequence information and allow hybridization of even highly mismatched oligonucleotides [47]. Even in the present case, the mismatched duplexes were actually stabilized more than the matched ones by the cyclopalladated benzylamine moiety. However, the difference in T_m between the matched and the most stable mismatched duplex (**ON1b-Pd•ON5a** and **ON1b-Pd•ON5g**, respectively) was still nearly 18 °C, translating into a 10^4-fold preference of **ON5a** over **ON5g** in hybridization with **ON1b-Pd**.

4. Materials and Methods

4.1. General Methods

NMR spectra were recorded on Bruker 500 NMR spectrometers (Bruker, Billerica, MA, USA) and chemical shifts (δ, ppm) are quoted relative to the residual solvent peak as an internal standard. Mass spectra were recorded on a Bruker Daltonics microTOF-Q mass spectrometer (Bruker, Billerica, MA, USA). The solvents for organic synthesis were of reagent grade and dried over 4 Å molecular sieves. For preparation of HPLC elution buffers, freshly distilled triethylamine was used. The other chemicals, including unmodified oligonucleotides, were commercial products that were used as received.

4.2. N-benzyl-2,2,2-trifluoro-N-[2-(2-hydroxyethoxy)ethyl]acetamide (3)

2-[2-(benzylamino)ethoxy]ethanol (**2**, 2.00 g, 10.2 mmol) was dissolved in MeOH (10 mL). Et₃N (2.90 mL, 20.4 mmol) was added and the resulting mixture stirred for 30 min at 25 °C. Ethyl trifluoroacetate (1.46 g, 10.2 mmol) was then added and the reaction mixture stirred for 16 h at 25 °C, after which it was evaporated to dryness to afford the desired product **3** as a mixture of two slowly interconverting rotamers (2.98 g, near quantitative yield). ¹H NMR (CDCl₃, 500 MHz, major rotamer): δ 7.19-7.48 (m, 5H), 4.80 (s, 2H), 3.40-3.70 (m, 8H), 2.82 (br, 1H). ¹H NMR (CDCl₃, 500 MHz, minor rotamer): δ 7.19-7.48 (m, 5H), 4.82 (s, 2H), 3.40-3.70 (m, 8H), 2.82 (br, 1H). ¹³C NMR (CDCl₃, 125 MHz, major rotamer): δ 157.1 (q, J = 35.8 Hz), 136.3, 128.7, 127.7, 127.6, 116.8 (q, J = 285.9 Hz), 72.5, 67.6, 60.9, 50.3, 46.8 (q, J = 2.9 Hz). (CDCl₃, 125 MHz, minor rotamer): δ 156.9 (q, J = 35.8 Hz), 135.8, 128.8, 127.9, 127.2, 116.8 (q, J = 285.9 Hz), 72.4, 68.9, 61.0, 51.4 (q, J = 3.0 Hz), 46.0. HRMS (ESI⁺) m/z calcd 314.0974 obsd 314.0972 [M + Na]⁺.

4.3. 2-[2-(N-benzyl-2,2,2-trifluoroacetamido)ethoxy]ethyl 2-cyanoethyl N,N-diisopropylphosphoramidite (1)

N-benzyl-2,2,2-trifluoro-N-(2-(2-hydroxyethoxy)ethyl)acetamide (**3**, 670 mg, 2.30 mmol) and Et₃N (1.93 mL, 13.8 mmol) were dissolved in anhydrous CH₂Cl₂ (7 mL) under N₂ atmosphere. 2-Cyanoethyl N,N-diisopropylchlorophosphoramidite (0.616 mL, 2.76 mmol) was added and the resulting mixture stirred for 3 h at 25 °C, after which the reaction was quenched by addition of saturated aq. NaHCO₃ (100 mL). The phases were separated and the aqueous phase was extracted with CH₂Cl₂ (100 mL). The combined organic phases were washed with saturated aq. NaHCO₃

(100 mL), dried over anhydrous Na$_2$SO$_4$ and evaporated to dryness. The crude product thus obtained was passed through a silica gel column eluting with a mixture of EtOAc and hexane (1:1, v/v) to afford the desired product **1** containing a major impurity of 2-Cyanoethyl *N,N*-diisopropylphosphonamidate (859 mg, 72% yield based on ^{31}P NMR). This material was used in oligonucleotide synthesis without further purification. ^{1}H NMR (CDCl$_3$, 500 MHz, major rotamer): δ 7.22-7.48 (m, 5H), 4.80 (s, 2H), 3.36-3.89 (m, 12H), 2.66 (m, 2H), 1.20 (d, *J* = 6.8 Hz, 6H), 1.19 (d, *J* = 6.8 Hz, 6H). ^{1}H NMR (CDCl$_3$, 500 MHz, minor rotamer): δ 7.22-7.48 (m, 5H), 4.82 (s, 2H), 3.36-3.89 (m, 12H), 2.66 (m, 2H), 1.20 (d, *J* = 6.8 Hz, 6H), 1.19 (d, *J* = 6.8 Hz, 6H). ^{13}C NMR (CDCl$_3$, 125 MHz, major rotamer): δ 157.1 (q, *J* = 36.0 Hz), 136.3, 128.7, 127.7, 127.6, 118.6, 116.8 (q, *J* = 288.2 Hz), 71.1 (d, *J* = 7.6 Hz), 67.8, 62.6 (d, *J* = 16.8 Hz), 58.4 (d, *J* = 19.2 Hz), 50.3, 46.7 (q, *J* = 2.9 Hz), 42.9 (d, *J* = 12.6 Hz), 24.0 (d, *J* = 7.2 Hz), 20.0. ^{13}C NMR (CDCl$_3$, 125 MHz, minor rotamer): δ 156.7 (q, *J* = 35.4 Hz), 135.7, 129.0, 127.9, 127.2, 118.0, 116.8 (q, *J* = 288.7 Hz), 71.0 (d, *J* = 7.7 Hz), 69.1, 62.7 (d, *J* = 17.1 Hz), 58.4 (d, *J* = 19.2 Hz), 51.5 (q, *J* = 3.2 Hz), 45.9, 42.9 (d, *J* = 12.6 Hz), 24.0 (d, *J* = 7.2 Hz), 20.0 (d, *J* = 6.9 Hz). ^{31}P NMR (CDCl$_3$, 202 MHz, major rotamer): δ 148.1. ^{31}P NMR (CDCl$_3$, 202 MHz, minor rotamer): δ 148.0. HRMS (ESI$^+$) *m/z* calcd 514.2053 obsd 514.2034 [M + Na]$^+$.

*4.4. Bis{N-[2-(2-hydroxyethoxy)ethyl]benzylaminato-C^2,N} bis(μ-chloro) dipalladium(II) (**4**)*

2-[2-(benzylamino)ethoxy]ethanol (**2**, 240 mg, 1.28 mmol) was dissolved in a mixture of H$_2$O (3 mL) and MeCN (3 mL). Li$_2$PdCl$_4$ (336 mg, 1.28 mmol) was dissolved in a mixture of H$_2$O (10 mL) and MeCN (10 mL) and the resulting solution was added to the solution of **2**. After stirring for 16 h at 25 °C, the reaction mixture was evaporated to dryness. The residue was purified by silica gel column chromatography eluting with a mixture of EtOAc and hexane (8:2, v/v), affording the desired product **4** as a mixture of *cisoid* and *transoid* stereoisomers (189 mg, 44% yield). ^{1}H NMR (CDCl$_3$, 500 MHz, major stereoisomer): δ 7.34-7.47 (m, 4H), 4.46 (ddd, *J*$_1$ = 10.6 Hz, *J*$_2$ = 10.4 Hz, *J*$_3$ = 2.5 Hz, 1H), 4.28 (dd, *J*$_1$ = 13.1 Hz, *J*$_2$ = 6.7 Hz, 1H), 4.03 (m, 1H), 3.44-3.80 (m, 6 H), 3.02 (m, 1H), 2.38 (m, 1H). ^{1}H NMR (CDCl$_3$, 500 MHz, minor stereoisomer): δ 7.34-7.47 (m, 4H), 4.33 (m, 1H), 4.24 (dd, *J*$_1$ = 13.0 Hz, *J*$_2$ = 7.6 Hz, 1H), 4.03 (m, 1H), 3.44-3.80 (m, 6 H), 3.02 (m, 1H), 2.38 (m, 1H). ^{13}C NMR (CDCl$_3$, 125 MHz, major stereoisomer): δ 147.5, 135.0, 129.9, 129.7, 128.9, 128.6, 72.8, 67.8, 61.6, 57.35, 50.8. ^{13}C NMR (CDCl$_3$, 125 MHz, minor stereoisomer): δ 142.7, 135.2, 129.9, 129.7, 128.9, 128.6, 72.8, 67.6, 61.4, 57.26, 51.5. HRMS (ESI$^+$) *m/z* calcd 300.0216 obsd 300.0152 [M/2 − Cl]$^+$.

4.5. Oligonucleotide Synthesis

The modified oligonucleotides **ON1b**, **ON2b**, **ON3b** and **ON4b** were assembled on an Applied Biosystems 3400 (Applied Biosystems, Waltham, MA, USA) automated DNA/RNA synthesizer using conventional phosphoramidite strategy. For the benzylamine building block **1**, an extended coupling time (600 s) was used. Removal of the base and phosphate protections and release of the oligonucleotides from the solid support was accomplished by treatment with 25% aq. NH$_3$ for 16 h at 55 °C. The cyclopalladated oligonucleotides **ON1b-Pd**, **ON2b-Pd**, **ON3b-Pd** and **ON4b-Pd** were prepared by incubating **ON1b**, **ON2b**, **ON3b** and **ON4b** (192, 260, 290 and 173 nmol, respectively) and Li$_2$PdCl$_4$ (384, 520, 580 and 346 nmol, respectively) in a mixture of H$_2$O (530 μL) and MeCN (30 μL) for 16 h at 25 °C. All modified oligonucleotides were purified by reversed-phase high performance liquid chromatography (RP-HPLC) on a Hypersil ODS C18 column (250 × 4.6 mm, 5 μm, Thermo Fisher Scientific, Waltham, MA, USA) eluting with a linear gradient (0 to 30% over 25 min) of MeCN in 50 mM aqueous triethylammonium acetate. The flow rate was 1.0 mL·min^{-1} and the detection wavelength 260 nm. The purified oligonucleotides were characterized by electrospray ionization mass spectrometry (ESI-MS) and quantified UV spectrophotometrically using molar absorptivities calculated by an implementation of the nearest-neighbors method. Molar absorptivity of both free and cyclopalladated benzylamine was assumed to be negligible.

4.6. Melting Temperature Measurements

Melting profiles were recorded on a PerkinElmer Lambda 35 UV-Vis spectrometer equipped with a Peltier temperature control unit (PerkinElmer, Waltham, MA, USA). Samples were prepared by mixing the appropriate oligonucleotides (3.0 µM) in 20 mM cacodylate buffer (pH 7.4), the ionic strength of which was adjusted to 0.10 M with $NaClO_4$. When applicable, 2-mercaptoethanol was used in 100 µM concentration and added after mixing of the oligonucleotides. Before measurement, the samples were annealed by heating to 90 °C and gradually cooling to room temperature. UV melting curves were acquired by monitoring the absorbance at $\lambda = 260$ nm over a temperature range of 10–90 °C, sampling at 10 °C intervals. The melting temperatures were determined as inflection points on the UV melting curves.

4.7. CD Measurements

CD spectra were recorded on an Applied Photophysics Chirascan spectropolarimeter equipped with a Peltier temperature control unit (Applied Photophysics, Leatherhead, UK). Samples used in the CD measurements were identical to those used in the UV melting temperature measurements. CD spectra were acquired between $\lambda = 200$ and 400 nm over a temperature range of 10–90 °C, sampling at 10 °C intervals. At each temperature, samples were allowed to equilibrate for 600 s before acquisition.

5. Conclusions

Cyclopalladated benzylamine, tethered at a terminal position of a short oligonucleotide by a flexible linker, promotes hybridization despite a significant entropy penalty for "freezing" the linker in an appropriate conformation. Sensitivity to the presence of competing ligands suggests direct coordination of Pd(II) to a nucleobase of the complementary strand as the origin of stabilization by the palladacyclic moiety. In light of these results, oligonucleotides furnished with palladacyclic "warheads" could prove useful as a sequence-selective alternative to current platinum-based anticancer agents.

Supplementary Materials: Supplementary materials can be found at http://www.mdpi.com/1422-0067/19/6/1588/s1.

Author Contributions: Conceptualization, T.L.; Methodology, T.L. and S.M.; Investigation, M.H. and S.M.; Resources, T.L.; Data Curation, T.L.; Writing-Original Draft Preparation, M.H. and T.L.; Writing-Review and Editing, T.L.; Visualization, T.L.; Supervision, T.L. and S.M.; Project Administration, T.L.; Funding Acquisition, T.L.

Funding: This project has received funding from the European Union's Horizon 2020 research and innovation programme under the Marie Skłodowska-Curie grant agreement No. 721613 and from the Academy of Finland (decisions No. 286478 and No. 294008).

Conflicts of Interest: The authors declare no conflict of interest.

Abbreviations

CD	circular dichroism
DNA	deoxyribonucleic acid
PNA	peptide nucleic acid
RP-HPLC	reversed-phase high performance liquid chromatography
ESI-MS	electrospray ionization mass spectrometry

References

1. Dasari, S.; Tchounwou, P.B. Cisplatin in cancer therapy: Molecular mechanisms of action. *Eur. J. Pharmacol.* **2014**, *740*, 364–378. [CrossRef] [PubMed]
2. Rosenberg, B.; Van Camp, L.; Krigas, T. Inhibition of cell division in *Escherichia coli* by electrolysis products from a platinum electrode. *Nature* **1965**, *205*, 698–699. [CrossRef] [PubMed]

3. Sanchez-Cano, C.; Hannon, M.J. Novel and emerging approaches for the delivery of metallo-drugs. *Dalton Trans.* **2009**, 10702–10711. [CrossRef] [PubMed]

4. Ott, I.; Gust, R. Non platinum metal complexes as anti-cancer drugs. *Arch. Pharm.* **2007**, *340*, 117–126. [CrossRef] [PubMed]

5. Bruijnincx, P.C.A.; Sadler, P.J. New trends for metal complexes with anticancer activity. *Curr. Opin. Chem. Biol.* **2008**, *12*, 197–206. [CrossRef] [PubMed]

6. Ott, I. On the medicinal chemistry of gold complexes as anticancer drugs. *Coord. Chem. Rev.* **2009**, *253*, 1670–1681. [CrossRef]

7. Levina, A.; Mitra, A.; Lay, P.A. Recent developments in ruthenium anticancer drugs. *Metallomics* **2009**, *1*, 458–470. [CrossRef] [PubMed]

8. Berners-Price, S.J.; Filipovska, A. Gold compounds as therapeutic agents for human diseases. *Metallomics* **2011**, *3*, 863–873. [CrossRef] [PubMed]

9. Cutillas, N.; Yellol, G.S.; de Haro, C.; Vicente, C.; Rodríguez, V.; Ruiz, J. Anticancer cyclometalated complexes of platinum group metals and gold. *Coord. Chem. Rev.* **2013**, *257*, 2784–2797. [CrossRef]

10. Johnstone, T.C.; Suntharalingam, K.; Lippard, S.J. The next generation of platinum drugs: Targeted Pt(II) agents, nanoparticle delivery, and Pt(IV) prodrugs. *Chem. Rev.* **2016**, *116*, 3436–3486. [CrossRef] [PubMed]

11. Bindoli, A.; Rigobello, M.P.; Scutari, G.; Gabbiani, C.; Casini, A.; Messori, L. Thioredoxin reductase: A target for gold compounds acting as potential anticancer drugs. *Coord. Chem. Rev.* **2009**, *253*, 1692–1707. [CrossRef]

12. Kelland, L. The resurgence of platinum-based cancer chemotherapy. *Nat. Rev. Cancer* **2007**, *7*, 573–584. [CrossRef] [PubMed]

13. Hartmann, J.T.; Lipp, H.-P. Toxicity of platinum compounds. *Expert Opin. Pharmacother.* **2003**, *4*, 889–901. [CrossRef] [PubMed]

14. Daugaard, G.; Abildgaard, U. Cisplatin nephrotoxicity. A review. *Cancer Chemother. Pharmacol.* **1989**, *25*, 1–9. [CrossRef] [PubMed]

15. Johnstone, T.C.; Suntharalingam, K.; Lippard, S.J. Third row transition metals for the treatment of cancer. *Philos. Trans. A Math. Phys. Eng. Sci.* **2015**, *373*, 20140185. [CrossRef] [PubMed]

16. Carreira, M.; Calvo-Sanjuán, R.; Sanaú, M.; Marzo, I.; Contel, M. Organometallic palladium complexes with a water-soluble iminophosphorane ligand as potential anticancer agents. *Organometallics* **2012**, *31*, 5772–5781. [CrossRef] [PubMed]

17. Caires, A.C.F. Recent advances involving palladium (II) complexes for the cancer therapy. *Anticancer Agents Med. Chem.* **2007**, *7*, 484–491. [CrossRef] [PubMed]

18. Taube, H. Rates and mechanisms of substitution in inorganic complexes in solution. *Chem. Rev.* **1952**, *50*, 69–126. [CrossRef]

19. Lippert, B.; Miguel, P.J.S. Comparing Pt^{II}- and Pd^{II}-nucleobase coordination chemistry: Why Pd^{II} not always is a good substitute for Pt^{II}. *Inorg. Chim. Acta* **2018**, *472*, 207–213. [CrossRef]

20. Serrano, F.A.; Matsuo, A.L.; Monteforte, P.T.; Bechara, A.; Smaili, S.S.; Santana, D.P.; Rodrigues, T.; Pereira, F.V.; Silva, L.S.; Machado, J.; et al. A cyclopalladated complex interacts with mitochondrial membrane thiol-groups and induces the apoptotic intrinsic pathway in murine and cisplatin-resistant human tumor cells. *BMC Cancer* **2011**, *11*, 296. [CrossRef] [PubMed]

21. Marchán, V.; Grandas, A. Platinated oligonucleotides: Synthesis and applications for the control of gene expression. In *Metal Complex–DNA Interactions*; John Wiley & Sons, Ltd.: Hoboken, NJ, USA, 2009; pp. 273–300. [CrossRef]

22. Colombier, C.; Lippert, B.; Leng, M. Interstrand cross-linking reaction in triplexes containing a monofunctional transplatin-adduct. *Nucleic Acids Res.* **1996**, *24*, 4519–4524. [CrossRef] [PubMed]

23. Brabec, V.; Reedijk, J.; Leng, M. Sequence-dependent distortions induced in DNA by monofunctional platinum(II) binding. *Biochemistry* **1992**, *31*, 12397–12402. [CrossRef] [PubMed]

24. Algueró, B.; Pedroso, E.; Marchán, V.; Grandas, A. Incorporation of two modified nucleosides allows selective platination of an oligonucleotide making it suitable for duplex cross-linking. *J. Biol. Inorg. Chem.* **2007**, *12*, 901–911. [CrossRef] [PubMed]

25. Algueró, B.; López de la Osa, J.; González, C.; Pedroso, E.; Marchán, V.; Grandas, A. Selective platination of modified oligonucleotides and duplex cross-links. *Angew. Chem. Int. Ed.* **2006**, *45*, 8194–8197. [CrossRef] [PubMed]

26. Schmidt, K.S.; Boudvillain, M.; Schwartz, A.; van der Marel, G.A.; van Boom, J.H.; Reedijk, J.; Lippert, B. Monofunctionally trans-diammine platinum(II)-modified peptide nucleic acid oligomers: A new generation of potential antisense drugs. *Chem. Eur. J.* **2002**, *8*, 5566–5570. [CrossRef]

27. Maity Sajal, K.; Lönnberg, T. Oligonucleotides incorporating palladacyclic nucleobase surrogates. *Chem. Eur. J.* **2018**, *24*, 1274–1277. [CrossRef] [PubMed]

28. Ruiz, J.; Cutillas, N.; Vicente, C.; Villa, M.D.; López, G.; Lorenzo, J.; Avilés, F.X.; Moreno, V.; Bautista, D. New palladium(II) and platinum(II) complexes with the model nucleobase 1-methylcytosine: Antitumor activity and interactions with DNA. *Inorg. Chem.* **2005**, *44*, 7365–7376. [CrossRef] [PubMed]

29. Ruiz, J.; Lorenzo, J.; Vicente, C.; López, G.; López-de-Luzuriaga, J.M.; Monge, M.; Avilés, F.X.; Bautista, D.; Moreno, V.; Laguna, A. New palladium(II) and platinum(II) complexes with 9-aminoacridine: Structures, luminescence, theoretical calculations, and antitumor activity. *Inorg. Chem.* **2008**, *47*, 6990–7001. [CrossRef] [PubMed]

30. Fuchita, Y.; Tsuchiya, H. Synthesis of ortho-palladated complexes of n-methylbenzylamine: The first example of ortho-palladation of α-unsubstituted secondary benzylamine. *Inorg. Chim. Acta* **1993**, *209*, 229–230. [CrossRef]

31. Fuchita, Y.; Tsuchiya, H.; Miyafuji, A. Cyclopalladation of secondary and primary benzylamines. *Inorg. Chim. Acta* **1995**, *233*, 91–96. [CrossRef]

32. Selvakumar, K.; Vancheesan, S. Synthesis and characterization of cyclopalladated complexes of secondary benzylamines. *Polyhedron* **1997**, *16*, 2405–2411. [CrossRef]

33. Mergny, J.-L.; Lacroix, L. Analysis of thermal melting curves. *Oligonucleotides* **2003**, *13*, 515–537. [CrossRef] [PubMed]

34. Golubev, O.; Lönnberg, T.; Lönnberg, H. Interaction of Pd^{2+} complexes of 2,6-disubstituted pyridines with nucleoside 5'-monophosphates. *J. Inorg. Biochem.* **2014**, *139*, 21–29. [CrossRef] [PubMed]

35. Martin, R.B. Nucleoside sites for transition metal ion binding. *Acc. Chem. Res.* **1985**, *18*, 32–38. [CrossRef]

36. Pneumatikakis, G.; Hadjiliadis, N.; Theophanides, T. Complexes of inosine, cytidine, and guanosine with palladium(II). *Inorg. Chem.* **1978**, *17*, 915–922. [CrossRef]

37. Scharf, P.; Müller, J. Nucleic acids with metal-mediated base pairs and their applications. *ChemPlusChem* **2013**, *78*, 20–34. [CrossRef]

38. Takezawa, Y.; Shionoya, M. Metal-mediated DNA base pairing: Alternatives to hydrogen-bonded Watson–Crick base pairs. *Acc. Chem. Res.* **2012**, *45*, 2066–2076. [CrossRef] [PubMed]

39. Clever, G.H.; Kaul, C.; Carell, T. DNA–metal base pairs. *Angew. Chem. Int. Ed.* **2007**, *46*, 6226–6236. [CrossRef] [PubMed]

40. Clever, G.H.; Shionoya, M. Metal–base pairing in DNA. *Coord. Chem. Rev.* **2010**, *254*, 2391–2402. [CrossRef]

41. Müller, J. Chemistry: Metals line up for DNA. *Nature* **2006**, *444*, 698. [CrossRef] [PubMed]

42. Müller, J. Metal-ion-mediated base pairs in nucleic acids. *Eur. J. Inorg. Chem.* **2008**, *2008*, 3749–3763. [CrossRef]

43. Mandal, S.; Müller, J. Metal-mediated DNA assembly with ligand-based nucleosides. *Curr. Opin. Chem. Biol.* **2017**, *37*, 71–79. [CrossRef] [PubMed]

44. Takezawa, Y.; Müller, J.; Shionoya, M. Artificial DNA base pairing mediated by diverse metal ions. *Chem. Lett.* **2017**, *46*, 622–633. [CrossRef]

45. Lippert, B.; Sanz Miguel, P.J. The renaissance of metal–pyrimidine nucleobase coordination chemistry. *Acc. Chem. Res.* **2016**, *49*, 1537–1545. [CrossRef] [PubMed]

46. Taherpour, S.; Golubev, O.; Lönnberg, T. On the feasibility of recognition of nucleic acid sequences by metal-ion-carrying oligonucleotides. *Inorg. Chim. Acta* **2016**, *452*, 43–49. [CrossRef]

47. Clever, G.H.; Söltl, Y.; Burks, H.; Spahl, W.; Carell, T. Metal–salen-base-pair complexes inside DNA: Complexation overrides sequence information. *Chem. Eur. J.* **2006**, *12*, 8708–8718. [CrossRef] [PubMed]

International Journal of
Molecular Sciences

MDPI

Article

Design, Synthesis, and Biological Evaluation of Benzimidazole-Derived Biocompatible Copper(II) and Zinc(II) Complexes as Anticancer Chemotherapeutics

Mohamed F. AlAjmi [1], Afzal Hussain [1,*], Md. Tabish Rehman [1], Azmat Ali Khan [2], Perwez Alam Shaikh [1] and Rais Ahmad Khan [3,*]

[1] Department of Pharmacognosy, College of Pharmacy, King Saud University, P.O. Box 2457, Riyadh 11451, KSA; malajmii@ksu.edu.sa (M.F.A.); m.tabish.rehman@gmail.com (M.T.R.); aperwez@ksu.edu.sa (P.A.S.)

[2] Department of Pharmaceutical Chemistry, College of Pharmacy, King Saud University, P.O. Box 2457, Riyadh 11451, KSA; azmatbiotech@gmail.com

[3] Department of Chemistry, College of Science, King Saud University, P.O. Box 2455, Riyadh 11451, KSA

* Correspondence: afzal.hussain.amu@gmail.com (A.H.); raischem@gmail.com (R.A.K.); Tel.: +966-504767847 (A.H.); +966-536745404 (R.A.K.)

Received: 22 April 2018; Accepted: 12 May 2018; Published: 16 May 2018

Abstract: Herein, we have synthesized and characterized a new benzimidazole-derived "BnI" ligand and its copper(II) complex, [Cu(BnI)₂], **1**, and zinc(II) complex, [Zn(BnI)₂], **2**, using elemental analysis and various spectroscopic techniques. Interaction of complexes **1** and **2** with the biomolecules viz. HSA (human serum albumin) and DNA were studied using absorption titration, fluorescence techniques, and in silico molecular docking studies. The results exhibited the significant binding propensity of both complexes **1** and **2**, but complex **1** showed more avid binding to HSA and DNA. Also, the nuclease activity of **1** and **2** was analyzed for pBR322 DNA, and the results obtained confirmed the potential of the complexes to cleave DNA. Moreover, the mechanistic pathway was studied in the presence of various radical scavengers, which revealed that ROS (reactive oxygen species) are responsible for the nuclease activity in complex **1**, whereas in complex **2**, the possibility of hydrolytic cleavage also exists. Furthermore, the cytotoxicity of the ligand and complexes **1** and **2** were studied on a panel of five different human cancer cells, namely: HepG2, SK-MEL-1, HT018, HeLa, and MDA-MB 231, and compared with the standard drug, cisplatin. The results are quite promising against MDA-MB 231 (breast cancer cell line of **1**), with an IC_{50} value that is nearly the same as the standard drug. Apoptosis was induced by complex **1** on MDA-MB 231 cells predominantly as studied by flow cytometry (FACS). The adhesion and migration of cancer cells were also examined upon treatment of complexes **1** and **2**. Furthermore, the in vivo chronic toxicity profile of complexes **1** and **2** was also studied on all of the major organs of the mice, and found them to be less toxic. Thus, the results warrant further investigations of complex **1**.

Keywords: metal complexes; interaction with biomolecules; nuclease activity; cytotoxicity; toxicity

1. Introduction

The serendipitous discovery of cisplatin [1] as an anticancer chemotherapeutic agent [2–4] swiftly marked significant breakthroughs in the field of metallodrugs as potential anticancer agents [5], and prompted several researchers and groups to explore the field of the metal-based chemotherapeutic [6]. Several potential metallodrugs were discovered since then and were used in clinics worldwide to treat cancers, starting from cisplatin, oxaliplatin, carboplatin, etc. However, the severe side effects and resistance by the cancer cells confine their clinical application widely [7]. To curb increasing resistance and the unbearable cost of treatment, it is a necessity to design potential alternatives.

Continual efforts have been made to design and develop non-classical metal-based chemotherapeutics. Among them, NAMI-A "(imidazole)RuCl$_4$(DMSO)(imidazole)" and KP1019 "(indazole)RuCl$_4$(DMSO)(indazole)" as ruthenium-centered potential anticancer drugs entered into the clinical trials, but still lead to side effects, and cost remains a hurdle. So, many scientists concentrated their attention on transition metals, particularly copper and zinc, to design novel metal-based anticancer agents. Both the metal ions are bio-essential elements that play a prominent role in the active site of many metalloproteins. Copper redox systems are involved in the catalytic process in the body [8,9]. Copper generates reactive oxygen species (ROS), endogenous DNA damage, and nucleobase affinity, and thus favors proliferative activity [10–12]. For growth and development, zinc ion is needed; it regulates the metabolism of the cells and is also a major regulatory ion in the metabolism of cells, which shrink the cardiotoxicity and hepatotoxicity that is associated with several anticancer drugs [13,14].

Furthermore, copper and zinc also possess affinity towards protein and DNA, although their propensity of binding mostly depends on the organic motifs. Thus, the role of the organic moiety/coordinated ligand is also pivotal. In this context, benzimidazole moieties are unique and interesting as they bear coordination potential for metal centers, planarity, and mimic the imidazole functions in proteins [15,16]. The benzimidazole core is also present inside the body as a component of Vitamin B$_{12}$ [17]. The importance of this core can be understood by its presence in active drugs such as mebendazole ("Vermox" anthelmintic agents) [18]. The presence of biologically active pharmacophore, an imidazole ring, gives it versatility in possessing broad biological activities, from antihistaminic, antitubercular, anti-HIV, NSAIDs, antihypertensive to anticancer, etc. [19–23]. Thus, benzimidazole-derived moieties' introduction to the drug design may enhance the biological activities.

Prompted by the impetus of the above findings, we have synthesized and characterized new copper(II) and zinc(II) complexes of the benzimidazole-derived organic motif. Further, we have studied their interaction with HSA (human serum albumin) and DNA using spectroscopic techniques and in silico molecular docking studies. Moreover, both complexes and the ligand were tested against five human cancer cell lines, and their chronic toxicity profile in vivo was studied.

2. Results and Discussion

2.1. Synthesis and Characterization

The tridentate biologically active pharmacophore, the benzimidazole-derived ligand, was synthesized by stirring the mixture of 2-aminobenzimidazole and 1-methyl-1*H*-imidazole-2-carbaldehyde in toluene as solvent and in the presence of glacial acetic acid as a catalyst at 80 °C for 12 h. The yellow precipitate was isolated via rotavapor and recrystallized in methanol as a yellow-crystalline material suitable for X-ray crystallography. This organic motif "BnI" was characterized by UV, Fourier transform infrared (FT-IR), NMR, and elemental analysis.

The synthesis of complexes [Cu(BnI)$_2$] **1** and [Zn(BnI)$_2$] **2** was straightforward as the reaction was carried out between the ligand BnI and Cu(acetate)$_2$·nH$_2$O (**1**)/Zn(acetate)$_2$·nH$_2$O (**2**) in methanol solution in a molar ratio of 2:1, respectively (Figure 1).

Both the complexes **1** and **2** were isolated as green/yellow precipitate in reasonably good yield (73–86%). Complexes **1** and **2** were characterized by elemental microanalysis and various spectroscopic techniques. After several attempts, we failed to get crystals suitable for single X-ray crystallography. The electrospray ionization mass spectra (ESI-MS) of the complexes **1** and **2** revealed the presence of a molecular ion in significantly good agreement of the calculated and experimental *m*/*z* values.

The FT-IR spectrum of the ligand showed a characteristic strong peak at 1613 cm^{-1}, which is associated with the C=N vibration of the Schiff base (Figure S1). Upon complexation with metal acetates, the FT-IR spectra of the complexes exhibited that both the signals of C=N vibrations got the significant shift to

1642 cm^{-1} for **1** (Figure S2) and 1644 cm^{-1} for **2** (Figure S3), thus confirming the coordination of the metal center with the 'BnI' ligand. The peak associated with M–N exhibited at 428 cm^{-1} and 438 cm^{-1} for complexes **1** and **2**, respectively (Figures S2 and S3).

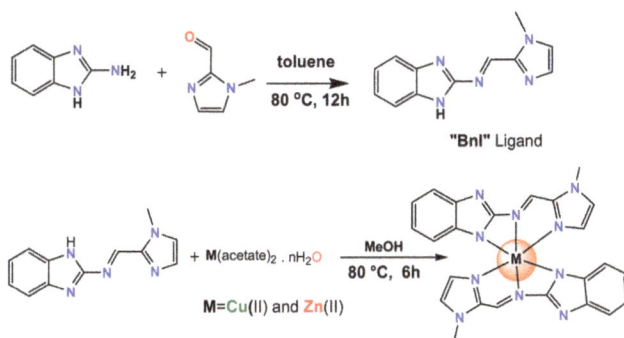

Figure 1. Schematic representation of the synthesis of the benzimidazole-derived ligand and its copper(II) complex [Cu(BnI)$_2$], **1** and zinc(II) complex, [Zn(BnI)$_2$], **2**.

The ^{1}H NMR spectrum of the ligand exhibited a broad signals singlet for N–H of the benzimidazole ring at δ = 12.67 ppm (Figure S4). The characteristic signal for the aldimine proton H<u>C</u>=N was observed at 9.21 ppm, and the aromatic proton peak appeared in the range of 7.57–7.16 ppm for all six protons. Also, the singlet peak that was associated with the N–<u>H</u>$_3$ protons appeared at 4.11 ppm. Upon complexation with Zn(OAc)$_2$ (**2**) (Figure S5), the ^{1}H NMR spectrum shows the absence of NH proton. Furthermore, the characteristic signal of aldimine protons gets shifted to 9.69 ppm and the aromatic protons also get shifted to a range of 7.24–6.84 ppm, which is suggestive of the coordination of the metal center with the nitrogens of the benzimidazole moiety and aldimine group. The protons of the imidazole ring appeared with a very little shift at 7.574 ppm from 7.571 ppm (0.003 ppm), thus confirming that the free movement of the imidazole ring may be because of the steric hindrance provided by the methyl group (N–CH$_3$) over it. The peak associated with N–C<u>H</u>$_3$ appeared at 3.92 ppm.

The ^{13}C NMR spectra of the ligand and complex **2** further ascertain our proposed structure. The peak attributed to the ligand appeared in the range of 155.35 ppm for H<u>C</u>=N and 154.49–110.89 ppm for aromatic carbons, and at 35.42 ppm for the N–CH$_3$ (Figure S6). Upon complexation with Zn(OAc), the organic moiety showed significant shifts in the signals. The characteristic signal associated with the aldimine carbon appears at 155.74 ppm (Figure S7). Other aromatic carbon peaks also get shifted and exhibited in the range of 143.29–120.44 ppm. The N–C<u>H</u>$_3$ signal appeared at 64.90 ppm. Thus, the NMR of the complex **2** (Zn(II) complex) further confirmed the above-proposed structure.

The stability of the complexes **1** and **2** using absorption spectroscopy was studied in a dimethyl sulfoxide (DMSO) and phosphate buffer mixture that was incubated at room temperature and studied for 24 h. Both the complexes were quite stable; however, complex **2** showed signs of hydrolysis after 24 h.

2.2. HSA–Metal Complex Interaction by Spectroscopy

2.2.1. Quenching Experiment

In this study, we followed quenching in human serum albumin (HSA) fluorescence emission in the presence of complexes **1** and **2** at different temperatures (298, 303, and 308 K). The emission intensity of HSA at 338 nm (λ_{max}) was decreased significantly to 70% (Figure 2A and inset) and 63% (Figure 2D and inset) respectively for complexes **1** and **2** when the molar ratio of HSA:complex was 1:10. Stern–Volmer plots (Figure 2B,F) and the results presented in Table 1 shows that the K_{SV} values of the two complexes (**1** and **2**) were of the order of 10^4 M^{-1}, thereby indicating a very strong quenching phenomenon. The linear dependency of emission quenching as a function of the concentrations of complexes **1** and **2** suggested that only one type of quenching mechanism (either static or dynamic) was operational, and there was only one type of equivalent binding site on HSA for complex **1** and **2**. We also calculated the bimolecular quenching constant (k_q) after taking the τ_o of HSA as 5.71×10^{-9} s [24]. The k_q values for **1** and **2** were of the order of 10^{12} M^{-1} s^{-1}, which were at least 100 times more than the maximum collision constant of 10^{10} M^{-1} s^{-1} [25]. These results clearly indicate that the quenching of HSA emission intensity in the presence of complexes **1** and **2** was due to the formation of a complex rather than merely a collision event. Further, the type of quenching process that was involved was determined by evaluating the dependence of emission intensity quenching at different temperatures. The decreased values of K_{SV} and k_q at elevated temperatures showed that a static quenching mechanism was involved in HSA and the systems of complexes **1** and **2** (Table 1). Furthermore, the slope of the modified Stern–Volmer plot indicated that both complexes **1** and **2** had only one type of binding site present on HSA (Figure 2C,F). Also, an intercept of the modified Stern–Volmer plot indicated that the binding constants of the two complexes **1** and **2** were of the magnitude of 10^4 M^{-1}, thus indicating a strong binding of complexes **1** and **2** with HSA (Table 1). Earlier studies also indicated that several ligands bind strongly to HSA with a binding constant ranging between 10^3 M^{-1} to 10^5 M^{-1} [26–29].

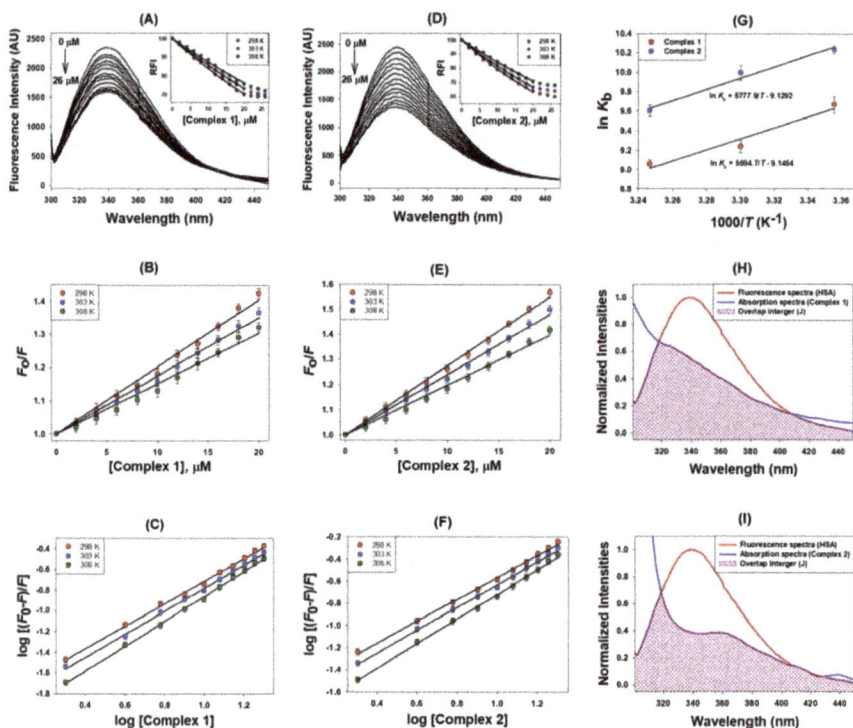

Figure 2. Molecular interaction between human serum albumin (HSA) and complexes **1** and **2**. (**A,D**) represent quenching in the fluorescence of HSA in the presence of complex **1** and **2**, respectively; (**B,E**) represent the Stern–Volmer plots of HSA-complex **1** and HSA-complex **2** respectively; (**C,F**) represent modified Stern–Volmer plots of HSA-complex **1** and HSA-complex **2** respectively; (**G**) represents the van't Hoff plots for determining thermodynamic parameters; (**H,I**) represent FRET between HSA and complex **1** and **2**, respectively.

Table 1. Quenching constants and binding parameters for HSA–metal complex interactions.

Complex	Temp	K_{SV}	k_q	n	K_b
	(K)	**$\times 10^4$ (M^{-1})**	**$\times 10^{12}$ (M^{-1} s^{-1})**		**$\times 10^4$ (M^{-1})**
			Binding Parameters		
1	298	2.02	3.54	1.09	1.58
	303	1.75	3.24	1.12	1.27
	308	1.51	2.64	1.20	0.86
2	298	2.74	4.79	0.99	2.78
	303	2.39	4.19	1.03	2.19
	308	1.97	3.45	1.11	1.48

Complex	Temp	ΔH	ΔS	$T\Delta S$ (kcal/mol)	ΔG
	(K)	**(kcal/mol)**	**(cal/mol/K)**		**(kcal/mol)**
			Thermodynamic Parameters		
1	298			−5.41	−5.71
	303	−11.12	−18.17	−5.50	−5.62
	308			−5.60	−5.52
2	298			−5.40	−6.08
	303	−11.48	−18.14	−5.49	−5.99
	308			−5.59	−5.89

Complex	J	R_o	R
	$\times 10^{-15}$ (M^{-1} cm^{-1})	**(nm)**	**(nm)**
		FRET Parameters	
1	3.26	2.04	2.35
2	4.63	2.16	2.36

The thermodynamic parameters associated with the interaction between complexes **1** and **2** and HSA were estimated by assuming that the change in enthalpy (ΔH) was not significant over the studied temperature range. Figure 3G shows the correlation between the binding constant (K_a) and $1/T$, the slope and intercept of which gives the values of $-\Delta H/R$ and $\Delta S/R$, respectively. We found that the values of ΔH and ΔS for the HSA-**1** interaction were −11.12 kcal/mol and −18.17 cal/mol/K, respectively. Similarly, for the HSA-**2** interaction, the ΔH and ΔS values were −11.48 kcal/mol and −18.14 cal/mol/K, respectively (Table 1). It is inferred from the above values that the enthalpy and entropy changes involved in the binding of complexes **1** and **2** to HSA were of a similar nature and magnitude. The negative values of ΔH and ΔS suggest that hydrogen bonding and van der Waal's interactions were predominantly responsible for the formation of a stable HSA-**1** and HSA-**2** complex. Moreover, the interaction of HSA with complexes **1** and **2** was spontaneous, as suggested by the negative ΔG values (Table 1). At 298 K, the changes in Gibb's free energy for HSA-**1** and HSA-**2** interactions were −5.71 kcal/mol and −6.08 kcal/mol, respectively.

2.2.2. Förster Resonance Energy Transfer (FRET) between HSA and Metal Complexes

FRET occurs when the emission spectrum of the donor overlaps with the absorption spectrum of the acceptor [30]. Figure 2H,I depicts the overlap between the HSA fluorescence spectrum and the absorption spectrum of complexes **1** and **2**, respectively. We found that the efficiency of the energy transfer between HSA and the metal complexes was in the range of 30–37%. The value of r (distance between HSA and the metal complexes) for complexes **1** and **2** was 2.35 nm and 2.36 nm, respectively. Likewise, the values of R_0 (distance at which the efficiency of energy transfer becomes 50%) for complexes **1** and **2** with HSA interaction were 2.04 nm and 2.16 nm, respectively (Table 1). The deduced r values satisfied the condition of FRET that it should be within the 2–8 nm range. Moreover, the r values were within the range of $0.5R_o < r < 1.5R_o$, thereby indicating that the quenching of HSA emissions was due to a complex formation between HSA and metal complexes (i.e., static quenching mechanism).

2.3. Molecular Docking Studies with HSA by Autodock

Prediction of Binding Sites of Metal Complexes on HSA

HSA is the most abundant protein in the plasma, and is responsible for the transportation of drugs. Three independent domains (I–III) are present in the structure of HSA. Domain I spans residues 1–195, domain II extends from residues 196–383, and domain III extends from residues 384–585; these are the basic HSA folds. Each domain has been divided into two subdomains, namely A and B [28]. The two principal binding sites of HSA are located in the hydrophobic cavities of subdomain IIA (Sudlow's site I) and subdomain IIIA (Sudlow's site II). A new binding site on HSA that is located at subdomain IB has been recently identified [31]. The binding site of metal complexes on HSA was predicted by an in silico approach using Autodock 4.2 software (The Scripps Research Institute, La Jolla, CA, USA). We found that both complexes **1** and **2** were bound to HSA at domain IIA near Sudlow's site I mainly through hydrogen bonding and hydrophobic interactions (Table 2, Figure 3). In the HSA-**1** complex, a hydrogen bond (3.71 Å) between the H-atom of metal complex **1** and the O-atom of Asp451 was observed along with six hydrophobic interactions with Lys195, Lys199, Trp214, Arg218, and His242. Also, other key residues involved in stabilizing the HSA-**1** metal complex through van der Waal's interactions were Gln196, Phe211, Arg218, Arg222, Arg257, Ala291, and Glu292. Similarly, complex **2** has been observed to interact with HSA by forming two hydrogen bonds with Ala291 and His242, and forming seven hydrophobic interactions with Lys195, Lys199, Trp214, Leu238, His242, and Ala291. Also, other residues involved in stabilizing the HSA-**1** complex through van der Waal's interactions were Gln196, Ala215, Arg218, Arg222, and Arg257 (Table 2, Figure 3). The binding energy and Gibb's free energy for complex **1** were 7.8×10^6 M^{-1} and -9.4 kcal/mol, while for metal complex **2**, the binding affinity and Gibb's free energy were 4.2×10^7 M^{-1} and -10.4 kcal/mol, respectively. It is clear from the Table 2 and Figure 3 that complexes **1** and **2** were bound near Trp214, which explains the quenching of HSA fluorescence in the presence of these metal complexes. However, it should be noted that the values of ΔG obtained from in vitro binding experiments and computational prediction differ significantly. The ΔG value obtained in the in vitro experiment accurately measured the interaction between HSA and metal complexes in a solvent mimicking physiological conditions (i.e., it is global phenomenon). Conversely, the ΔG value acquired from the in silico docking experiment measured the interaction of metal complexes at the binding site of the protein only (i.e., it is a local phenomenon).

Table 2. List of residues of HSA interaction with different complexes.

Interaction	Nature of Interaction	Bond Length (Å)	Binding Affinity, K_d (M^{-1})	ΔG (kcal/mol)
		Complex 1		
Unk:C—Asp451:O$^{\delta 2}$	Hydrogen Bond	3.71		
Lys199:C$^{\beta}$—Unk	Hydrophobic (π-σ)	3.89		
Trp214—Unk	Hydrophobic (π-π)	4.79		
His242—Unk	Hydrophobic (π-π)	5.25	7.8×10^6	-9.4
Unk—Trp214	Hydrophobic (π-π)	4.91		
Unk—Arg218	Hydrophobic (π-alkyl)	5.23		
Unk—Lys195	Hydrophobic (π-alkyl)	4.92		
		Complex 2		
Unk:C—His242:N$^{\epsilon 2}$	Hydrogen Bond	3.25		
Unk:C—Ala291:O	Hydrogen Bond	3.62		
Lys199:C$^{\beta}$—Unk	Hydrophobic (π-σ)	3.68		
Ala291:C$^{\beta}$—Unk	Hydrophobic (π-σ)	3.59		
His242—Unk	Hydrophobic (π-π)	4.86	4.2×10^7	-10.4
His242—Unk	Hydrophobic (π-π)	4.69		
Unk—Trp214	Hydrophobic (π-π)	5.05		
Unk—Lys195	Hydrophobic (π-alkyl)	5.02		
Unk—Leu238	Hydrophobic (π-alkyl)	5.15		

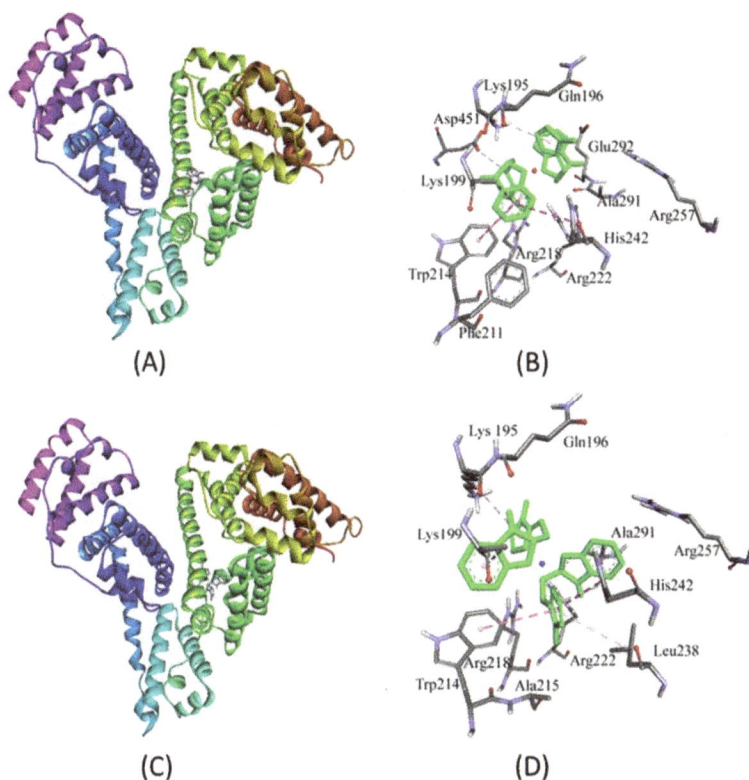

Figure 3. Molecular docking of complexes **1** and **2** with HSA. Panels (**A,C**) show the binding of complexes **1** and **2** near Sudlow's site I in subdomain IIA respectively, and panels (**B,D**) represent the involvement of different amino acid residues of HSA in stabilizing complex with **1** and **2**, respectively. The amino acid residues are color coded according to element properties.

2.4. DNA Binding/Nuclease Activity

DNA is thought to be one of the primary targets of the metallodrugs after cisplatin, which is an unanticipated finding for anticancer chemotherapeutics. Hence, we studied the binding affinity of complexes **1** and **2** with CT-DNA using spectroscopic studies such as absorption titration and an ethidium-bromide displacement assay. In the absorption studies, with the incremental addition of CT-DNA, the spectra exhibited hyperchromism with significant 6–9 nm blue-shifts, which is attributed to the binding of complexes **1** and **2** with CT-DNA via electrostatic mode (external groove) or partial intercalation (Figure 4A,B). Furthermore, the binding strengths were evaluated in the form of an intrinsic binding constant 'K_b' using Wolfe–Shimer equation [32]. The K_b values were found to be 2.67×10^4 M^{-1} and 1.07×10^4 M^{-1} for complexes **1** and **2**, respectively.

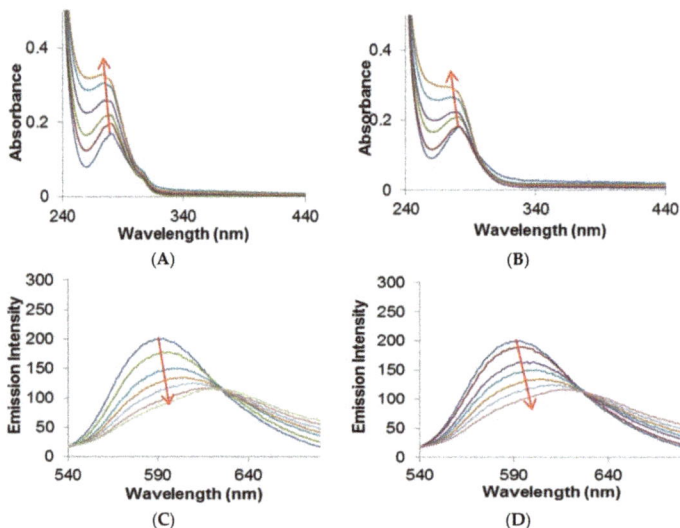

Figure 4. Absorption spectra of (**A**) complex **1**; and (**B**) complex **2** with CT-DNA. Fluorescence quenching spectra of (**C**) complex **1**; and (**D**) complex **2** with EthBr-DNA adduct. Both experiments were carried out in 5 mM Tris-HCl/50mM NaCl buffer, pH = 7.5, at room temperature. The spectra in different colors represent the effect of increasing concentrations (symbolized by an arrow) of complexes **1** and **2** on the studied spectroscopic properties.

A further mode of interaction of complexes **1** and **2** with CT-DNA was studied via ethidium-bromide (EtBr) displacement assay, which involved competitive binding between a standard intercalator and complexes **1** and **2** against CT-DNA (Figure 4C,D). In this experiment, the pre-treated CT-DNA with EtBr was used in the standard ratio (EtBr is weakly emissive, whereas the DNA+EtBr adduct exhibits high emission intensity). The DNA-EtBr adduct was treated with an increasing concentration of complexes **1** and **2**; the results give rise to a significant reduction in the fluorescence intensity but don't diminish the emission intensity 100%. These results ascertain that complexes **1** and **2** are not typical intercalators, but the possibilities of partial intercalation cannot be ruled out. The extent of quenching was evaluated by using the Stern–Volmer equation, and the K_{SV} values were calculated and were found to be 2.03 M^{-1} and 1.21 M^{-1} for complexes **1** and **2**, respectively.

We have further evaluated the nuclease activity of complexes **1** and **2** against pBR322 DNA. The concentration-dependent cleavage pattern was obtained with an increasing concentration of complexes **1** and **2** (0.5–2.5 µM) with pBR322 DNA (100 µM) (Figure 5).

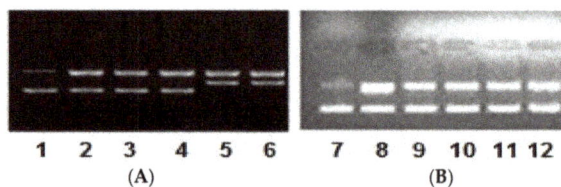

Figure 5. Electrophoretic pattern of pBR322 DNA (100 ng) by (**A**) complex **1** and (**B**) complex **2** (0.5–2.5 µM) in 50 mM Tris-HCl/NaCl buffer pH, 7.4 after 45 min of incubation at various concentrations. Lane 1: DNA alone (control); Lane 2: DNA + 0.5 µM **1**; Lane 3: DNA + 1.0 µM **1**; Lane 4: DNA + 1.5 µM **1**; Lane 5: DNA + 2.0 µM **1**; Lane 6: DNA + 2.5 µM **1**; Lane 7: DNA alone (control); Lane 8: DNA + 0.5 µM **2**; Lane 9: DNA + 1.0 µM **2**; Lane 10: DNA + 1.5 µM **2**; Lane 11: DNA + 2.0 µM **2**; Lane 12: DNA + 2.5 µM **2**.

The electrophoretic pattern that was observed exhibited significantly good cleavage with complex **1**, and was exhibited by the appearance of form III (linear form from the supercoiled circular 'native form'), which ascertained the double-stranded incision of the DNA. However, complex **2** displayed moderate cleavage activity, giving rise to form II (nicked circular form), which resulted due to the single-stranded incision of the DNA. However, the appearance of the broad band swelling of bands in the presence of complex **2** ascertains the hydrolysis of the DNA. Furthermore, the mechanistic pathway of the DNA cleavage by complexes **1** and **2** was also examined, and the results are presented in Figure 6. In this experiment, the DNA and complex reaction mixture were treated with radical scavengers (hydroxyl radical scavengers viz., DMSO, EtOH), a singlet oxygen radical scavenger (viz., NaN$_3$) and a superoxide radical scavenger (viz., SOD). The electrophoretic pattern observed confirms that the generation of hydroxyl radical, singlet oxygen radical, and superoxide radicals are the responsible species for the cleavage of the DNA. Thus, both complexes **1** and **2** cleave the DNA by the freely diffusible radical mechanism. At higher concentrations, that complex **2** exhibits hydrolytic cleavage is quite possible. To get into the insight of the DNA groove-binding propensity of complexes **1** and **2**, the DNA cleavage was also carried out in the presence of 4',6-diamidino-2-phenylindole (DAPI, minor groove binder) and methyl green (MG, major groove binder) standards. The mixture of DNA and known groove binders were treated with the complexes and the results exhibited that complexes **1** and **2** possess the minor groove-binding tendency (Figure 6).

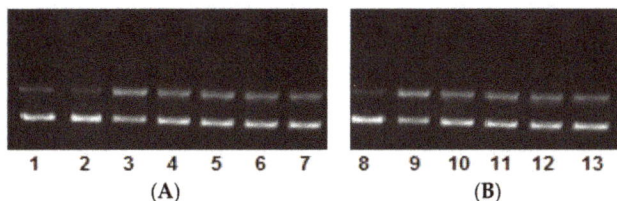

Figure 6. Mechanistic electrophoretic pattern of pBR322 DNA (100 ng) by (**A**) complex **1** and (**B**) complex **2** (0.5–2.5 µM) in 50 mM of Tris-HCl/NaCl buffer pH 7.4 after 45 min of incubation at various concentrations. Lane 1: DNA alone (control); Lane 2: DNA + DAPI + **1**; Lane 3: DNA + MG + **1**; Lane 4: DNA + DMSO + **1**; Lane 5: DNA + EtOH + **1**; Lane 6: DNA + NaN$_3$ + **1**; Lane 7: DNA + SOD+ **1**; Lane 8: DNA + DAPI + **2**; Lane 9: DNA + MG + **2**; Lane 10: DNA + DMSO + **2**; Lane 11: DNA + EtOH + **2**; Lane 12: DNA + NaN$_3$ + **2**; Lane 13: DNA + SOD + **2**.

2.5. Molecular Docking Studies with DNA by Hex

Prediction of Binding Sites of Metal Complexes on DNA

In the present study, we have used Hex 8.0 software (Dave Ritchie, Capsid research team at the LORIA/Inria, Nancy, France) to probe the binding site of complexes **1** and **2** on the dodecameric DNA molecule, and the molecular interactions involved in stabilizing the complex (Table 3, Figure 7).

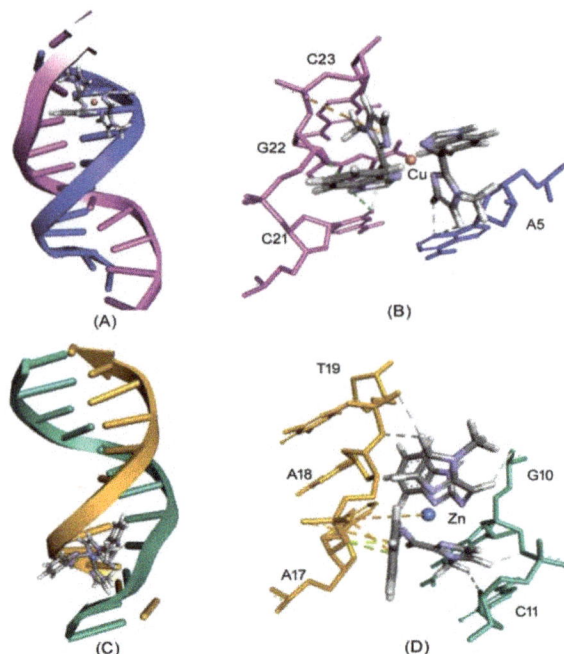

Figure 7. Molecular docking of complexes **1** and **2** with DNA. Panels (**A,C**) represent the binding of complexes **1** and **2** at the major groove of DNA, respectively; and panels (**B,D**) depict the residues of DNA making contacts with complexes **1** and **2**, respectively. Different strands and nucleotides of DNA are represented by separate color (Magenta and Blue in (**A,B**), and Green and Yellow in (**C,D**)). Metal complexes **1** and **2** are represented according to element composition.

Table 3. Molecular interactions between DNA and complexes **1** and **2**.

Interaction	Bond Distance (Å)	Nature of Interaction	E_{total} Value
Complex 1			
Unk:H—B:d21:O2	2.57	Hydrogen Bond	
Unk:H—B:dG22:O4′	2.25	Hydrogen Bond	
Unk:H—A:dA5:N3	2.34	Hydrogen Bond	−238.14
Unk:H—A:dA5:N3	2.98	Hydrogen Bond	
Unk:H—B:d21:O2	2.79	Hydrogen Bond	
B:d23:OP1—Unk	4.63	Electrostatic (π-Anion)	
Complex 2			
Zn—B:dA18:OP1	5.41	Electrostatic	
Unk:H—A:dG10:O3′	2.01	Hydrogen Bond	
Unk:H—A:d11:O4′	2.89	Hydrogen Bond	
Unk:H—A:dG10:OP1	2.93	Hydrogen Bond	
Unk:H—B:dA18:O3′	2.62	Hydrogen Bond	−248.34
Unk:H—B:dT19:OP1	2.81	Hydrogen Bond	
B:dA18:OP1—Unk	2.88	Electrostatic (π-Anion)	
B:dA18:OP2—Unk	4.88	Electrostatic (π-Anion)	
B:dA17:O3′—Unk	2.87	π-Lone Pair	
B:dA17:O3′—Unk	2.55	π-Lone Pair	

Findings of the docking results further support our in vitro results that the most preferred binding site of complexes **1** and **2** on DNA was located at the minor groove of DNA, with an overall binding score of −238.14 and −248.34, respectively. It was bound in a GC-rich region in such a fashion that allowed the planar part of the molecule to form favorable interactions with the nitrogenous bases of the DNA molecule (Table 3).

The DNA-complex **1** adduct was stabilized by five hydrogen bonds (dA5:N3 of chain A and d21:O2 of chain B formed two hydrogen bonds each with complex **1**, while dG22:O4′ formed one hydrogen bond with complex **1**). Moreover, the DNA-complex **1** adduct was stabilized by one electrostatic interaction with d23:OP1 of chain B. Similarly, complex **2** formed three electrostatic interactions and five hydrogen bonds along with two π-lone pair interactions with DNA molecules (Table 3, Figure 7). The metal ligand Zn was electrostatically involved with dA18:OP1 of chain B. Also, dA18:OP1 of chain B formed two electrostatic interactions, while dA17:O3′ of chain B formed two π-lone pair interactions. Moreover, the overall DNA-complex **2** adduct was further stabilized by five hydrogen bonds with d10:O3′, dG10:O4′, and d11:O4′ of chain A, and dA18:O3′ and dT19:OP1 of chain B.

2.6. In Vitro Cytotoxicity Assays

2.6.1. Analysis of Growth Inhibition Using 5-Diphenyltetrazolium Bromide (MTT) Assay

In view of the above findings, the two novel complexes **1** and **2** were investigated for their anticancer efficacy in vitro. The cytotoxic effect of the synthesized complexes was examined on a panel of cancer cells. The cytotoxic efficacy was expressed in terms of IC_{50} values, as shown in Table 4. The newly synthesized complex **1** inhibited the cell growth of all the five tested cell lines in a dose-dependent manner. MDA-MB 231 was observed to be the most sensitive to complex **1**, followed by HeLa > HepG2 > HT108 > SK-MEL-1. On the other hand, the novel complex **2** inhibited the cell growth of four out of five tested cell lines in a dose-dependent manner. Complex **2** was most effective on SK-MEL-1 followed by HepG2 > HeLa > HT108. However, complex **2** showed the least effect on MDA-MB 231 (Table 4). In comparison with complex **2**, complex **1** showed potential in vitro anticancer activity against all of the tested cancer cell lines. However, no activity against particular cancer cell lines was observed with free ligands. Nevertheless, the anticancer potential of complex **1** was due to the presence of its copper moiety, which efficiently inhibited the growth of cancer cells. Considering this, we continued our anticancer study further with complex **1** only. Complex **1** showed significant activity when compared with standard drugs and some of the previously reported copper(II) complexes. For example, the copper(II) complex of 4′-methoxy-5,7-dihydroxy-isoflavone exhibited cytotoxic effects against cancer cell lines viz., A549, HeLa, HepG2, SW620, and MDA-MB-435), with IC_{50} values in the 10−50 μM range [33]. Also, the copper complexes [Cu(L)Cl(H$_2$O)]Cl·3H$_2$O of 2-methyl-1*H*-benzimidazole-5-carbohydrazide (L) displayed cytotoxicity against A549 cancer cells (IC_{50} = 20 μM) [34].

Table 4. IC_{50} values of complexes **1**, **2**, the ligand, and cisplatin against five human cancer cell lines.

Complex	HepG2 (Liver) (μM)	SK-MEL-1 (Skin) (μM)	HT018 (Colon) (μM)	HeLa (Cervical) (μM)	MDA-MB 231 (Breast) (μM)
1	14 ± 2.2	17.8 ± 2.7	15 ± 2.1	13 ± 2.2	3.5 ± 2.4
2	19 ± 2.6	18 ± 2.3	25 ± 4.0	24.5 ± 2.2	26.7 ± 4.1
Ligand	NA	NA	NA	NA	NA
Vehicle control (0.1% DMSO)	NA	NA	NA	NA	NA
Cisplatin (Positive control)	6 ± 0.4	5.6 ± 0.8	5.7 ± 0.2	6 ± 0.6	3.1 ± 0.2

NA stands for Not Active.

2.6.2. Effect on Cancer Cell Adhesion/Migration

Cancer metastasis is the spread of cancer cells to tissues and organs far from where a tumor originated. Cancer cells must be able to stick together, move, and migrate in order to spread. The properties of cell adhesion and cell migration are fundamental in regulating cell movement and cancer metastasis, and are necessary for the cells to move into the bloodstream. To further characterize the anticancer activity of the potent complex **1**, its inhibitory effect was investigated on the adhesion and migration of cancer cells at an IC_{50} concentration for the respective cancer cell lines. Complex **1** was tested at concentrations of 14 µM, 17.8 µM, 15 µM, 13 µM, and 3.5 µM against HepG2, SK-MEL-1, HT 018, HeLa, and MDA-MB 231, respectively. We found that complex **1** inhibited cell adhesion by about 16.4% (14 µM), 21.3% (17.8 µM), 16.5% (15 µM), 12.2% (13 µM), and 35.2% (3.5 µM) of HepG2, SK-MEL-1, HT 018, HeLa, and MDA-MB 231, respectively (Figure 8A). As far as migration is concerned, complex **1** inhibited the migration phenomenon of the tested cancer cells. However, the rate of inhibition varied with the respective cell lines, respectively (Figure 8B). Therefore, we conclude that complex **1** was able to affect the metastatic potential of cancer cells by inhibiting their adhesion as well as migration properties.

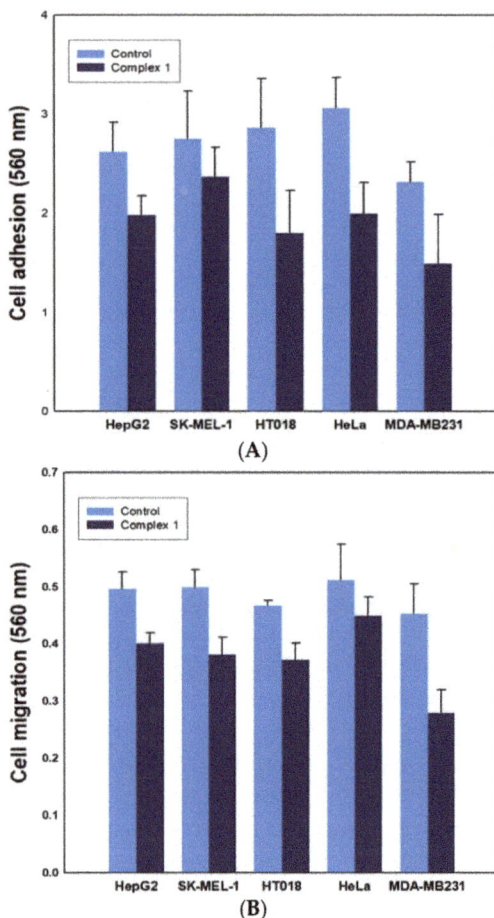

Figure 8. Effect of complex **1** on (**A**) cell adhesion and (**B**) cell migration against five cancer cell lines. Assays for adhesion and migration were performed with a cytoselect 24-well plate, and the absorbance of extracted samples was read at 560 nm.

2.6.3. Annexin V Apoptosis Detection Assay

Apoptosis and necrosis are two primary mechanisms of cell death. Cells that are damaged by external injury undergo necrosis, while cells that are induced to commit programmed suicide because of internal or external stimuli undergo apoptosis. An increasing number of chemopreventive agents have been shown to stimulate apoptosis in pre-malignant and malignant cells in vitro or in vivo [35]. A gross majority of classical apoptotic attributes can be quantitatively examined by flow cytometry (FACS). The apoptotic effect of complex **1** was evaluated using annexin-V staining. All of the tested cancer cells were harvested after treatment by complex **1** for 48 h and incubated with annexin V-FITC and PI. First, 10,000 cells were analyzed per determination. Dot plots show annexin V–FITC binding on the *X*-axis and propidium iodide (PI) staining on the *Y*-axis (Figure 9).

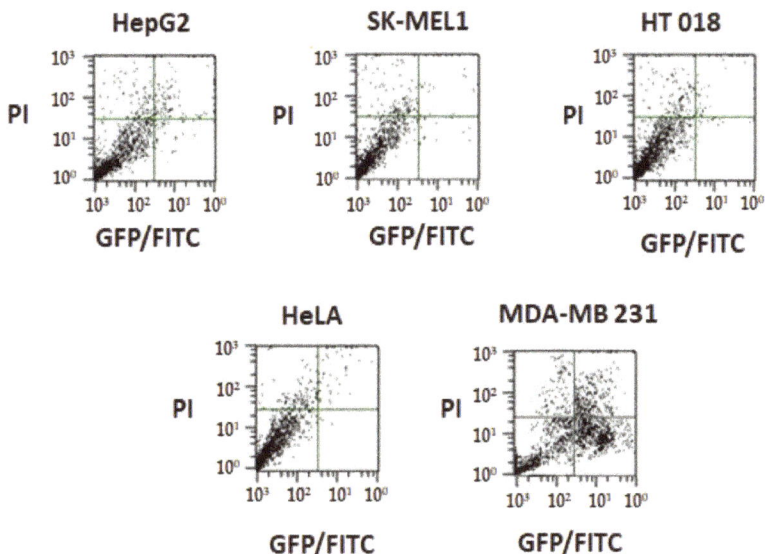

Figure 9. The apoptotic effect of complex **1** using annexin-V staining of tested cancer cell lines. Dots represent cells as follows: lower left quadrant, normal cells (FITC$^-$/PI$^-$); lower right quadrant, early apoptotic cells (FITC$^+$/PI$^-$); upper left quadrant, necrotic cells (FITC$^-$/PI$^+$); upper right quadrant, late apoptotic cells (FITC$^+$/PI$^+$).

Data of all of the cell lines after treatment with complex **1** showed a decrease in viable cancer cells. Distinct 3.9% (HepG2), 1.9% (SK-MEL-1), 1.0% (HT018), 0.2% (HeLa), and 32.6% (MDA-MB 231) increases in apoptotic cell population was recorded. After 48 h of exposure time, due to late apoptosis, the cell population increased to 7.2%, 2.7%, 1.2%, and 1.5%, and 26.1% for HepG2, SK-MEL-1, HT018, HeLa, and MDA-MB 231, respectively. We found that complex **1** was the most effective on MDA-MB-231 cells. The cytotoxic effect of the complex was not correlated to the increase of the necrotic cell population.

2.7. In Vivo Chronic Toxicity Studies of Complex 1 vs. 2

To evaluate the safety of complexes **1** and **2**, we studied their chronic toxicity in male as well as female mice. The hematology studies revealed that both complexes **1** and **2** increased the count of red blood cells (RBC) and white blood cells (WBC) in male mice, as well as hemoglobin, platelets, neutrophils, and lymphocytes, with neutrophils and lymphocytes being the most significantly affected cells (Figure 10A). Complex **2** was more potent in increasing the count of these cells than complex **1**. This might be due to the stimulation of bone marrow or increased protection of it from the deleterious effects, leading to the production of cells. In female mice, complex **1** showed weak activity against the above-mentioned cells except for neutrophils and lymphocytes, where it possessed significantly increased initial counts. Although both complexes significantly increased the clotting time in both male and female mice, complex **2** seemed to be more potent. Interestingly, both complexes increased fertility in mice. Both had reduced dead fetus percentages at the end of term, with complex **2** having a far lower dead fetus percentage.

In liver function tests (Figure 10B), both complexes did not alter liver functions in male animals except SGOT (Serum Glutamic Oxaloacetic Transaminase), which was significantly elevated by both complexes. Both complexes significantly lowered blood glucose levels, with complex **2** possessing stronger blood sugar lowering activity. However, both complexes significantly elevated SGOT, SGPT (Serum Glutamic Pyruvic Transaminase), GGT (Gamma-Glutamyl Transferase), ALP (Alkaline Phosphatase), and bilirubin in female animals. Moreover, both complexes significantly decreased blood glucose levels.

In renal function tests (Figure 10C), both complexes significantly increased sodium and potassium levels, but decreased calcium, urea, uric acid, and blood creatinine in male mice to significant (and potent) levels. Both complexes possessed the same activity in female mice, but decreased potassium blood levels.

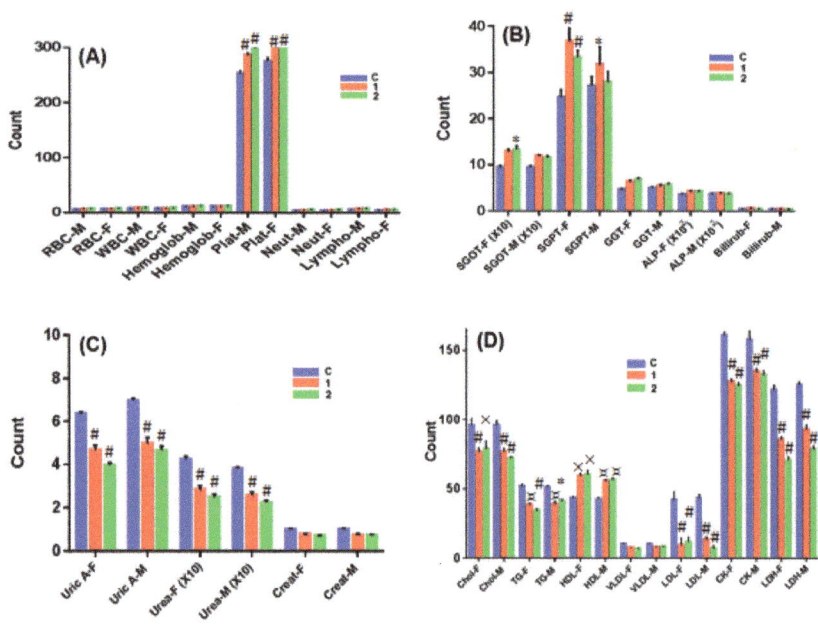

Figure 10. Chronic toxicity profile of complexes **1** and **2**. Effect of complexes **1** and **2** on (**A**) blood components; (**B**) biochemical parameters of liver function; (**C**) biochemical parameters of kidney function; and (**D**) biochemical parameters of lipids and heart function. M and F denote male and female mice, respectively. Statistical analysis was performed by two-way ANOVA, as well as the Dunnett test as the post-hoc test, compared with control (**C**), $n = 4$. (* $p < 0.05$; ¤$p < 0.01$; × $p < 0.005$; # $p < 0.001$).

The lipid profile and protein tests (Figure 10D) confirmed that both the complexes decreased significant levels of blood triglycerides, cholesterol, LDL (Low Density Lipoproteins), and VLDL (Very Low Density Lipoproteins), but significantly increased the blood levels of HDL and total proteins in both male and female mice. In cardiac function tests (Figure 10D), both complexes significantly lowered the blood levels of LDH and creatine kinase in both male and female mice. Complex **2** showed stronger activity on both enzymes.

3. Experimental Section

3.1. Material

First, 2-aminobenzimidazole, 1-methyl-1H-imidazole-2-carbaldehyde, Cu(CH$_3$COO)$_2$·H$_2$O, Zn(CH$_3$COO)$_2$·2H$_2$O, pBR322 DNA, sodium azide, superoxide dismutase, H$_2$O$_2$, DAPI, methyl green (MG), the sodium salt of CT-DNA, and HSA essentially fatty acid-free ($\geq$98%) were purchased from Sigma-Aldrich (St. Louis, MO, USA). *Tris*(hydroxymethyl)aminomethane hydrochloride (Tris-HCl) was of analytical grade and also obtained from Sigma-Aldrich (St. Louis, MO, USA). Rosewell Park Memorial Institute (RPMI)-1640 medium, Dulbecco's minimum essential medium (DMEM), fetal calf serum (FCS), dimethyl sulphoxide (DMSO), trypsin-EDTA solution, dithiothreitol (DTT); 3-4,5-dimethylthiazol-2-yl-2, 5-diphenyltetrazolium bromide (MTT), and cisplatin were procured from Sigma-Aldrich (St. Louis, MO, USA). All of the other chemicals used were also of the highest purity. Cell adhesion and cell migration kits were obtained from Cell Biolabs, Inc (San Diego, CA, USA). A Vybrant Apoptosis Assay Kit was procured from Molecular Probe (Eugene, OR, USA). Solvents were used as received.

3.2. Synthesis Procedure of Schiff Base

A mixture of 2-aminobenzimidazole (5.0 mmol) and 1-aminoimizole-2-carboxyaldehyde (5.0 mmol) in 25 mL of toluene by adding a few drops of acetic acid was stirred at 80 °C for 12 h. The yellow coloration of the mixture deepened, and after 12 h of stirring, the solvent was removed by rotavapor. A yellow-colored compound was isolated. To recrystallize the compound, it was dissolved in MeOH, and the resulting solution was kept for slow evaporation. Yellow single crystals were isolated after two days. Yield: (68%). M.p. 265 °C. Anal. Calc. for 1$_2$H$_{11}$N$_5$: C, 63.99; H, 4.92; N, 31.09. Found: C, 63.86; H, 4.89; N, 31.05%. FT-IR (KBr pellet, cm^{-1}): 3384, 3050, 2979, 1612, 1565, 1463, 1427, 1380, 1279, 1159, 740. ^{1}H NMR (CDCl$_3$, δ, 293 K): 12.67 (s, 1H, 1NH), 9.21 (s, 1H, $\underline{H}$C=N), 7.51–7.16 (ArH, 6H), 4.11 (s, 3H, N–C$\underline{H_3}$), ESI-MS: 226.1.

3.3. Copper Complex: [Cu(BnI)$_2$], (1)

Yield: (86%). M.p.198 °C. Anal. Calc. for C$_{24}$H$_{20}$N$_{10}$Cu: C, 56.30; H, 3.94; N, 27.36. Found: C, 56.46; H, 4.03; N, 27.29%. ESI MS (+ve) DMSO, m/z: 515.0 for C$_{24}$H$_{20}$N$_{10}$Cu + 3H$^+$. FT IR (KBr pellet, cm^{-1}): 3408, 3062, 1642, 1597, 1469, 1390, 1272, 742, 672, 662, 507, 438. Λ$_M$ (1 × 10^{-3} M, DMSO): 15.90 Ω^{-1} cm^2 mol^{-1} (non-electrolyte in nature). UV-vis (1 × 10^{-3} M, DMSO, λ$_{nm}$): 656 nm. μ$_{eff}$ (B.M.) = 1.89. EPR (solid state, g$_{av}$, 298 K) = 2.09.

3.4. Zinc Complex, [Zn(BnI)$_2$], (2)

Yield: (73%). M.p. 174 °C. Anal. Calc. for C$_{24}$H$_{20}$N$_{10}$Zn: C, 56.09; H, 3.92; N, 27.26. Found: C, 55.86; H, 3.89; N, 27.21%. ESI-MS (+ve) DMSO, m/z: 516.2 for C$_{24}$H$_{20}$N$_{10}$Zn + 2H$^+$. FT IR (KBr pellet, cm^{-1}): 3408, 3055, 1644, 1582, 1466, 1392, 1271, 741, 668, 647, 512, 428. ^{1}H NMR (DMSO, δ, 293 K): 9.68 (s, 2H, $\underline{H}$C=N), 7.58-6.84 (ArH, 12H), 3.92 (s, 6H, N–C$\underline{H_3}$).

3.5. Fluorescence Quenching Measurements

Samples of HSA were prepared in 20 mM of sodium phosphate buffer (pH 7.4), while the 1-mM stock solution of complexes **1** and **2** in 10% DMSO was prepared and further diluted in 20 mM of sodium phosphate buffer to reach the desired concentration. In all of the samples, the final concentration of DMSO was below 1%. The concentration of HSA was determined from Beer–Lambert's law using the molar extinction coefficient of 36,500 M^{-1} cm^{-1} at 280 nm [36].

Quenching in the fluorescence emission intensity of HSA was measured on a JASCO spectrofluorometer (FP-8300) fitted with a thermostatically controlled cell holder attached to a water bath. To the HSA sample, (2 μM, 3 mL) 2 μM of complexes **1** and **2** was successively added in such a manner that the total volume of metal complexes added was not more than 30 µL. Fluorescence quenching was measured by exciting Trp214 of HSA at 295 nm, and the fluorescence emission spectrum was collected between 300–450 nm [27]. Excitation and emission slits were kept at 5 nm. All of the fluorescence emission intensities were corrected for the inner filter effect using the following equation:

$$F_{Corr} = F_{Obs} \times e^{(A_{ex}+A_{em})/2}$$

(1)

The quenching parameters were deduced using the following Stern–Volmer (Equation (2)) and modified Stern–Volmer (Equation (3)) equations:

$$\frac{F_o}{F} = 1 + K_{SV}[Q] = 1 + k_q \tau_0[Q]$$

(2)

where F_o and F are the fluorescence emission intensities of HSA before and after metal complex binding; K_{SV} is the Stern–Volmer constant; $[Q]$ is the molar concentration of the quencher, i.e., metal complex; k_q is the bimolecular quenching rate constant, and τ_0 is the lifetime of HSA fluorescence in the absence of any quencher.

$$\log \frac{(F_o - F)}{F} = \log K_a + n \log[Q]$$

(3)

where F_o and F are the fluorescence intensities of HSA before and after metal complex binding; K_a is the binding constant; n is the number of binding sites, and $[Q]$ is the molar concentration of the quencher.

The thermodynamic parameters for the interactions between HSA and complexes **1** and **2** were determined by measuring emission quenching at three different temperatures (298, 303, and 308 K). The following van't Hoff and thermodynamic equations (equations (4) and (5)) were used to deduce the change in enthalpy (ΔH), entropy (ΔS), and Gibb's free energy (ΔG).

$$\ln K_a = \frac{\Delta S}{R} - \frac{\Delta H}{RT}$$

(4)

$$\Delta G = \Delta H - T\Delta S$$

(5)

3.6. FRET Measurements

The FRET between HSA and complexes **1** and **2** were observed by measuring the absorption spectra of complexes **1** and **2**, and the emission spectra of HSA in the 300–450 nm range. All of the intensities were normalized as described earlier [30]. The distance (r) between Trp214 of HSA and bound complexes **1** and **2** was measured using the following equations:

$$E = \frac{R_0^6}{R_0^6 + r^6} = 1 - \frac{F}{F_o}$$

(6)

$$R_0^6 = 8.79 \times 10^{-25} K^2 n^{-4} \phi J$$

(7)

where E is the efficiency of the energy transfer; R_o is the distance at which the efficiency of energy transfer becomes 50%; r is the distance between HSA (donor) and the metal complex (acceptor); F and F_o are the emission intensities of HSA in the presence and absence of the metal complex (quencher); K^2 is the geometry of the dipoles ($K^2 = 2/3$ for HSA); n is the refractive index of the medium (here, it is 1.33); φ is the emission quantum yield of HSA in the absence of quencher ($\varphi = 0.118$), and J is the overlap integral of the HSA emission spectrum and the absorption spectrum of the metal complex.

The overlap integral (J) is determined from the following equation:

$$J = \frac{\int_0^\infty F_\lambda \varepsilon_\lambda \lambda^4 d\lambda}{\int_0^\infty F_\lambda d\lambda}$$

(8)

where F_λ is the fluorescence intensity of HSA at the wavelength λ; and ε_λ is the molar extinction coefficient of the metal complex at the wavelength λ.

3.7. DNA Binding and Nuclease Activity

The standard protocol [32] was adopted for both the experiments, with the slight modifications that we reported earlier [36–40].

3.8. Molecular Docking Studies

3.8.1. Preparation of Ligands and Receptors

The three-dimensional structures of HSA (PDB Id:1AO6) and B-DNA dodecamer d(CGCGAATTCGCG)$_2$ (PDB Id:1BNA) that were used for docking were retrieved from the RCSB PDB database bank [41]. Structures of the ligands (complexes **1** and **2**) were drawn in CHEMSKETCH (Available online: http://www.acdlabs.com) and converted to a PDB file using OPENBABEL (Available online: http://www.vcclab.org/lab/babel). The energies of complexes **1** and **2** were minimized using a MMFF94 forcefield. The water molecules and any heteroatoms were deleted from the receptor files (i.e., HSA and DNA) before setting up the docking program.

3.8.2. Molecular Docking of Complexes **1** and **2** with DNA

HEX 8.0.0 software was used to study the binding of complexes **1** and **2** with a DNA molecule with default settings. The binding site of the ligand on the receptor was searched on the basis of shape as well as electrostatics, and the post-processing was done by optimized potentials for liquid simulations (OPLS) minimization. The GRID dimension was set at 0.6, and 10,000 solutions were computed. The results were analyzed in Discovery Studio 4.0 (Accelrys Software Inc., San Diago, USA, 2013) [42].

3.8.3. Molecular Docking of Complexes **1** and **2** with HSA

Autodock4.2 was used for the docking of complexes 1 and 2 with HSA as described previously [27]. Briefly, Gasteiger partial charges were added to the atoms of complexes 1 and 2. Non-polar hydrogen atoms were merged, and rotatable bonds were defined. Essential hydrogen atoms, Kollman united-atom charges, and solvation parameters were added to HSA with the aid of the AutoDockTool. Affinity grid maps of 120 × 120 × 120 Å grid points and 0.7 Å spacing were generated using the AutoGrid program. Docking simulations were performed using the Lamarckian genetic algorithm (LGA) and the Solis and Wets local search methods. Initial positions, orientations, and torsions of the ligand molecules were set randomly. All of the rotatable torsions were released during docking. Each run of the docking experiment was set to terminate after a maximum of 2.5×10^6 energy evaluations. The population size was set to 150. During the search, a translational step of 0.2 Å and quaternion and torsion steps of five were applied. The docking figures for publication were prepared in Discovery Studio 4.0 (Accelrys Software Inc., 2013) [42].

3.9. Cytotoxicity

3.9.1. Cell Lines and Culture Conditions

Human cancer cell lines HepG2 (Liver), SK-MEL-1 (Skin), HT 018 (Colon), HeLa (Cervical), and MDA-MB 231 (Breast) were procured from the American Type Culture Collection (Rockville, MD, USA). MDA-MB-231 and HepG2 cell lines were maintained in DMEM; whereas SK-MEL-1, HeLa, and HT-018 were maintained in RPMI medium. The complete growth medium was supplemented with 10% (v/v) heat-inactivated FCS, 2 mM L-glutamine, and antimycotic-antibiotic solution. All of the cells were maintained in a standard culture condition of 37 °C temperature and 95% humidified atmosphere containing 5% CO_2. Cells were screened periodically for mycoplasma contamination.

3.9.2. MTT Assay

The novel complexes **1** and **2** were examined for their cytotoxicity against five different types of cancer cell lines viz., HepG2, SK-MEL-1, HT 018, HeLa, and MDA-MB 231 using a standard MTT reduction assay, according to Khan et al. (2012) [35]. Briefly, cells at a concentration of 1×10^6 cells/200 mL/well were plated in 96-well plates and further grown in their respective medium containing 10% FCS. After 24 h, cells were incubated with 0–25 μM concentrations of test complexes or respective free ligands. Cells with 0.1% DMSO (vehicle control) and cisplatin (positive control) were also cultured under the same conditions. Cells only were used as negative control. Ensuing 48 h of incubation, old medium from treated cells was replaced with fresh medium. MTT reagent (5 mg/mL in PBS) was added to each well, and cells were further incubated for 2–3 h at 37 °C. After treatment, the supernatants were carefully removed, and 100 μL of DMSO was added to each well. Absorbance was measured at 620 nm in a multi-well plate reader, and drug sensitivity was expressed in terms of the concentration of drug required for a 50% reduction of cell viability (IC_{50}).

3.9.3. Measurement of Cancer Cell Adhesion

Assay for adhesion was performed with cytoselect 24-well cell adhesion as per protocol [43]. For quantitative analysis, the IC_{50} concentration of respective complexes (in triplicate) was tested on HepG2, SK-MEL-1, HT 018, HeLa, and MDA-MB 231 cancer cells. Cancer cell ($0.1–1.0 \times 10^6$ cells/mL) suspension in serum-free medium was added to the inside of each well of a pre-warmed adhesion plate. The plates were incubated for 30–90 min in a CO_2 incubator and treated as per the protocol [43]. After air-drying the wells, stained adhered cells were extracted using an extraction solution. Then, the absorbance of the extracted sample (adhered cells) was read at 560 nm in a microtiter plate reader.

3.9.4. Measurement of Cancer Cell Migration

The cell migration assay was performed with a cytoselect 24-well cell adhesion assay according to the protocol (Khan et al., 2013) [43] on HepG2, SK-MEL-1, HT 018, HeLa, and MDA-MB 231 cancer cell lines. For quantitative analysis, the IC_{50} concentration of the respective complexes (in triplicate) was tested on HepG2 (Liver), SK-MEL-1 (Skin), HT 018 (Colon), Hela (Cervical), and MDA-MB 231 (Breast) cancer cells. The test complexes were supplemented with medium containing 10% FBS in the lower well of the migration plate. To the inside of each insert, 100 μL of $0.5–1.0 \times 10^6$ cells/mL of cell suspension was added. The plates were then incubated for 8 h at 37 °C in a humidified CO_2 incubator. Finally, the absorbance of 100 μL of each sample was then read at 560 nm.

3.9.5. Analysis of Annexin-V Binding by Flow Cytometry

Annexin-V staining was performed according to the protocol (Khan et al., 2017) [44]. For quantitative analysis, the IC_{50} concentration of respective complexes (in triplicate) was tested on HepG2 (Liver), SK-MEL-1 (Skin), HT 018 (Colon), Hela (Cervical), and MDA-MB 231 (Breast) cancer cells. Cancer cell ($0.1–1.0 \times 10^7$ cells/mL) suspension in serum-free medium was incubated with respective complexes in six-well plates in a CO_2 incubator. After treating with complexes for 48 h, the cancer cells were harvested and incubated with annexin V-FITC and propidium iodide (PI). The fluorescence emission of Annexin-V stained cells was measured at 530–575 nm in a flow cytometer (MACSQuant, Bergisch Gladbach, Germany). Dots represent cells as follows: lower left quadrant, normal cells ($FITC^-/PI^-$); lower right quadrant, early apoptotic cells ($FITC^+/PI^-$); upper left quadrant, necrotic cells ($FITC^-/PI^+$); upper right quadrant, late apoptotic cells ($FITC^+/PI^+$).

3.10. In Vivo Toxicity

3.10.1. Toxicity Study Design

Mice (males) were randomly divided into different groups (n = 6–10). Different doses of complexes **1** and **2** (0.5, 1, 2, 5, 8 and 10 g/kg) were administered intraperitoneally. The complexes were suspended in 0.2% aqueous Tween 80 or 0.25% carboxymethylcellulose. The animals were observed for 72 h for signs of toxicity and mortality, and LD_{50} was calculated according to a published method [45].

3.10.2. Chronic Toxicity Study

A total of 40 male and 40 female Swiss albino mice were randomly allocated to the control and test groups. The complexes 1 and 2 in each case were mixed with drinking water for the feasibility of administration due to long treatment duration. The dose selected was $1/10^{th}$ of the LD_{50}. The treatment was continued for a period of 12 weeks [46] (WHO Scientific Group, Geneva, 1967). The animals were then observed for all external general symptoms of toxicity, body weight changes, and mortality. Ten male and 10 female rats were used in each group, having one control and two treated groups. One group of treated female rats was mated with treated males, and pregnancy outcomes were studied. Urine was collected 1–2 days before the end of the treatment. The treated animals were fasted for 12 h and then anesthetized with ether. Blood samples were collected via heart puncture and centrifuged at 3000 rpm for 10 min. The plasma was then stored at −20 °C pending for analysis of the biochemical parameters. Vital organs were removed, weighed, and investigated for apparent signs of toxicity, and stored in 10% formalin for histological studies. The percentage of each organ relative to the body weight of the animal was calculated.

3.10.3. Hematological Studies

Whole non-centrifuged blood was used for the determination of some hematological parameters. The blood was analyzed for WBC and RBC count, hemoglobin, platelets, neutrophils, and lymphocytes measurements using Contraves Digicell 3100H (Zurich, Switzerland) [47].

3.10.4. Serum Analysis of Biochemical Parameters

A colorimetric method was used for the determination of the biochemical parameters (SGPT, SGOT, GGT, ALP, bilirubin, and lipid profile) in plasma. The enzyme activity was quantified spectrophotometrically using commercial enzymatic kits (Crescent diagnostics test kits, Jeddah, KSA).

3.10.5. Statistical Analysis

The results were presented as a mean ± standard error of the mean (S.E.M). Statistical differences were analyzed using ANOVA with the Dunnett test as a post-hoc test. A value of $p < 0.05$ was considered statistically significant [48].

4. Conclusions

The new copper(II) (1) and zinc(II) (2) complexes have been designed and prepared with the benzimidazole-derived ligand as a biologically active moiety. The focus of our work was on its biological significance, in particular its anticancer property. Thus, complexes 1 and 2 were studied to examine the propensity of binding with HSA and DNA, which confirmed the avid binding. DNA cleavage experiments showed the evident nuclease activity of complex 1 with a double-stranded cleavage mechanism, and further revealed that ROS were responsible for the cleavage activity. Interestingly, the cytotoxicity of both the complexes and the ligand was examined on the five different cancer cells line, and the results showed that the activity of complex 1 was considerably good on breast cancer cells, with the IC_{50} values comparable to the standard drug cisplatin. Furthermore, we have studied the cell adhesion and cell migration properties of different cancer cell line in the presence of complex 1. Our findings exhibited that complex 1 showed significant anti-metastatic potential by inhibiting the adhesion and migration property of cancer cells. Moreover, the in vivo chronic toxicity of both complexes 1 and 2 revealed that they were safe and could be developed as a potential anticancer drug for human consumption. Overall, the results of this study ascertain that copper complex (1) can be a promising chemotherapeutic intervention for future application in cancer therapy, but warrants further investigations.

Supplementary Materials: The following are available online at www.mdpi.com/link.

Author Contributions: R.A.K., A.H., M.F.A., and M.T.R. carried out design, synthesis, characterization, analyzed the data and prepared the manuscript. R.A.K. and M.T.R performed interaction studies and molecular modeling. A.A.K. conducted the in vitro anticancer studies. All the authors, R.A.K., A.H., M.F.A., M.T.R., A.A.K., and P.A.S., have discussed all the results and helped in writing, language editing and preparing the manuscript.

Acknowledgments: The authors would like to extend their sincere thanks to King Abdul-Aziz City for Science and Technology, Kingdom of Saudi Arabia for funding this project (KACST Project Number- ARP-35-227).

Conflicts of Interest: The authors declare no conflict of interest.

References

1. Rosenberg, B.; VanCamp, L.; Grimley, E.B.; Thomson, A.J. The inhibition of growth or cell division in *Escherichia coli* by different ionic species of platinum (IV) complexes. *J. Biol. Chem.* **1967**, *242*, 1347–1352. [PubMed]

2. Rosenberg, B.; VanCamp, L.; Trosko, J.E.; Mansour, V.H. Platinum compounds: A new class of potent antitumour agents. *Nature* **1969**, *222*, 385–386. [CrossRef] [PubMed]

3. Rosenberg, B. Noble metal complexes in cancer chemotherapy. *Adv. Exp. Med. Biol.* **1977**, *91*, 129–150. [PubMed]

4. Rosenberg, B. Platinum Complexes for the Treatment of Cancer. *Interdiscip. Sci. Rev.* **1978**, *3*, 134–147. [CrossRef]

5. Wang, C.H.; Shih, W.C.; Chang, H.C.; Kuo, Y.Y.; Hung, W.C.; Ong, T.G.; Li, W.S. Preparation and Characterization of Amino-Linked Heterocyclic Carbene Palladium, Gold, and Silver Complexes and Their Use as Anticancer Agents That Act by Triggering Apoptotic Cell Death. *J. Med. Chem.* **2011**, *54*, 5245–5249. [CrossRef] [PubMed]

6. Rehman, M.T.; Khan, A.U. Understanding the interaction between human serum albumin and anti-bacterial/anti-cancer compounds. *Curr. Pharm. Des.* **2015**, *21*, 1785–1799. [CrossRef] [PubMed]

7. Muhammad, N.; Guo, Z. Metal-based anticancer chemotherapeutic agents. *Curr. Opin. Chem. Biol.* **2014**, *19*, 144–153. [CrossRef] [PubMed]

8. Qiao, X.; Ma, Z.Y.; Xie, C.Z.; Xue, F.; Zhang, Y.W.; Xu, J.Y.; Qiang, Z.Y.; Lou, J.S.; Chen, G.J.; Yan, S.P. Study on potential antitumor mechanism of a novel Schiff base copper(II)complex: Synthesis, crystal structure, DNA binding, cytotoxicity and apoptosis induction activity. *J. Inorg. Biochem.* **2011**, *105*, 728–737. [CrossRef] [PubMed]

9. Zhou, W.; Wang, X.Y.; Hu, M.; Zhu, C.C.; Guo, Z.J. A mitochondrion-targeting copper complex exhibits potent cytotoxicity against cisplatin-resistant tumor cells through a multiple mechanism of action. *Chem. Sci.* **2014**, *5*, 2761–2770. [CrossRef]

10. Khodade, V.S.; Dharmaraja, A.T.; Chakrapani, H. Synthesis, reactive oxygen species generation and copper-mediated nuclease activity profiles of 2-aryl-3-amino-1, 4-naphthoquinones. *Bioorg. Med. Chem. Lett.* **2012**, *22*, 3766–3769. [CrossRef] [PubMed]

11. Palanimuthu, D.; Shinde, S.V.; Somasundaram, K.; Samuelson, A.G. In Vitro and in Vivo Anticancer Activity of Copper Bis(thiosemicarbazone) Complexes. *J. Med. Chem.* **2013**, *56*, 722–734. [CrossRef] [PubMed]

12. Santini, C.; Pellei, M.; Gandin, V.; Porchia, M.; Tisato, F.; Marzano, C. Advances in Copper Complexes as Anticancer Agents. *Chem. Rev.* **2014**, *114*, 815–862. [CrossRef] [PubMed]

13. Ali, M.M.; Frei, E.; Straubb, J.; Breuerb, A.; Wiesslerb, M. Induction of metallothionein by zinc protects from daunorubicin toxicity in rats. *Toxicology* **2002**, *85*, 85–93. [CrossRef]

14. Pushie, M.J.; Pickering, I.J.; Korbas, M.; Hackett, M.J.; George, G.N. Elemental and Chemically Specific X-ray Fluorescence Imaging of Biological Systems. *Chem. Rev.* **2014**, *114*, 8499. [CrossRef] [PubMed]

15. Gupta, R.K.; Sharma, G.; Pandey, R.A.; Kumar, A.; Koch, B.; Li, P.-Z.; Xu, Q.; Pandey, D.S. DNA/Protein Binding, Molecular Docking, and in Vitro Anticancer Activity of Some Thioether-Dipyrrinato Complexes. *Inorg. Chem.* **2013**, *52*, 13984–13996. [CrossRef] [PubMed]

16. Zhao, J.A.; Li, S.S.; Zhao, D.D.; Chen, S.F.; Hu, J.Y. Metal and structure tuned in vitro antitumor activity of benzimidazole-based copper and zinc complexes. *J. Coord. Chem.* **2013**, *66*, 1650–1660. [CrossRef]

17. Overington, J.P.; Al-Lazikani, B.; Hopkins, A.L. How many drug targets are there? *Nat. Rev. Drug Discov.* **2006**, *5*, 993–996. [CrossRef] [PubMed]

18. Mentese, E.; Yilmaz, F.; Ülker, S.; Kahveci, B. Microwave-assisted synthesis of some new coumarin derivatives including 1,2,4-triazol-3-one and investigation of their biological activities. *Chem. Heterocycl. Compd.* **2015**, *50*, 447–456.

19. Sridevi, C.H.; Balaji, K.; Naidu, A.; Sudhakaran, R. Synthesis of Some Phenylpyrazolo Benzimidazolo Quinoxaline Derivatives as Potent Antihistaminic Agents. *E-J. Chem.* **2010**, *7*, 234–238. [CrossRef]

20. Pan, T.; He, X.; Chen, B.; Chen, H.; Geng, G.; Luo, H.; Zhang, H.; Bai, C. Development of benzimidazole derivatives to inhibit HIV-1 replication through protecting APOBEC3G protein. *Eur. J. Med. Chem.* **2015**, *95*, 500–513. [CrossRef] [PubMed]

21. Keri, R.S.; Hiremathad, A.; Budagumpi, S.; Nagaraja, B.M. Comprehensive Review in Current Developments of Benzimidazole-Based Medicinal Chemistry. *Chem. Biol. Drug Des.* **2015**, *86*, 19–65. [CrossRef] [PubMed]

22. Bansal, Y.; Silakari, O. The therapeutic journey of benzimidazoles: A review. *Bioorg. Med. Chem.* **2012**, *20*, 6208–6236. [CrossRef] [PubMed]

23. Yadav, S.; Narasimhan, B.; Kaur, H. Perspectives of Benzimidazole Derivatives as Anticancer Agents in the New Era. *Anticancer Agents Med. Chem.* **2016**, *16*, 1403–1425. [CrossRef] [PubMed]

24. Ware, W.R. Oxygen quenching of fluorescence in solution: AN experimental study of the diffusion process. *J. Phys. Chem.* **1962**, *66*, 455–458. [CrossRef]

25. Kang, J.; Liu, Y.; Xie, M.X.; Li, S.; Jiang, M.; Wang, Y.D. Interactions of human serum albumin with chlorogenic acid and ferulic acid. *Biochim. Biophys. Acta* **2004**, *1674*, 205–214. [CrossRef] [PubMed]

26. Agudelo, D.; Bourassa, P.; Bruneau, J.; Bérubé, G.; Asselin, E.; Tajmir-Riahi, H.A. Probing the binding sites of antibiotic drugs doxorubicin and N-(trifluoroacetyl) doxorubicin with human and bovine serum albumins. *PLoS ONE* **2012**, *7*, e43814. [CrossRef] [PubMed]

27. Rehman, M.T.; Shamsi, H.; Khan, A.U. Insight into the Binding Mechanism of Imipenem to Human Serum Albumin by Spectroscopic and Computational Approaches. *Mol. Pharm.* **2014**, *11*, 1785–1797. [CrossRef] [PubMed]

28. Rehman, M.T.; Ahmed, S.; Khan, A.U. Interaction of meropenem with 'N' and 'B' isoforms of human serum albumin: A spectroscopic and molecular docking study. *J. Biomol. Struct. Dyn.* **2016**, *34*, 1849–1864. [CrossRef] [PubMed]

29. Kamtekar, N.; Pandey, A.; Agrawal, N.; Pissurlenkar, R.R.S.; Borana, M.; Ahmad, B. Interaction of multimicrobial synthetic inhibitor 1,2-bis(2-benzimidazolyl)-1,2-ethanediol with serum albumin: Spectroscopic and computational studies. *PLoS ONE* **2013**, *8*, e53499. [CrossRef] [PubMed]

30. Förster, T. Intermolecular energy migration and fluorescence. *Ann. Phys.* **1948**, *437*, 55–75. [CrossRef]

31. Zsila, F. Subdomain IB Is the Third Major Drug Binding Region of Human Serum Albumin: Toward the Three-Sites Model. *Mol. Pharm.* **2013**, *10*, 1668–1682. [CrossRef] [PubMed]

32. Wolfe, A.; Shimer, G.H.; Meehan, T. Polycyclic aromatic hydrocarbons physically intercalate into duplex regions of denatured DNA. *Biochemistry* **1987**, *26*, 6392–6396. [CrossRef] [PubMed]

33. Chen, X.; Tang, L.-J.; Sun, Y.-N.; Qiu, P.-H.; Liang, G. Syntheses, characterization and antitumor activities of transition metal complexes with isoflavone. *J. Inorg. Biochem.* **2010**, *104*, 379–384. [CrossRef] [PubMed]

34. Galal, S.A.; Hegab, K.H.; Kassab, A.S.; Rodriguez, M.L.; Kerwin, S.M.; El-Khamry, A.-M.A.; El Diwani, H.I. New transition metal ion complexes with benzimidazole-5-carboxylic acid hydrazides with antitumor activity. *Eur. J. Med. Chem.* **2009**, *44*, 1500–1508. [CrossRef] [PubMed]

35. Khan, A.A.; Jabeen, M.; Chauhan, A.; Owais, M. Synthesis and characterization of novel n-9 fatty acid conjugates possessing antineoplastic properties. *Lipids* **2012**, *47*, 973–86. [CrossRef] [PubMed]

36. Rehman, M.T.; Faheem, M.; Khan, A.U. Insignificant β-lactamase activity of human serum albumin: No panic to nonmicrobial-based drug resistance. *Lett. Appl. Microbiol.* **2013**, *57*, 325–329. [CrossRef] [PubMed]

37. Usman, M.; Zaki, M.; Khan, R.A.; Alsalme, A.; Ahmad, M.; Tabassum, S. Coumarin centered copper(II) complex with appended-imidazole as cancer chemotherapeutic agents against lung cancer: Molecular insight via DFT-based vibrational analysis. *RSC Adv.* **2017**, *7*, 36056–36071. [CrossRef]

38. Tabassum, S.; Asim, A.; Khan, R.A.; Arjmand, F.; Divya, R.; Balaji, P.; Akbarsha, M.A. A multifunctional molecular entity CuII–SnIV heterobimetallic complex as a potential cancer chemotherapeutic agent: DNA binding/cleavage, SOD mimetic, topoisomerase Iα inhibitory and in vitro cytotoxic activities. *RSC Adv.* **2015**, *5*, 47439–47450. [CrossRef]

39. Usman, M.; Arjmand, F.; Khan, R.A.; Alsalme, A.; Ahmad, M.; Tabassum, S. Biological evaluation of dinuclear copper complex/dichloroacetic acid cocrystal against human breast cancer: Design, synthesis, characterization, DFT studies and cytotoxicity assays. *RSC Adv.* **2017**, *7*, 47920–47932. [CrossRef]

40. Khan, R.A.; Usman, M.; Rajakumar, D.; Balaji, P.; Alsalme, A.; Arjmand, F.; AlFarhan, K.; Akbarsha, M.A.; Marchetti, F.; Pettinari, C.; et al. Heteroleptic Copper(I) Complexes of "Scorpionate" Bis-pyrazolyl Carboxylate Ligand with Auxiliary Phosphine as Potential Anticancer Agents: An Insight into Cytotoxic Mode. *Sci. Rep.* **2017**, *7*, 45229. [CrossRef] [PubMed]

41. Berman, H.M.; Westbrook, J.; Feng, Z.; Gilliland, G.; Bhat, T.N.; Weissig, H.; Shindyalov, I.N.; Bourne, P.E. The Protein Data Bank. *Nucleic Acids Res.* **2000**, *28*, 235–242. [CrossRef] [PubMed]

42. Accelrys Software Inc. *Discovery Studio Modeling Environment, Release 4.0*; Accelrys Software Inc.: San Diego, CA, USA, 2013.

43. Khan, A.A.; Alanazi, A.M.; Jabeen, M.; Chauhan, A.; Abdelhameed, A.S. Design, Synthesis and In Vitro Anticancer Evaluation of a Stearic Acid-based Ester Conjugate. *Anticancer Res.* **2013**, *33*, 1217–1224.

44. Khan, A.A. Pro-apoptotic activity of nano-escheriosome based oleic acid conjugate against 7,12-dimethylbenz(a)anthracene (DMBA) induced cutaneous carcinogenesis. *Biomed. Pharmacother.* **2017**, *90*, 295–302. [CrossRef] [PubMed]

45. Shah, A.H.; Qureshi, S.; Tariq, M.; Ageel, A.M. Toxicity studies on six plants used in the traditional Arab system of medicine. *Phytother. Res.* **1989**, *3*, 25–29. [CrossRef]

46. WHO Scientific Group. *Principles for Pre-clinical Testing of Drugs Safety*; Technical Report Series; World Health Organization: Geneva, Switzerland, 1967; Volume 341, pp. 9–11.

47. Edwards, C.R.W.; Bouchier, I.A.D. *Davidson's Principles, and Practice Medicine*; Churchill Livingstone Press: London, UK, 1991; p. 492.

48. Daniel, W.W. *Biostatistics: A Foundation for Analysis in the Health Sciences*, 6th ed.; Wiley: New York, NY, USA, 1995; pp. 273–303.

International Journal of
Molecular Sciences

MDPI

Article

Binding Interactions of Zinc Cationic Porphyrin with Duplex DNA: From B-DNA to Z-DNA

Tingxiao Qin [1,2,3], Kunhui Liu [2], Di Song [1,*], Chunfan Yang [2], Hongmei Zhao [1] and Hongmei Su [2,*]

[1] Beijing National Laboratory for Molecular Sciences (BNLMS), Institute of Chemistry, Chinese Academy of Sciences, Beijing 100190, China; qtx0521@iccas.ac.cn (T.Q.); hmzhao@iccas.ac.cn (H.Z.)
[2] College of Chemistry, Beijing Normal University, Beijing 100875, China; kunhui@bnu.edu.cn (K.L.); yangchunfan@bnu.edu.cn (C.Y.)
[3] University of Chinese Academy of Sciences, Beijing 100049, China
* Correspondence: songdi@iccas.ac.cn (D.S.); hongmei@bnu.edu.cn (H.S.); Tel.: +86-10-6255-2723 (D.S.)

Received: 28 February 2018; Accepted: 27 March 2018; Published: 4 April 2018

Abstract: Recognition of unusual left-handed Z-DNA by specific binding of small molecules is crucial for understanding biological functions in which this particular structure participates. Recent investigations indicate that zinc cationic porphyrin (ZnTMPyP4) is promising as a probe for recognizing Z-DNA due to its characteristic chiroptical properties upon binding with Z-DNA. However, binding mechanisms of the ZnTMPyP4/Z-DNA complex remain unclear. By employing time-resolved UV-visible absorption spectroscopy in conjunction with induced circular dichroism (ICD), UV-vis, and fluorescence measurements, we examined the binding interactions of ZnTMPyP4 towards B-DNA and Z-DNA. For the ZnTMPyP4/Z-DNA complex, two coexisting binding modes were identified as the electrostatic interaction between pyridyl groups and phosphate backbones, and the major groove binding by zinc(II) coordinating with the exposed guanine N_7. The respective contribution of each mode is assessed, allowing a complete scenario of binding modes revealed for the ZnTMPyP4/Z-DNA. These interaction modes are quite different from those (intercalation and partial intercalation modes) for the ZnTMPyP4/B-DNA complex, thereby resulting in explicit differentiation between B-DNA and Z-DNA. Additionally, the binding interactions of planar TMPyP4 to DNA were also investigated as a comparison. It is shown that without available virtual orbitals to coordinate, TMPyP4 binds with Z-DNA solely in the intercalation mode, as with B-DNA, and the intercalation results in a structural transition from Z-DNA to B-ZNA. These results provide mechanistic insights for understanding ZnTMPyP4 as a probe of recognizing Z-DNA and afford a possible strategy for designing new porphyrin derivatives with available virtual orbitals for the discrimination of B-DNA and Z-DNA.

Keywords: Z-DNA; ZnTMPyP4; transient absorption spectroscopy; binding mode

1. Introduction

DNA molecules are highly polymorphic and can form multiple conformations under different physiological conditions [1,2]. The most common structure of DNA is known as B-DNA, a right-handed double helix with the negatively charged deoxyribose-phosphate backbone outside and stacked base pairs inside. In some conditions, e.g., high ionic strengths, in the presence of highly charged cations or some small molecules, certain B-DNA sequences can transform to left-handed helical conformations called Z-DNA with the zig-zag backbone and the stacked base pairs partially exposed to the outside [2–4]. With the unique structure and higher energy, Z-DNA is less common than B-DNA [5,6]. Nevertheless, a considerable number of studies have demonstrated that Z-DNA is relevant to many biological processes; for example, it has been found that Z-DNA sequences participate in chromatin-dependent activation of the CSF1 (the human colony-stimulating factor 1 gene) promoter [7].

Differentiating B-DNA and Z-DNA is of fundamental importance for investigating the role played by Z-DNA in many biological processes [1]. Generally, recognition of Z-DNA is achieved by utilizing Z-DNA binding small molecules, which have particular binding modes to Z-DNA that are distinct from those to B-DNA, thus allowing the specific DNA form to be determined [8,9]. Although a large number of small molecules such as porphyrin molecules and its analogues, have been reported to bind with B-DNA and Z-DNA, only a few of them can recognize Z-DNA via detection of particular binding modes with Z-DNA [9–12]. This indicates that selecting an appropriate small molecule as a probe for Z-DNA requires a deep mechanistic understanding for the binding modes of small molecules to B-DNA and Z-DNA.

As a derivative of meso-tetrakis[4-(N-methylpyridiumyl)]-21H,23H-porphyrin (TMPyP4), zinc(II) cationic porphyrin (ZnTMPyP4) has received much attention, because zinc(II) can additionally afford an axial coordination in the center of porphyrin, which may enable ZnTMPyP4 to become a promising probe for recognizing Z-DNA through different DNA binding behaviors. Balaz et al. investigated the binding of ZnTMPyP4 to B-DNA and Z-DNA (i.e., B-poly(dG-dC)$_2$ and Z-poly(dG-dC)$_2$) by using induced circular dichrosim (ICD) spectroscopy [9,13], and obtained distinct ICD signals in the Soret region of ZnTMPyP4 with a small negative peak for B-DNA and a very intense bisignate curve for Z-DNA. They assumed that such a dramatic change in ICD spectra might be ascribed to different interaction modes between the binding of ZnTMPyP4 with B-DNA and Z-DNA. The small negative ICD peak suggests ZnTMPyP4 should be partially intercalated between GC pairs of B-DNA, whereas the intense bisignate curve reflects the electronic interaction between two bound ZnTMPyP4, and each ZnTMPyP4 may adopt its central metal zinc(II) to bind with guanine N$_7$ of Z-DNA exposed externally (Scheme 1) [14].

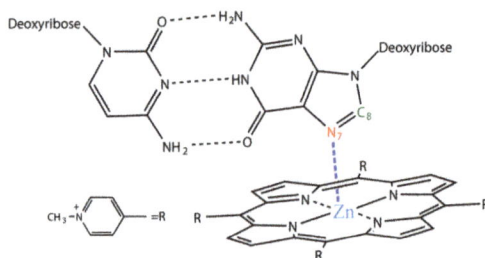

Scheme 1. Schematic diagram for the central metal zinc(II) of ZnTMPyP4 coordinating with guanine N$_7$ of Z-DNA.

Recently, Gong et al. studied the interactions of ZnTMPyP4 with Z-DNA (i.e., Z-poly(dG-dC)$_2$) and B-DNA (i.e., B-poly(dG-dC)$_2$) by means of linear dichroism (LD) spectroscopy [15]. They observed a strong positive LD peak in the Soret region for the ZnTMPyP4/Z-DNA complex, which is quite different from the LD spectrum of the ZnTMPyP4/B-DNA complex, which shows a very weak negative peak in the same region. By calculating angles of the B$_x$ and B$_y$ transitions of ZnTMPyP4 relative to the DNA helix axis, they proposed two possible ways for ZnTMPyP4 to bind with the major groove of Z-DNA: the molecular plane of porphyrin either parallel or perpendicular to the DNA helix axis.

To further understand the binding mechanisms underlying these spectroscopic changes, this work attempts to examine the binding of ZnTMPyP4 towards B-DNA and Z-DNA (B-poly(dG-dC)$_2$ and Z-poly(dG-dC)$_2$) respectively by employing time-resolved UV-visible absorption spectroscopy in conjunction with multiple steady-state spectroscopic techniques of ICD, UV-vis and fluorescence. The steady-state spectroscopic results first indicate a partial intercalation binding mode for the ZnTMPyP4/B-DNA complex and an external groove binding mode for ZnTMPyP4/Z-DNA complex.

Further, time-resolved UV-visible absorption spectroscopy monitors the triplet-state decay kinetics of the bound ZnTMPyP4 with B-DNA and Z-DNA, and reveals that there are two coexisting binding modes for ZnTMPyP4 with each DNA. For ZnTMPyP4 interacting with B-DNA, the two lifetime components of 30.4 ± 0.1 μs (58%) and 11.9 ± 0.2 μs (42%) are separately assigned to the intercalation mode and partial intercalation mode, according to the degree of shielding of triplet states from oxygen quenching. For the binding of ZnTMPyP4 with Z-DNA, the shorter-lived component (3.4 ± 0.3 μs) arises from the electrostatic interaction between pyridyl groups of ZnTMPyP4 and phosphate backbones of Z-DNA, allowing two sides of the ZnTMPyP4 macrocycle to be almost fully exposed to oxygen quenching. The exposed guanine N_7 atom in Z-DNA provides a site for axial coordination with the central zinc(II) of ZnTMPyP4. The longer-lived component (14.8 ± 0.2 μs) thereby corresponds to a major groove binding mode in which the central metal zinc(II) of ZnTMPyP4 coordinates with guanine N_7 of Z-DNA. This assignment validates the existence of the axially coordinated binding between ZnTMPyP4 and Z-DNA. In addition, the binding interactions of planar TMPyP4 to DNA are also investigated. The comparison of TMPyP4 with ZnTMPyP4 reveals the key roles of the coordination ability of Zn(II) in altering the binding behavior of porphyrin. These results provide clear pictures of binding modes for ZnTMPyP4/DNA interactions and shed light on the mechanistic understanding of ZnTMPyP4 as a probe for recognizing the structural change form B-DNA and Z-DNA.

2. Results and Discussion

2.1. Characterizing the Formation of B-DNA and Z-DNA Conformations

The alternating purine–pyrimidine sequence poly(dG-dC)$_2$ (poly(deoxyguanylicdeoxycytidylic) acid; approximate average length in base pairs is 800) forms a typical B-DNA structure [9] and is subject to favorable transition to the Z-form conformation in the presence of micromolar concentrations of protonated spermine [13]. Figure 1 shows the Circular dichroism (CD) spectrum of B-poly(dG-dC)$_2$ (denoted by B-DNA hereafter), which features a complex positive CD band centered at 274 nm and a negative CD band at 253 nm. Under the experimental conditions of spermine (12 μM) and Na-cacodylate buffer (1 mM, pH = 7.0), the CD spectrum presents a negative CD band at 293 nm and a positive CD band at 263 nm (Figure 1), which is characteristic for Z-DNA, indicating that the B-DNA has been successfully induced to the Z-poly(dG-dC)$_2$ (denoted by Z-DNA hereafter).

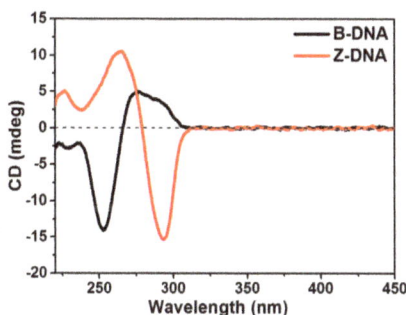

Figure 1. The CD spectra of B-DNA (50 μM) and Z-DNA (50 μM) in the presence of spermine (12 μM) in Na-cacodylate buffer (1 mM, pH = 7.0).

2.2. Interaction of TMPyP4 with B-DNA and Z-DNA

For comparison with ZnTMPyP4/DNA interactions, we first investigated the binding behaviors of planar porphyrin TMPyP4 with B-DNA and Z-DNA. Figure 2a shows the absorption spectra of TMPyP4 (2 μM) in the absence and presence of B-DNA (50 μM) in Na-cacodylate buffer. Compared to free TMPyP4, large hypochromic effects (H = 38%) and red-shifts (22 nm, from 422.0 to 444.0 nm) of the

Soret band of TMPyP4 are observed in the presence of B-DNA, indicating the intercalative binding of TMPyP4 with B-DNA [16,17]. The ICD spectrum of TMPyP4 further confirms this intercalation binding mode, which displays an induced negative ICD band at 438 nm (−2.8 mdeg) in the presence of B-DNA, as shown in Figure 3a [16,18,19]. In addition, Figure S1a shows the splitting of the emission into two peaks and the decreased fluorescence intensity of the TMPyP4/B-DNA complex relative to free TMPyP4. This quenching may originate from a good π-π stacking between TMPyP4 and GC pair of B-DNA, implying the existence of the intercalation mode [20]. Interestingly, identical hypochromicity and bathochromic shift, similar negative ICD bands at 436 nm and fluorescence quenching also appear in the absorption, ICD and fluorescence spectra of TMPyP4 (Figure 2b, Figure 3b and Figure S1b) when B-DNA is replaced by Z-DNA. These spectroscopic results suggest that TMPyP4 also intercalates into Z-DNA, which is in good agreement with previous reports [10,11].

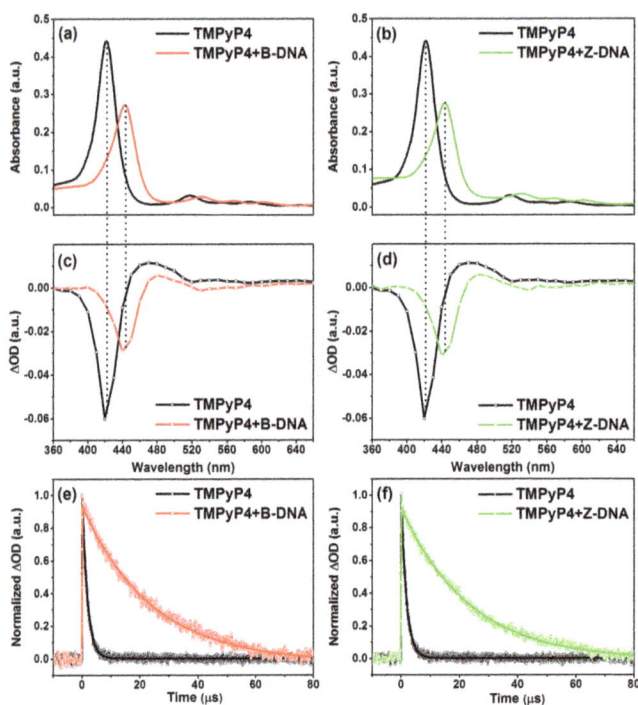

Figure 2. (**a**,**b**) Steady-state UV/Vis absorption spectra and (**c**,**d**) transient UV/Vis absorption spectra recorded instantly (50 ns) after laser flash photolysis upon 355 nm excitation for free TMPyP4 (2 μM) and its complexes with each of the two DNA (50 μM) in Na-cacodylate buffer (1 mM, pH = 7.0). Normalized triplet decay signals after laser flash photolysis of TMPyP4 (2 μM) upon 355 nm excitation in the absence (black) and presence of B-DNA (50 μM) (**e**) and Z-DNA (50 μM) (**f**). Fitted curves are shown by solid lines. The fits are all obtained from complete decay traces.

Second, transient absorption spectroscopy was performed to examine the binding of TMPyP4 to B-DNA and Z-DNA by monitoring the triplet state of TMPyP4. In the transient absorption spectrum of free TMPyP4, a positive peak at 470 nm due to triplet excited-state formation and the negative peak of ground-state depletion at 420 nm are observed, as shown in Figure 2c,d. Accompanying the hypochromicity, the transient absorption of the triplet state is red-shifted to 480 nm when the TMPyP4 is bound with either B-DNA or Z-DNA. Therefore, the triplet excited-state decay kinetics were monitored for free TMPyP4 at 470 nm and DNA/TMPyP4 complexes at 480 nm.

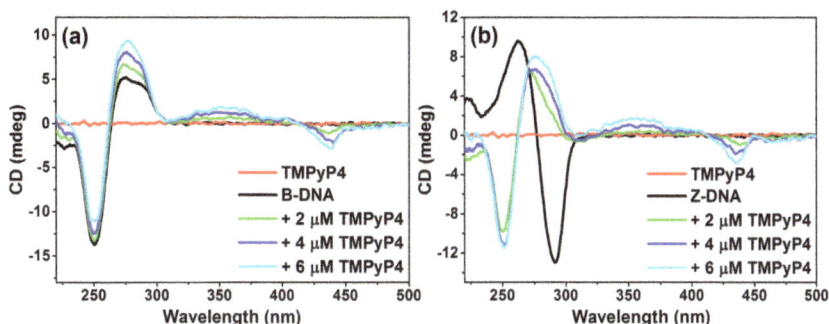

Figure 3. CD spectra of (**a**) B-DNA (50 µM) and (**b**) Z-DNA (50 µM) in the presence of different TMPyP4 concentrations from 2 to 6 µM in Na-cacodylate buffer (1 mM, pH = 7.0).

Figure 2e,f displays the triplet decay curves for TMPyP4 bound, respectively, with B-DNA and Z-DNA, in comparison with free TMPyP4. As can be seen, the triplet decay of free TMPyP4 follows a monoexponential law with a lifetime of 1.7 ± 0.02 µs, which matches early reports [21–23]. When TMPyP4 is bound with B-DNA or Z-DNA, the triplet decays become pronouncedly slower and exhibit first-order exponential laws separately with a lifetime of 27.8 ± 0.3 µs for B-DNA and a lifetime of 28.6 ± 0.2 µs for Z-DNA. Similar lifetimes might correspond to similar binding sites, since the triplet lifetime reflects the microenvironment of TMPyP4 bound with the duplex DNA, which could screen the triplet state from being quenched by oxygen molecules in bulk solution. According to the literature [24], the lifetime components of 27.8 ± 0.3 µs for B-DNA and 28.6 ± 0.2 µs for Z-DNA can both be assigned to the population in intercalation mode, which is consistent with the above steady-state experimental results.

From the steady-state and transient experimental results, it can be concluded that an identical binding mode (intercalation mode) is observed for TMPyP4 interacting with B-DNA or Z-DNA. Furthermore, intercalative TMPyP4 can cause the structural transition of Z-DNA to B-DNA, as evidenced by the disappearance of the marked CD peaks of Z-DNA and the appearance of the characteristic bands of B-DNA (Figure 3b) [10]. These results indicate that the planar TMPyP4 cannot serve as a probe for recognizing Z-DNA.

2.3. Interaction of ZnTMPyP4 with B-DNA and Z-DNA

2.3.1. Interaction of ZnTMPyP4 with Right-Handed B-DNA

Parallel spectroscopic studies were carried out for the binding of ZnTMPyP4 to B-DNA. Similar to TMPyP4, the ICD spectrum of ZnTMPyP4 displays an induced negative CD band at 448.0 nm (−2.7 mdeg) (Figure 4a), and the fluorescence spectrum shows reduced intensity (Figure S2a). However, the absorption spectrum of ZnTMPyP4 shows a similar hypochromic effect (H = 37%), but with a relative small bathochromic shift (10 nm) (Figure 5a) compared to TMPyP4. From these observations, an atypical intercalation mode—namely, partial intercalation—is suggested for the pentacoordinated ZnTMPyP4 with B-DNA, and this may be caused by the steric hindrance of the axially coordinated water molecule on zinc(II), which is in agreement with general knowledge of ZnTMPyP4 [13,23].

The interaction of ZnTMPyP4 with B-DNA was further examined using transient absorption spectroscopy by monitoring the triplet excited-state decay kinetics of ZnTMPyP4. As shown in Figure 5c, a positive peak (triplet excited-state formation) and negative peak (ground-state depletion) are observed at 480 nm and 440 nm in the transient absorption spectrum of free ZnTMPyP4. When ZnTMPyP4 binds to B-DNA, large hypochromicity and small redshifts (10 nm) are observed for both triplet excited-state formation and ground-state depletion, which is in accordance with the steady-state absorption spectra. Therefore, the triplet decay kinetics of free ZnTMPyP4 and the ZnTMPyP4/B-DNA complex were measured at 480 nm and at 490 nm, respectively.

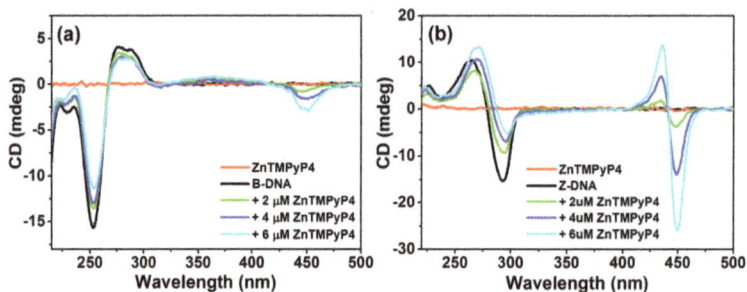

Figure 4. CD spectra of (**a**) B-DNA (50 µM) and (**b**) Z-DNA (50 µM) in the presence of different ZnTMPyP4 concentrations from 2 to 6 µM in Na-cacodylate buffer (1 mM, pH = 7.0).

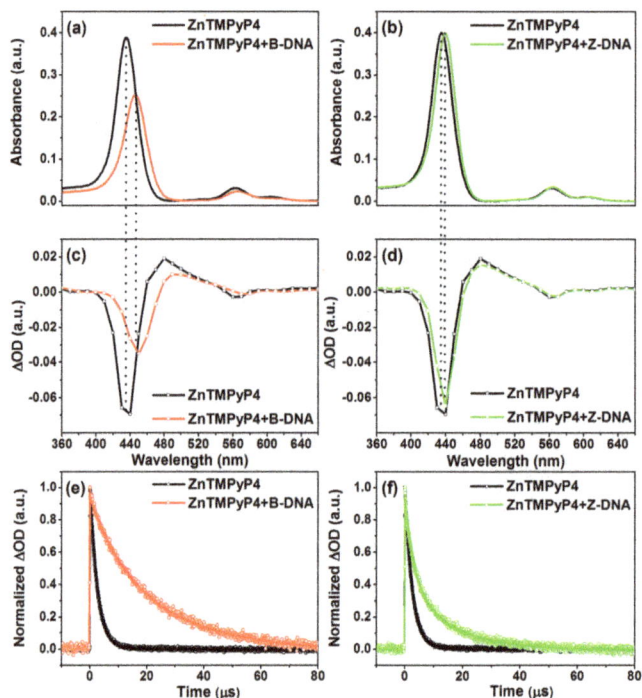

Figure 5. (**a,b**) Steady-state UV/Vis absorption spectra and (**c,d**) transient UV/Vis absorption spectra recorded instantly (50 ns) after laser flash photolysis upon 355 nm excitation for free ZnTMPyP4 (2 µM) and its complexes with each of the two DNA (50 µM)) in Na-cacodylate buffer (1 mM, pH = 7.0). Normalized triplet decay signals after laser flash photolysis of ZnTMPyP4 (2 µM) upon 355 nm excitation in the absence (black) and presence of B-DNA (50 µM) (**e**) and Z-DNA (50 µM) (**f**). Fitted curves are shown by solid lines. The fits are all obtained from complete decay traces.

As shown in Figure 5e, a monoexponential decay behavior with a triplet lifetime of 2.6 ± 0.01 µs is exhibited for free ZnTMPyP4, and this inherent triplet lifetime is longer than that of TMPyP4. In contrast, the triplet state decay kinetics of the ZnTMPyP4/B-DNA complex becomes much slower and can be well fitted by two exponential components with lifetimes of 11.9 ± 0.2 μs and 30.4 ± 0.1 μs (Figure 5e, Table 1). As mentioned above, the triplet state lifetime reflects the degree of shielding of the triplet ligand from oxygen quenching; thus, the biexponential decay for the bound ZnTMPyP4 strongly implies the simultaneous existence of two binding modes. One of the two coexisting binding modes should be the partial intercalation mode, which is also suggested by the steady-state experiments. Since the intercalation of ZnTMPyP4 between GC base pairs of the duplex was characterized in previous work [20], and the axial water was assumed to be released by gaining enough energy during its intercalation process, the other binding mode may possibly be the intercalation mode. Obviously, the intercalation mode much better protects the porphyrin macrocycle from molecular oxygen access, which quenches the triplet state, than the partial intercalation mode. The currently observed longer lifetime component of 30.4 ± 0.1 μs should thus correspond to the population bound in the intercalation mode, whereas the shorter lifetime component of 11.9 ± 0.2 μs is assigned to the population bound in the partial intercalation mode. The 30.4 ± 0.1 μs lifetime is comparable with the intercalative mode of TMPyP4 (27.8 ± 0.3 μs or 28.6 ± 0.2 μs) in poly(dG-dC)$_2$, indicating the reasonability of this assignment.

Table 1. Triplet decay lifetimes of TMPyP4 and ZnTMPyP4 in air-saturated solution and in their complexes with B-DNA and Z-DNA. Pre-exponential factors of the two lifetime components in the biexponential fitting $I = I_0 + A_1 e^{-t/\tau_1} + A_2 e^{-t/\tau_2}$ yield the respective percentages of two binding modes (values shown in brackets).

Sample	τ_1 (μs)	τ_2 (μs)
TMPyP4	1.7 ± 0.02	
TMPyP4+B-DNA	27.8 ± 0.3	
TMPyP4+Z-DNA	28.6 ± 0.2	
ZnTMPyP4	2.6 ± 0.01	
ZnTMPyP4+B-DNA	11.9 ± 0.2 (42%)	30.4 ± 0.1 (58%)
ZnTMPyP4+Z-DNA	3.4 ± 0.3 (53%)	14.8 ± 0.2 (47%)

In addition, the triplet reporter method also assesses the contribution of different binding modes. With the biexponential fitting, the pre-exponential factors obtained for the two lifetime components actually correspond to the respective percentages of the binding modes, and these values are also listed in Table 1. From Table 1, it is shown that intercalation is the main binding mode for nearly 58%, while the partial intercalation mode accounts for ~42%.

2.3.2. Interaction of ZnTMPyP4 with Left-Handed Z-DNA

The binding of ZnTMPyP4 with Z-DNA was further examined, and is different from that with B-DNA, interestingly. As shown in Figure 5b, addition of ZnTMPyP4 to the spermine-induced Z-form of poly(dG–dC)$_2$ results in a bathochromic shift ($\Delta\lambda = 4$ nm, from 435 to 439 nm) and no hypochromicity of the Soret band, which is quite different from the absorption spectrum of ZnTMPyP4/B-DNA, which displays large hypochromic effects ($H = 37\%$) and red-shifts (10 nm). Normally, for the interaction of porphyrin with duplexes, the intercalation mode results in a strong bathochromic shift (≥ 15 nm) and marked hypochromicity ($H \geq 35\%$) in the Soret band [16,17], whereas the external groove binding mode exhibits a smaller bathochromic shift (≤ 8 nm) and weaker hypochromicity ($H \leq 10\%$) [16,19]. Therefore, in the current work, such a small red-shift and absence of hypochromicity suggest an external groove binding mode for the ZnTMPyP4/Z-DNA complex (Scheme 2), instead of the partial intercalation and intercalation modes that are detected for the ZnTMPyP4/B-DNA complex.

The different binding behavior can also be seen from the fluorescence spectrum of the ZnTMPyP4/Z-DNA complex, which is shown in Figure S2b. Unlike TMPyP4, the fluorescence of ZnTMPyP4 is not quenched upon binding to Z-DNA, and the emission at 635 nm is obviously blue-shifted to 627 nm. This is clearly indicative of the absence of π-π interactions between GC base pair and ZnTMPyP4 core and the existence of an external groove binding mode in which guanine N_7 atom of Z-DNA is available for coordination with Zn(II), as concluded in previous fluorescence studies [9].

Figure 4b shows the CD spectrum of ZnTMPyP4 in the presence of Z-DNA, which reveals a bisignate signal in the Soret region with a negative CD band at 450 nm (-26 mdeg) and a positive CD band at 436 nm (+14 mdeg). As discussed by previous reports, the induced bisignate CD signals may originate from long-distance through-space, electric dipole–dipole, and exciton coupling of the two ZnTMPyP4 chromophores [9,25], which might adopt a groove binding mode with substitution of the axial water molecule in ZnTMPyP4 by guanine N_7 of Z-DNA.

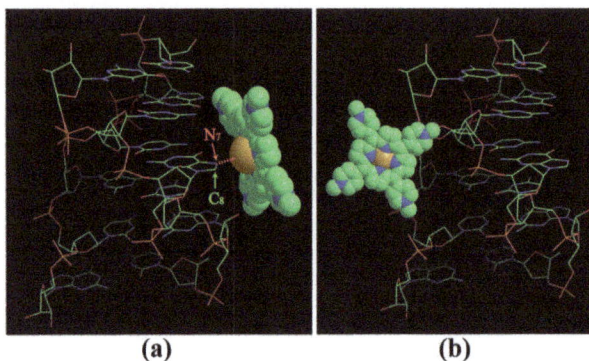

(a)   **(b)**

Scheme 2. Depiction of the possible binding modes of ZnTMPyP4 to the Z-DNA: (a) the major groove coordination binding mode; (b) the electrostatic interaction binding mode. (Crystal structure of the Z-DNA with about half a turn was extracted from RCSB, PDB entry: 5EBI [26]).

To further reveal the binding modes of ZnTMPyP4 with Z-DNA, transient absorption spectroscopy measurements were performed. In the transient absorption spectrum of the bound ZnTMPyP4 with Z-DNA, no hypochromicity and slight redshifts are observed for triplet state formation and ground-state depletion (Figure 5d). The triplet state decay kinetics of ZnTMPyP4/Z-DNA complex was measured at 480 nm. Figure 5f shows its triplet decay behavior, which is obviously distinct from that of the ZnTMPyP4/B-DNA complex. The ZnTMPyP4 triplet state follows a second-order exponential decay with two lifetime components of 3.4 ± 0.3 μs and 14.8 ± 0.2 μs, which suggests the existence of two coexisting binding modes with their respective percentages of 53% and 47%.

The long lifetime of 14.8 ± 0.2 μs is much shorter than the typical intercalative lifetime ($\sim$30 μs) [24], which first excludes the possibility of the intercalation mode for ZnTMPyP4/Z-DNA complex. The possibility of a partial intercalation mode for ZnTMPyP4/Z-DNA complex can be also ruled out, according to the structural features of Z-DNA. Previous structural inspections of Z-DNA have indicated that the alternate *anti* and *syn* conformation of nucleobases leads to exposure of guanine atoms N_7 and C_8, which are both shielded in the B form [2]. In Z-DNA, the nitrogen atom N_7 is available for coordination with transition-metal ions, as shown previously by IR study [27,28], and thus can provide a site for axial coordination with the central Zn(II) of ZnTMPyP4 (Scheme 2a). It is well known that the nitrogen atom N is less electronegative and more easily renders electron pairs than the oxygen atom O, resulting in its stronger coordination ability. Therefore, the central Zn(II) of ZnTMPyP4 should more readily coordinate with the guanine atom N_7 of Z-DNA than the water molecule in the solution, which facilitates the coordination binding of ZnTMPyP4 with the guanine N_7 of Z-DNA, and rules out the possibility of partial intercalation binding.

Additionally, it should also be noted that the minor groove of Z-DNA is narrow and deep, and spermine is located here [29,30], preventing the binding of the ZnTMPyP4 in the minor groove. In this case, the most plausible binding mode for the long lifetime of 14.8 ± 0.2 μs is coordination binding in the major groove with substitution of the axial water molecule in ZnTMPyP4 by guanine N_7 of Z-DNA, as shown in Scheme 2a. In the geometry associated with this binding mode, one side of the prophyrin macrocycle faces the major groove of Z-DNA, while the other side is exposed to the solution. It follows that only one side of ZnTMPyP4 can receive efficient protection from oxygen access, leading to a moderate triplet state lifetime. This assignment of the coordination binding in the major groove is corroborated by the CD research results of the ZnTMPyP4/Z-DNA complex, in which induced bisignate CD signals due to the electronic coupling of the two ZnTMPyP4 chromophores are observed.

As for the short lifetime of 3.4 ± 0.3 µs of Z-DNA/ZnTMPyP4 complex, it is close to the typical triplet lifetime corresponding to electrostatic interaction binding mode, which was obtained for ZnTMPyP4 in B-poly(dG-dC)$_2$ (3.5 µs) at high ionic strength by Chirvony, V.S. et al. [20]. Consequently, the 3.4 ± 0.3 µs lifetime component can be assigned to the population in the electrostatic interaction binding mode. In this binding geometry (Scheme 2b), ZnTMPyP4 is bound electrostatically by the positively charged pyridyl groups to negatively charged phosphate backbones on the outside with minimal interaction between the ZnTMPyP4 macrocycle and Z-DNA, allowing the exposure of two sides of ZnTMPyP4 to oxygen quenching and thus causing a short triplet state lifetime of ~3.4 µs. This assignment is supported by LD study, which proposed that the ZnTMPyP4 molecule also binds across the groove by electrostatic interaction, with the molecular plane of porphyrin perpendicular to the DNA helix axis [15].

Notably, the short lifetime of 3.4 ± 0.3 µs of ZnTMPyP4/Z-DNA complex is only 31% longer than that of free ZnTMPyP4 (2.6 ± 0.01 µs). Should this lifetime component also be ascribed to the unbound ZnTMPyP4 in the solution? To scrutinize this possibility, additional CD spectra of ZnTMPyP4 were measured at different [ZnTMPyP4]/[Z-DNA] molar ratios ranging from 0.04 to 0.16. As shown in Figure 4b, the intensity of the bisignate peak at the soret region increases with increasing [ZnTMPyP4]/[Z-DNA] ratio. The saturation of the ICD signal is observed at the [ZnTMPyP4]/[Z-DNA] of 0.12 with the ICD of −26 mdeg (Figure S3). In our triplet state experiment, the concentration of the Z-poly(dG–dC)$_2$ (per base pair) is 50 µM and the concentration of the ZnTMPyP4 is 2 µM. Therefore, the [ZnTMPyP4]/[Z-DNA] ratio used in this work is 0.04, which is much lower than the binding saturation ratio 0.12. This indicates that each ZnTMPyP4 molecule is bound to the Z-DNA, thus ruling out the possibility that the short lifetime component of 3.4 ± 0.3 µs is caused by free ZnTMPyP4.

In brief, the triplet state kinetics results, together with the steady-state experimental results, reveal the simultaneous existence of major groove binding by forming N_7-Zn coordination bond and an electrostatic binding for the ZnTMPyP4/Z-DNA complex, which are quite different from the case of the ZnTMPyP4/B-DNA complex. In addition, compared with planar TMPyP4, ZnTMPyP4 provides a virtual orbital to coordinate with the exposed guanine N_7 atom of Z-DNA, leading to the unusual binding mode of ZnTMPyP4 distinguishing Z-DNA from B-DNA. These results not only provide mechanistic insights for ZnTMPyP4 acting as a probe for recognizing Z-DNA, but also afford a possible strategy for designing new porphyrin derivatives with available virtual orbitals for the discrimination of B-DNA and Z-DNA.

3. Materials and Methods

3.1. Materials

The porphyrin derivative meso-tetrakis[4-(*N*-methylpyridiumyl)]-21*H*,23*H*-porphyrin (TMPyP4) in the form of tetra-p-tosylate salt was purchased from Tokyo Chemical Industry (TCI, Tokyo, Japan) and used as received. Zinc(II)-meso-tetrakis(4-(*N*-methylpyridiumyl))-porphyrin (ZnTMPyP4) was purchased from Frontier Scientific (Logan, UT, USA) and used without any further purification. Poly(deoxyguanylic–deoxycytidylic) acid sodium salt (poly(dG–dC)$_2$), sodium cacodylate, sodium chloride, spermine tetrahydrochloride ($C_{10}H_{26}N_4\cdot4HCl$) were purchased from Sigma-Aldrich (St. Louis, MO, USA). Deionized water was obtained from a Milli-Q system with a resistivity of

18.2 MΩ·cm. DNA samples were prepared in a sodium cacodylate buffer (1 mM, pH 7.0), annealed at 70 °C for 20 min, and then cooled to 10 °C at 1 °C/min, and kept at 4 °C for 12 h. The concentration of the poly(dG–dC)$_2$ (per base pair) was quantified by UV-vis spectroscopy using the extinction coefficient $\varepsilon = 8.8 \times 10^3$ M^{-1}·cm^{-1} at 260 nm [13]. The porphyrin stock solutions were prepared in a sodium cacodylate buffer (1 mM, pH 7.0), and the concentration was determined by UV-vis spectroscopy using the following extinction coefficients $\varepsilon = 2.26 \times 10^5$ M^{-1}·cm^{-1} at 424 nm for TMPyP4 [22] and $\varepsilon = 2.04 \times 10^5$ M^{-1}·cm^{-1} at 437 nm for ZnTMPyP4 [16].

3.2. Steady-State Spectroscopy

CD spectra were recorded at room temperature on a Jasco J-815 spectropolarimeter (JASCO, Oklahoma City, OK, USA). Each sample was collected from 550 to 200 nm at a scan speed of 200 nm/min with a response time of 0.5 s in a 1 cm quartz cell. The reported spectroscopy was the average of three scans. The spectrum from a blank sample containing only buffer was used as the background that was subtracted from the averaged data. Steady-state absorption spectroscopy was recorded with a UV-vis spectrometer (model U-3010, Hitachi, Tokyo, Japan). Quartz cuvettes of 1 cm path length were used for all absorption measurement. Fluorescence spectroscopy was measured with a fluorescence spectrometer (F4600, Hitachi). All samples were excited at 355 nm.

3.3. Laser Flash Photolysis

Transient UV-visible spectroscopy and triplet state kinetics were measured by a nanosecond time-resolved laser flash photolysis (LFP) setup that has been described previously [22]. Briefly, the instrument comprises an Edinburgh LP920 spectrometer (Edinburgh Instrument Ltd., Livingstone, UK) combined with an Nd:YAG laser (Surelite II, Continuum Inc., Christiansburg, VI, USA). The excitation wavelength is a 355 nm laser pulse from Q-switched Nd:YAG laser (1 Hz, fwhm ≈ 7 ns, 10 mJ/pulse). The analyzing light is from a 450 W pulsed xenon lamp. A monochromator equipped with a photomultiplier for collecting the spectroscopy range from 350 to 700 nm was used to analyze transient absorption spectroscopy. The signals from the photomultiplier were displayed and recorded as a function of time on a 100 MHz (1.25 Gs/s sampling rate) oscilloscope (Tektronix, Beaverton, OR, USA, TDS 3012B), and the data were transferred to a personal computer. Data were analyzed by the online software of the LP920 spectrophotometer. The fitting quality was judged by weighted residuals and reduced χ^2 value. Sample solutions were freshly prepared for each measurement.

3.4. Structural Transition from B-DNA to Z-DNA

Micromolar concentration of spermine (a fully protonated tetraamine at pH = 7.0) was used to induce the B-DNA to Z-DNA transition in vitro. Z-DNA was formed by incubating at 60 °C with 12 μM spermine for 10 min, then slowly cooled down to room temperature. The successful transition from B-DNA to Z-DNA was ensured by CD spectroscopic results showing the characteristics of left-handed Z-DNA: negative CD bands at 293 nm, and a positive CD band at 263 nm [13].

4. Conclusions

In this work, we comprehensively studied the binding interactions of ZnTMPyP4 with two different duplex DNA, B-DNA and Z-DNA, by combining transient UV-vis absorption spectroscopy with steady-state measurements (UV-vis, fluorescence, and ICD). For the ZnTMPyP4/B-DNA complex, it is found that the partial intercalation and intercalation modes through π-π stacking of ZnTMPyP4 macrocycle with GC base pairs coexist. In contrast to this type of π-π stacking interaction mode, ZnTMPyP4 binds with Z-DNA by means of another two interaction modes, i.e., the electrostatic interaction between pyridyl groups and phosphate backbones, and major groove binding by Zn(II) coordinating with the exposed guanine N$_7$. The unusual binding behaviors of ZnTMPyP4 with Z-DNA thus make it possible to utilize ZnTMPyP4 as a probe for discerning Z-DNA from B-DNA. In addition, we also investigated the binding of planar TMPyP4 to B-DNA and Z-DNA as a comparison. It is

Int. J. Mol. Sci. **2018**, *19*, 1071

shown that without available virtual orbitals to coordinate, TMPyP4 binds with Z-DNA solely in the intercalation mode, which is identical to B-DNA. The intercalation of TMPyP4 results in structural transition from Z-DNA to B-ZNA. These results elaborate the binding mechanisms of ZnTMPyP4 with Z-DNA in comparison with B-DNA and reveal the key roles of the coordination ability of the Zn(II) cation in altering the binding behavior of porphyrin, thus providing valuable guidance for the design and selection of molecular probes in recognizing the important Z-DNA structure.

Supplementary Materials: Supplementary materials can be found at http://www.mdpi.com/1422-0067/19/4/1071/s1.

Acknowledgments: This work was financially supported by the National Natural Science Foundation of China (Grant No. 21773257, No. 21703011, No. 21373233, and No. 91441108).

Author Contributions: Hongmei Su conceived the project and designed the experiments; Tingxiao Qin performed experiments; Tingxiao Qin, Kunhui Liu, Di Song, Chunfan Yang, Hongmei Zhao and Hongmei Su discussed the results and analyzed the data. Tingxiao Qin, Di Song and Hongmei Su wrote the manuscript. All authors reviewed the manuscript.

Conflicts of Interest: The authors declare no conflict of interest.

References

1. Rich, A.; Nordheim, A.; Wang, A.H.-J. The chemistry and biology of left-handed Z-DNA. *Annu. Rev. Biochem.* **1984**, *53*, 791–846. [CrossRef] [PubMed]

2. Belmont, P.; Constant, J.-F.; Demeunynck, M. Nucleic acid conformation diversity: From structure to function and regulation. *Chem. Soc. Rev.* **2001**, *30*, 70–81. [CrossRef]

3. Pohl, F.M.; Jovin, T.M. Salt-induced co-operative conformational change of a synthetic DNA: Equilibrium and kinetic studies with poly (dG-dC). *J. Mol. Biol.* **1972**, *67*, 375–396. [CrossRef]

4. Lafer, E.M.; Möller, A.; Nordheim, A.; Stollar, B.D.; Rich, A. Antibodies specific for left-handed Z-DNA. *Proc. Natl. Acad. Sci. USA* **1981**, *78*, 3546–3550. [CrossRef] [PubMed]

5. Jovin, T.M.; Soumpasis, D.M.; McIntosh, L.P. The transition between B-DNA and Z-DNA. *Annu. Rev. Phys. Chem.* **1987**, *38*, 521–558. [CrossRef]

6. Rich, A.; Zhang, S. Z-DNA: The long road to biological function. *Nat. Rev. Genet.* **2003**, *4*, 566. [CrossRef] [PubMed]

7. Liu, R.; Liu, H.; Chen, X.; Kirby, M.; Brown, P.O.; Zhao, K. Regulation of CSF1 promoter by the SWI/SNF-like BAF complex. *Cell* **2001**, *106*, 309–318. [CrossRef]

8. Xu, Y.; Zhang, Y.X.; Sugiyama, H.; Umano, T.; Osuga, H.; Tanaka, K. (*P*)-helicene displays chiral selection in binding to Z-DNA. *J. Am. Chem. Soc.* **2004**, *126*, 6566–6567. [CrossRef] [PubMed]

9. Balaz, M.; De Napoli, M.; Holmes, A.E.; Mammana, A.; Nakanishi, K.; Berova, N.; Purrello, R. A cationic zinc porphyrin as a chiroptical probe for Z-DNA. *Angew. Chem. Int. Ed.* **2005**, *44*, 4006–4009. [CrossRef] [PubMed]

10. Pasternack, R.F.; Sidney, D.; Hunt, P.A.; Snowden, E.A.; Gibbs, J. Interactions of water soluble porphyrins with Z-poly (dG-dC). *Nucleic Acids Res.* **1986**, *14*, 3927–3943. [CrossRef] [PubMed]

11. McKinnie, R.E.; Choi, J.D.; Bell, J.W.; Pasternack, R.F.; Gibbs, E.J. Porphyrin induced Z to B conversion of poly (dG-dC)₂ in ethanol. *J. Inorg. Biochem.* **1988**, *32*, 207–224. [CrossRef]

12. D'Urso, A.; Mammana, A.; Balaz, M.; Holmes, A.E.; Berova, N.; Lauceri, R.; Purrello, R. Interactions of a tetraanionic porphyrin with DNA: From a Z-DNA sensor to a versatile supramolecular device. *J. Am. Chem. Soc.* **2009**, *131*, 2046–2047. [CrossRef] [PubMed]

13. Choi, J.K.; D'Urso, A.; Balaz, M. Chiroptical properties of anionic and cationic porphyrins and metalloporphyrins in complex with left-handed Z-DNA and right-handed B-DNA. *J. Inorg. Biochem.* **2013**, *127*, 1–6. [CrossRef] [PubMed]

14. Nejdl, L.; Ruttkay-Nedecky, B.; Kudr, J.; Krizkova, S.; Smerkova, K.; Dostalova, S.; Vaculovicova, M.; Kopel, P.; Zehnalek, J.; Trnkova, L.; et al. DNA interaction with zinc(II) ions. *Int. J. Biol. Macromol.* **2014**, *64*, 281–287. [CrossRef] [PubMed]

15. Gong, L.; Jang, Y.J.; Kim, J.; Kim, S.K. Z-form DNA specific binding geometry of Zn(II) *meso*-tetrakis (*N*-methylpyridinium-4-yl) porphyrin probed by linear dichroism spectroscopy. *J. Phys. Chem. B* **2012**, *116*, 9619–9626. [CrossRef] [PubMed]

16. Pasternack, R.F.; Gibbs, E.J.; Villafranca, J.J. Interactions of porphyrins with nucleic-acids. *Biochemistry* **1983**, *22*, 2406–2414. [CrossRef] [PubMed]

17. McMillin, D.R.; Shelton, A.H.; Bejune, S.A.; Fanwick, P.E.; Wall, R.K. Understanding binding interactions of cationic porphyrins with B-form DNA. *Coord. Chem. Rev.* **2005**, *249*, 1451–1459. [CrossRef]

18. Pasternack, R.F. Circular dichroism and the interactions of water soluble porphyrins with DNA—A minireview. *Chirality* **2003**, *15*, 329–332. [CrossRef] [PubMed]

19. D'Urso, A.; Nardis, S.; Pomarico, G.; Fragalà, M.E.; Paolesse, R.; Purrello, R. Interaction of tricationic corroles with single/double helix of homopolymeric nucleic acids and DNA. *J. Am. Chem. Soc.* **2013**, *135*, 8632–8638. [CrossRef] [PubMed]

20. Chirvony, V.S.; Galievsky, V.A.; Terekhov, S.N.; Dzhagarov, B.M.; Ermolenkov, V.V.; Turpin, P.Y. Binding of the cationic 5-coordinate Zn(II)-5, 10, 15, 20-tetrakis (4-*N*-methylpyridyl) porphyrin to DNA and model polynucleotides: Ionic-strength dependent intercalation in [poly (dG-dC)]$_2$. *Biospectroscopy* **1999**, *5*, 302–312. [CrossRef]

21. Song, D.; Yang, W.; Qin, T.; Wu, L.; Liu, K.; Su, H. Explicit differentiation of G-quadruplex/ligand interactions: Triplet excited states as sensitive reporters. *J. Phys. Chem. Lett.* **2014**, *5*, 2259–2266. [CrossRef] [PubMed]

22. Qin, T.; Liu, K.; Song, D.; Yang, C.; Su, H. Porphyrin bound to i-motifs: Intercalation versus external groove binding. *Chemistry-An Asian Journal* **2017**, *12*, 1578–1586. [CrossRef] [PubMed]

23. Yao, X.; Song, D.; Qin, T.; Yang, C.; Yu, Z.; Li, X.; Liu, K.; Su, H. Interaction between G-quadruplex and zinc cationic porphyrin: The role of the axial water. *Sci. Rep.* **2017**, *7*, 10951. [CrossRef] [PubMed]

24. Kruk, N.N.; Dzhagarov, B.M.; Galievsky, V.A.; Chirvony, V.S.; Turpin, P.-Y. Photophysics of the cationic 5, 10, 15,20-tetrakis (4-*N*-methylpyridyl) porphyrin bound to DNA, [poly(dA-dT)]$_2$ and [poly(dG-dC)]$_2$: Interaction with molecular oxygen studied by porphyrin triplet-triplet absorption and singlet oxygen luminescence. *J. Photochem. Photobiol. B Biol.* **1998**, *42*, 181–190. [CrossRef]

25. Mammana, A.; Pescitelli, G.; Asakawa, T.; Jockusch, S.; Petrovic, A.G.; Monaco, R.R.; Purrello, R.; Turro, N.J.; Nakanishi, K.; Ellestad, G.A.; et al. Role of environmental factors on the structure and spectroscopic response of 5'-DNA-porphyrin conjugates caused by changes in the porphyrin–porphyrin interactions. *Chemistry-An European Journal* **2009**, *15*, 11853–11866. [CrossRef] [PubMed]

26. Gilski, M.; Drozdzal, P.; Kierzek, R.; Jaskolski, M. Atomic resolution structure of a chimeric DNA-RNA Z-type duplex in complex with Ba^{2+} ions: A case of complicated multi-domain twinning. *Acta Crystallogr. Sect. D Struct. Biol.* **2016**, *72*, 211–223. [CrossRef] [PubMed]

27. Taboury, J.; Bourtayre, P.; Liquier, J.; Taillandier, E. Interaction of Z form poly (dG-dC). Poly (dG-dC) with divalent metal ions: Localization of the binding sites by IR spectroscopy. *Nucleic Acids Res.* **1984**, *12*, 4247–4258. [CrossRef] [PubMed]

28. Loprete, D.; Hartman, K. Conditions for the stability of the B, C, and Z structural forms of poly (dG-dC) in the presence of lithium, potassium, magnesium, calcium, and zinc cations. *Biochemistry* **1993**, *32*, 4077–4082. [CrossRef] [PubMed]

29. Egli, M.; Williams, L.D.; Gao, Q.; Rich, A. Structure of the pure-spermine form of Z-DNA (magnesium free) at 1-Å Resolution. *Biochemistry* **1991**, *30*, 11388–11402. [CrossRef] [PubMed]

30. Bancroft, D.; Williams, L.D.; Rich, A.; Egli, M. The low-temperature crystal structure of the pure-spermine form of Z-DNA reveals binding of a spermine molecule in the minor groove. *Biochemistry* **1994**, *33*, 1073–1086. [CrossRef] [PubMed]

International Journal of
Molecular Sciences

MDPI

Review

DNA Modified with Boron–Metal Cluster Complexes [M(C₂B₉H₁₁)₂]—Synthesis, Properties, and Applications

Agnieszka B. Olejniczak [1], Barbara Nawrot [2] and Zbigniew J. Leśnikowski [3,*]

[1] Screening Laboratory, Institute of Medical Biology, Polish Academy of Sciences, 106 Lodowa St.,
     93-232 Lodz, Poland; aolejniczak@ibmpan.pl
[2] Centre of Molecular and Macromolecular Studies, Polish Academy of Sciences, Sienkiewicza 112,
     90-363 Lodz, Poland; bnawrot@cbmm.lodz.pl
[3] Laboratory of Molecular Virology and Biological Chemistry; Institute of Medical Biology, Polish Academy of
     Sciences, 106 Lodowa St., 93-232 Lodz, Poland
*   Correspondence: zlesnikowski@cbm.pan.pl; Tel.: +48-42-272-3629

Received: 27 September 2018; Accepted: 4 November 2018; Published: 7 November 2018

Abstract: Together with tremendous progress in biotechnology, nucleic acids, while retaining their status as "molecules of life", are becoming "molecular wires", materials for the construction of molecular structures at the junction between the biological and abiotic worlds. Herein, we present an overview of the approaches for incorporating metal centers into nucleic acids based on metal–boron cluster complexes (metallacarboranes) as the metal carriers. The methods are modular and versatile, allowing practical access to innovative metal-containing DNA for various applications, such as nucleic acid therapeutics, electrochemical biosensors, infrared-sensitive probes, and building blocks for nanoconstruction.

Keywords: DNA; metallacarboranes; boron clusters; oligonucleotide probes; therapeutic nucleic acids

1. Introduction

Metal complexes, due to their versatility in redox and photophysical features, together with well-defined, though diverse coordination geometries, represent a class of advantageous functional components with broad applicability for nucleic acids modification. The incorporation of metal centers into nucleic acids offers difficult to compete opportunities to create a new type of material that merges the informational properties of nucleic with the electronic, magnetic, and optical properties characteristic of metals and their complexes. As a result, materials possessing physical and chemical features that differ significantly from those accessible in bio-organic or inorganic compounds only can be fabricated [1,2].

One can distinguish between two different goals of nucleic acid modification. The first focuses on modulation and improvement of the natural properties of nucleic acids such as specificity and selectivity of binding towards complementary sequences or stability in a biological environment in order to improve the designed, specific activities. This approach is mainly used if modified nucleic acids are applied as therapeutic agents such as antisense oligonucleotides, ribozymes, or interfering RNAs, and perform their natural or nearly natural functions in biological systems [3].

The second goal of nucleic acids modification focuses on incorporation of new, "unnatural" properties into their structures such as fluorescence, luminescence, specific redox activity, radioactivity, and specific affinity-bearing residues. The purpose of these modifications is to equip a nucleic acid with constituents that provide detectable signals under designed conditions. In these technological rather than biological applications, a modified nucleic acid is often used as a material for the construction of diagnostic tests and biosensors. Modified DNA-oligomers used as primers for polymerase chain reaction (PCR) also find broad and diverse applications in molecular biology and molecular medical diagnostics. New applications of nucleic acids as materials for nanoconstruction are still emerging [4,5].

There is a wide variety of entities designed for nucleic acids modification, including the following: (1) fluorescent and affinity labels, (2) redox labels, and (3) modified nucleobases with different properties and applications. There are also different types of metal complexes and organometallic compounds used for that purpose. Some of them have been employed in molecular medicine, diagnostics, and technology, providing a range of useful advantages and practical applications [6–8]. The metal-containing nucleic acids can be divided into the following types: (1) nucleic acids with a metal incorporated in the form of a complex with chelate, (2) metallated-nucleic acids containing metal ions that form a complex directly with double-stranded DNA, and (3) metal-labeled nucleic acids containing metal complexes covalently bound to the DNA strand. The latter type is a major focus of the present outline. Detailed discussion of the specific topics of chemistry and applications of metal-containing nucleic acids is available in several general reviews and books published recently [4,5,9,10]. The present overview is focused on the modification of nucleic acids with metallacarboranes, a unique class of metal complexes consisting of metal ion(s) and boron clusters as metal ligands.

2. Metallacarboranes

Polyhedral boron-rich compounds (boron clusters: *closo*-borates and carboranes) [11] have a broad spectrum of practical applications, from catalysis, optoelectronics, and new materials to medical chemistry and drug design [12,13]. One of the important features of a boron-rich system, such as the dicarbaborate anion (*nido*-7,8-$C_2B_9H_{12}^{2-}$), is the ability to serve as a ligand to form stable complexes with metal ions, that is, metallacarboranes [14]. Metallacarboranes are a large family of metallocene-type complexes containing at least one carborane polyhedron and one or more metallocenters (M = Co, Fe, Ni, Cr, Re, Al, Au, Ir, Mn, Pt, etc.) (Figure 1).

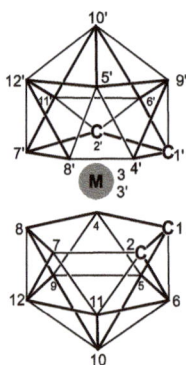

Figure 1. Structure and atom numbering of 3-metal-bis(1,2-dicarbollide) ate $[M(C_2B_9H_{11})_2]^-$. M = Co, Fe, Ni, Cr, Re, Al, Au, Ir, Mn, Pt, etc. Hydrogen atoms of C–H groups and B–H groups of the boron clusters, symbolized by the polygen vertex, were omitted for the clarity.

Metallacarboranes can be viewed as metal sandwich complexes in which a metal ion is facially coordinated to a carborane ligand analogous to metal–cyclopentadiene interactions in metallocenes. In most cases, the metal–carborane binding is strong, and separation of the ligand from the metal is rarely observed. The electron-delocalized bonding in these compounds has a strong stabilizing influence. Thus, both high and low formal metal oxidation states are stabilized as the effects of oxidation or reduction at the metal center are moderated via distribution over the cage skeleton [15].

Two general approaches have been developed for the modification of nucleic acids with carboranes and metallacarboranes: (1) de novo, step-by-step chemical synthesis of DNA oligomers with the use of metallacarborane-modified nucleoside monomers [16] for incorporation of the metallacarborane at specific locations; and (2) post-synthetic attachment of the metallacarborane to the suitably functionalized DNA oligonucleotides [17,18]. Recently, an enzymatic method for DNA modification with carboranes was also developed [19]. The expansion of the enzymatic approach for the incorporation of metallacarboranes into the nucleic acid chain seems feasible.

DNA containing metallacarborane groups via de novo synthesis was obtained using the following methods: (A) solid-phase synthesis employing H-phosphonate chemistry or (B) solid-phase synthesis employing phosphoramidite chemistry. The post-synthetic modification approach (C) is based on a one-step procedure employing copper(I)-catalyzed Huisgen–Meldal–Sharpless 1,3-dipolar cycloaddition of azides and alkynes [20] (Scheme 1).

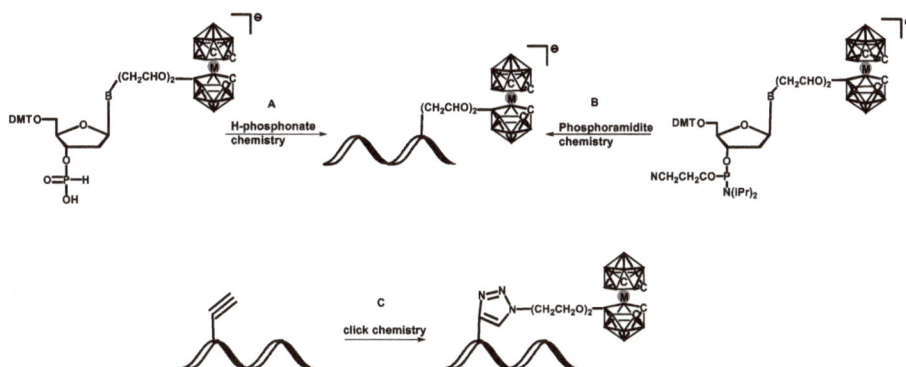

Scheme 1. Methods for the synthesis of oligonucleotides modified with metallacarborane: (A) de novo solid-phase, step-by-step synthesis employing H-phosphonate chemistry; (B) de novo solid-phase, step-by-step synthesis employing phosphoramidite chemistry; (C) post-synthetic one-step procedure based on copper(I)-catalyzed Huisgen–Meldal–Sharpless 1,3-dipolar cycloaddition of azides and alkynes. Hydrogen atoms of C–H groups and B–H groups of the boron clusters, symbolized by the polygen vertex, were omitted for clarity.

3. Synthesis of DNA Modified with Metallacarborane Complexes via de novo, Step-by-Step Chemical Synthesis

Originally, H-phosphonate chemistry was applied to synthesize oligonucleotides modified with carborane and metallacarborane of a sequence complementary to the fragment of the UL55 gene of human cytomegalovirus (HCMV) [21]. The solid-phase synthesis of modified oligonucleotide containing the 3,3′-iron-1,2,1′,2′-dicarbollide (BEFeC) at the 5′ end of 5′-d(BEFeCACGAGTCTTCGGCTCCGGA)-3′ oligomer was performed using an automated DNA synthesizer. The β-cyanoethyl phosphoramidite chemistry was used to synthesize the unmodified part of the oligonucleotide, and H-phosphonate chemistry was used to add the modified monomer bearing the metallacarborane moiety (Scheme 2). The expected nucleoside composition and the presence of the metallacarborane modification in the obtained oligomer was confirmed by enzymatic digestion [21]. Metallacarborane-containing dinucleotides bearing metallacarborane complexes were also synthesized using H-phosphonate chemistry and the solution-phase synthesis mode [22]. The same oligonucleotide, 5′-d(BEFeCACGAGTCTTCGGCTCCGGA)-3′, has been used as a target for biosensor devices [23]. The aqueous solution of the modified oligonucleotide was placed onto the electrode to allow hybridization with the unmodified oligonucleotide. Electrochemical signals were observed from the metallacarborane group, especially when a 3,3′-iron-1,2-1′,2′-dicarbollide complex was located far away from the electrode after hybridization. These studies confirmed the usefulness of the metallacarborane-containing Fe (III) ion as an effective redox marker.

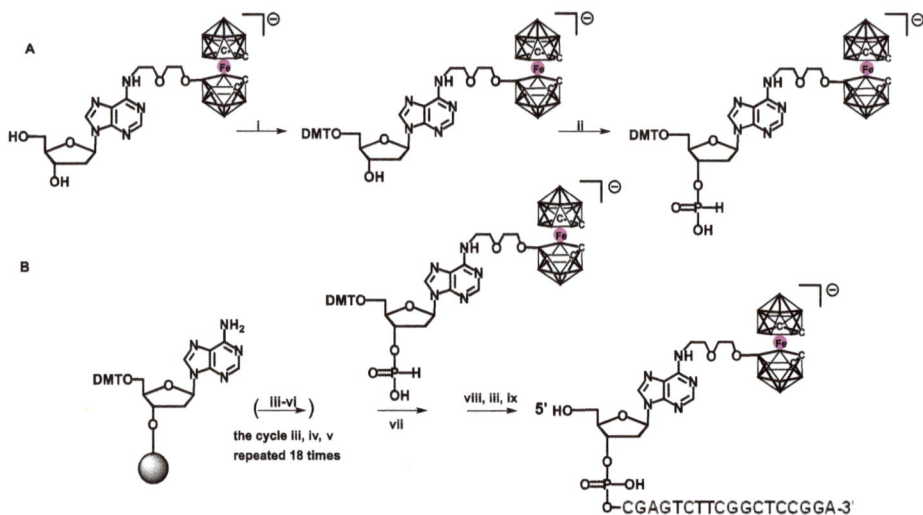

Scheme 2. Solid-phase synthesis of DNA modified with metallacarborane using H-phosphonate chemistry. (**A**) Synthesis of the H-phosphonate monomer: (i) dimethoxytrityl chloride (DMTCl)l and pyridine; (ii) PCl₃, imidazole, Et₃N, CH₂CH₂, and H₂O, −40 °C. (**B**) Synthesis of DNA oligonucleotides: (iii) detritylation: 3% dichloroacetic acid (DCA) in CH₂Cl₂; (iv) coupling: unmodified monomers, phosphoramidite chemistry; (v) oxidation: 0.1 mol/L I₂ in THF/pyridine/H₂O (13:6:1); (vi) capping: 1 mol/L (CH₃CO)₂O in THF/pyridine (1:8) and 0.5 mol/L 4-dimethylaminopyridine (DMAP) in THF; (vii) coupling: modified monomer, H-phosphonate chemistry; (viii) oxidation: 10% H₂O in CCl₄, Et₃N, *N*-methylimidazole (9.0:0.5:0.5, *v/v/v*); (ix) cleavage from the support: 30% NH₃ aq., 1 h, RT; base deprotection: 30% NH₃ aq. 2 h, 50 °C. Hydrogen atoms of C–H groups and B–H groups of the boron clusters, symbolized by the polygen vertex, were omitted for clarity.

DNA oligomers modified simultaneously with carborane (2′-*O*-(*o*-carboran-1-yl)methyluridine, 2′-CBM) and metallacarborane (2-*N*-{5-[3-cobalt bis(1,2-dicarbollide)-8-yl]-3-oxa-pentoxy}-2′-*O*-deoxyguanosine, BEMC) nucleosides complementary to the insulin receptor substrate (IRS-1) gene were also synthesized using a phosphoramidite approach for the introduction of carborane or metallacarborane groups-containing nucleoside monomers to the oligonucleotide [24]. Unmodified parts of the oligonucleotides were prepared by automated solid-phase synthesis (Scheme 3). The manual step was used for the insertion of phosphoramidites of 2′-CBM and BEMC bearing carborane or metallacarborane modification, respectively. It should be noted that because of the treatment of the oligonucleotide product with a concentrated solution of ammonia in water during the cleavage from the solid support and deprotection, the electroneutral *closo*-form of the carborane in 2′-CBM is transformed into a negatively charged *nido*-counterpart.

The physicochemical and biological properties of modified DNA were studied. The lipophilicity and resistance to enzymatic degradation by snake venom phosphodiesterase (SVPDE) of the modified oligonucleotides was higher than the unmodified counterparts, especially in the case of simultaneous modification at the 3′ and 5′ ends with 2′-CBM and BEMC groups, respectively. An effective formation of the doubled stranded structure with the complementary DNA-oligomers was shown [24].

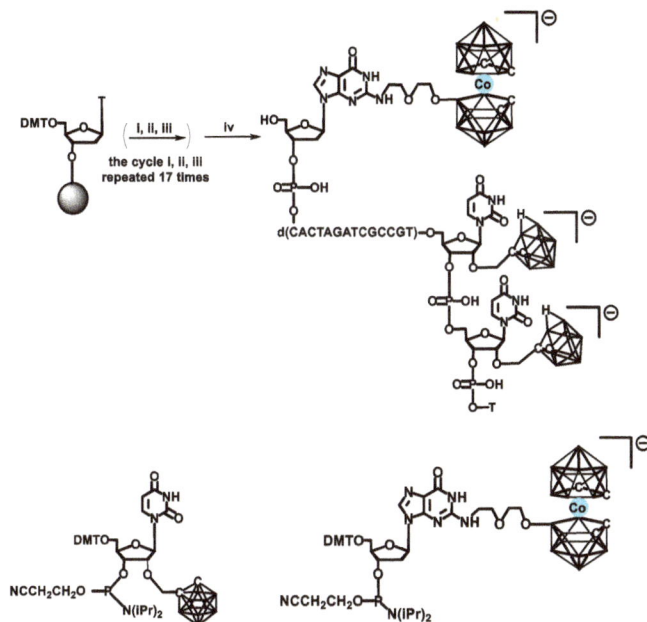

Scheme 3. Solid-phase synthesis of DNA modified with carborane and metallacarborane using phosphoramidite chemistry: (i) detritylation, (ii) coupling, (iii) oxidation and capping, and (iv) cleavage from the solid support and deprotection [25]. Hydrogen atoms of C–H groups and B–H groups of the boron clusters, symbolized by the polygen vertex, were omitted for clarity.

5′-*O*-Monomethoxytrityl-4-*O*-diethyleneoxy-8-[3-cobalt bis(1,2-dicarbollide)-8-yl]-thymidine-3′-*O*-(*N,N*-diisopropyl-β-cyanoethyl)phosphoramidite was used to synthesize DNA containing metallacarborane, 5′-d(^BEMCTGCTGGTTTGGCTG)-3′, using a β-cyanoethyl phosphoramidite cycle and an automated DNA synthesizer. The metallacarborane-modified nucleoside was attached to the 5′ end of the oligomer. In the oxidation step, *t*-BuOOH was applied instead of I₂, which is routinely used as the oxidizing agent. The synthesized oligonucleotide labeled with a metallacarborane cage was used as one of two primers in polymerase chain reaction (PCR) amplification of a fragment of HCMV DNA, strain AD 169. The other primer was unmodified oligonucleotide P12 (5′-AAACGGCGCAGCCACATAAGG-3′). The susceptibility of the primer with metallacarborane modification at its 5′ end to an extension by Taq DNA polymerase was proved by PCR amplification of the HCMV genome fragment on the junction of US14 and US13 gene. The 812-bp amplification product was obtained as expected [26].

Expanding these works, we have developed a general method for the synthesis of four canonical nucleosides T, dC, dA, and dG modified with 1,2-dicarba-*closo*-dodecaborane (C₂B₁₀H₁₁) cage, and their phosphoramidites, suitable for automated synthesis of DNA [27]. The modified nucleoside monomers bearing boron cluster were prepared from suitable 5-ethynyl derivatives in one-step reaction using cooper(I)-catalyzed Huisgen–Meldal–Sharpless 1,3-dipolar cycloaddition of azides and alkynes to give triazoles. The 5′-*O*-(4,4′-dimethoxytrityl)-3′-*O*-(*N,N*-diisopropyl-β-cyanoethyl)-*N*³-{[(*O*-carboran-1-yl)propyl)]-1*N*-1,2,3-triazol-4-yl}methylenethymidine phosphoramidite was used to synthesize four modified DNA oligonucleotides of various lengths and sequences of nucleobases. All the oligomers contained one carborane modification at the 5′-end, and two of them were additionally equipped with an amino linker C6 (Scheme 4). The oligonucleotides were obtained by solid-phase synthesis using a standard β-cyanoethyl cycle [25].

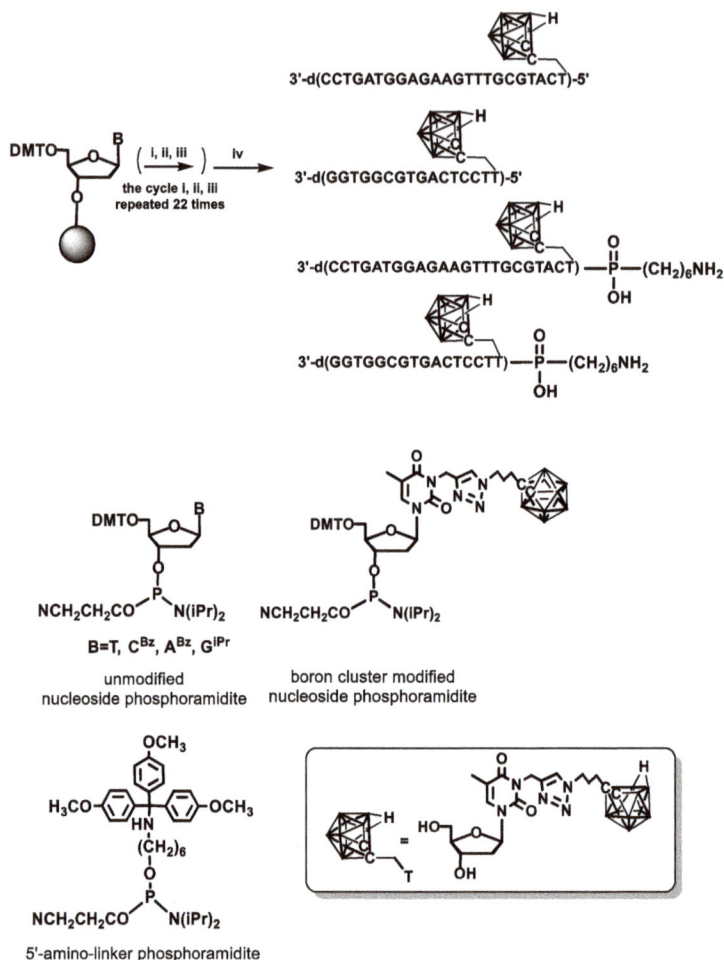

3'-d(CCTGATGGAGAAGTTTGCGTACT)-5'

3'-d(GGTGGCGTGACTCCTT)-5'

3'-d(CCTGATGGAGAAGTTTGCGTACT)—P—(CH₂)₆NH₂

3'-d(GGTGGCGTGACTCCTT)—P—(CH₂)₆NH₂

the cycle i, ii, iii
repeated 22 times

B=T, C^{Bz}, A^{Bz}, G^{iPr}

unmodified
nucleoside phosphoramidite

boron cluster modified
nucleoside phosphoramidite

5'-amino-linker phosphoramidite

Scheme 4. Synthesis of oligonucleotides containing carborane cluster at the 5′-end of the oligonucleotide chain: (i) detritylation; (ii) coupling; (iii) oxidation and capping; (iv) cleavage from the solid support and deprotection. Hydrogen atoms of C–H groups and B–H groups of the boron clusters, symbolized by the polygen vertex, were omitted for clarity.

Mass spectrometry analysis proved that the carborane cage is, as expected, in *nido*-form (7,8-dicarba-*nido*-undecaborate ion). The conversion of the boron cluster from *closo*- to *nido*-form takes place during the cleavage of the synthesized oligomer from the support and nucleic bases' deblocking under treatment with concentrated ammonia solution in water [28]. The modified oligonucleotides containing negatively charged, redox active 7,8-dicarba-*nido*-undecaborate have been used as a probe for electrochemical detection of specific DNA sequences derived from the cytomegalovirus (HCMV) [29]. Alternatively, they can be used as precursors for post-synthetic modification of the oligonucleotides with metallacarborane unit based on the 3,3,3-tricarbonyl-3-metal complex formation with 7,8-dicarba-*closo*-dodecaboran-1-yl ligand [30], according to the procedure shown in Scheme 5.

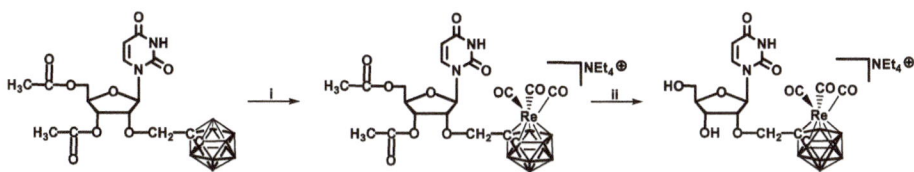

Scheme 5. Synthesis of a uridine-bearing 3,3,3-(CO)$_3$-*closo*-3,1,2-ReC$_2$B$_9$H$_{10}$ complex: (i) TEAF$_{aq.}$, EtOH, [Net$_4$]$_2$[ReBr$_3$(CO)$_3$], 100 °C, 30 h; (ii) 25% NH$_{3aq.}$, MeOH, 35 °C, 30 min. Hydrogen atoms of C–H groups and B–H groups of the boron clusters, symbolized by the polygen vertex, were omitted for clarity.

Using this approach, synthesis of a new type of nucleoside–metallacarborane conjugate, 2′-*O*-(3,3,3-tricarbonyl-3-rhenium-7,8-dicarba-*closo*-dodecaboran-1-yl) methyluridine, was recently achieved [31]. The method was based on the de novo formation of a metallacarborane complex via the reaction of [Net$_4$]$_2$[ReBr$_3$(CO)$_3$] with the uridine-bearing carborane as a metal ligand (Scheme 5).

4. Post-Synthetic Labeling of DNA with Metallacarboranes via a "Click Chemistry" Approach

The CuI-catalyzed 1,3-dipolar Huisgen–Meldal–Sharpless cycloaddition (HMSC) of azide and alkyne leads to the formation of a triazole moiety. This process, often called "click chemistry" or "chemical ligation", became an important tool for chemical modification of biomolecules [20,32]. The azide and alkyne components have several advantages. They can be conveniently introduced in an independent manner, are relatively stable and do not react with common organic reagents nor with a majority of functional groups present in biomolecules (are orthogonal). The formation of triazoles is irreversible and usually proceeds in high yield. In addition, this reaction is remarkably versatile because of the extremely mild and chemoselective copper(I) catalyst system, which is surprisingly insensitive to many solvents and operate well in a wide pH range. The advantages of the "click chemistry" approach, based not only on HMSC but also on other types of bioorthogonal chemistries such as Diels–Alder "click" reactions, allow for its application in areas such as synthesis on solid supports, combinatorial chemistry, or "organic chemistry in vivo" [33]. The click chemistry approach was used to label a DNA oligomer with the 3,3′-iron-1,2,1′,2′-dicarbollide group [34,35] using an alkyne derivative of DNA (with an acetylene moiety located close to the 5′ end) and an azide component bearing the metallacarborane group at the end of a diethylenoxy linker (Scheme 6).

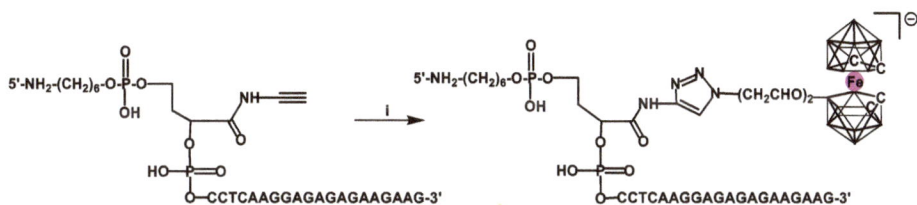

Scheme 6. Synthesis of DNA containing a metallacarborane moiety using Huisgen–Meldal–Sharpless cycloaddition (HMSC): (i) tris[(1-benzyl-1*H*-1,2,3-triazol-4-yl)methyl]amine (TBTA), sodium ascorbate, CuSO$_4$ × 5 H$_2$O, [8-N$_3$-(CH$_2$CH$_2$O)$_2$-3-iron bis(dicarbollide)], H$_2$O. Hydrogen atoms of C–H groups and B–H groups of the boron clusters, symbolized by the polygen vertex were omitted for clarity.

The obtained oligonucleotide deposited on a gold electrode acted as a main part of a novel electrochemical genosensor capable of electrochemical determination of the DNA sequences derived from avian influenza virus (AIV), type H5N1. The oligonucleotide probes containing an NH$_2$ group close to the 5′-end nucleotide were covalently attached to the electrode via coupling of *N*-hydroxysuccinimide/*N*-(3-dimethyl-aminopropyl)-*N*-ethylcarbodiimide hydrochloride (NHS/EDC) and 3-mercaptopropionic acid. The latter was previously deposited on the surface of gold electrode as a self-assembled monolayer (SAM). For the measurement, an Osteryoung square-wave voltammetry method was applied. The changes in the redox activity of the Fe(III) center of the metallacarborane complex before and after the hybridization process were found to be diagnostic and displayed good selectivity and sensitivity towards several complementary targets derived from AIV, H5N1. The calculated detection limits recorded for an ssDNA 20-mer and PCR products having complementary sequences at the 3′ end and in the middle of the oligonucleotide based on experimental data ranged from 0.03 to 0.08 fM, respectively; thus the genosensor was superior to many others previously reported. Its additional advantage was the ability to distinguish PCR products containing complementary sequences in different positions [34].

Another type of HMSC driven labeling of nucleic acids with metallacarboranes (shown in Schemes 7 and 8) consisted of two steps [35]. First, oligonucleotides containing, at a given position, a thymidine unit modified with 4-pentyn-1-yl substituent at the position 3-N were obtained by the use of corresponding phosphoramidite monomer. Also, oligomers bearing a uridine unit modified with a 2′-*O*-propargyl group were synthesized using commercially available monomer. In the second step, the alkyne-modified oligomers were post-synthetically conjugated to alkyl azide derivatives of boron clusters via HMSC [20,36].

For the proof of concept, a 22-mer antisense oligodeoxyribonucleotide (ASO) of the sequence 5′-d(TTT CTT TTC CTC CAG AGC CCGA)-3′ was chosen, which recognizes the epidermal growth factor receptor (EGFR) mRNA, known for being overexpressed in cancerous cells. The synthesized ASO oligonucleotides contained one or two thymidine or uridine metallacarboranes conjugates (of thymidine (TB)-type or uridine (UB)-type, respectively; Scheme 7) in positions 7 or/and 11 of the oligomer chain [35]. The metallacarborane moieties were anchored at position 3-*N* of thymidine (TB) or alternatively at 2′-oxygen atom of uridine (UB) unit. In addition, two UB-type ASOs (heavily loaded with metallacarborane, Scheme 8) [37] were also obtained according to the above procedure (see Table 1 for the sequences and modification type and position). The obtained models were used for evaluation of physicochemical and biological properties of DNA–metallacarborane conjugates.

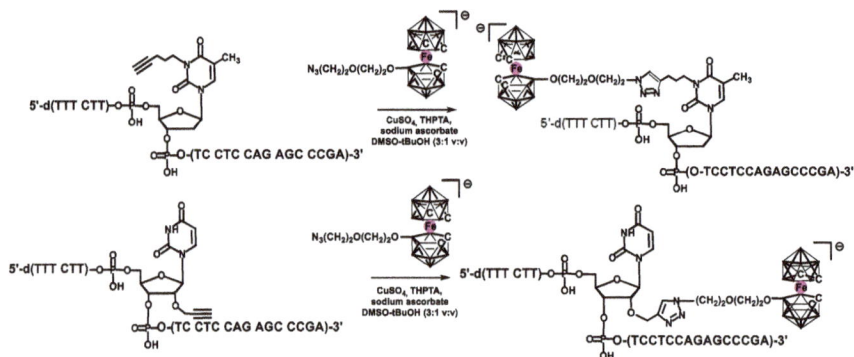

Scheme 7. Example synthesis of boron cluster-modified anti-epidermal growth factor receptor (EGFR) antisense oligonucleotides via post-synthetic modification using HMSC. DNA oligonucleotides bearing 3-*N*-metallacarborane-modified thymidine unit(s) (as in the upper reaction scheme) are assigned thymidine (TB)-type, while those bearing 2′-*O*-metallacarborane modified uridine units (a lower panel) are assigned uridine (UB)-type. Adapted in part from the literature [35]. Hydrogen atoms of C–H groups and B–H groups of the boron clusters, symbolized by the polygen vertex, were omitted for clarity.

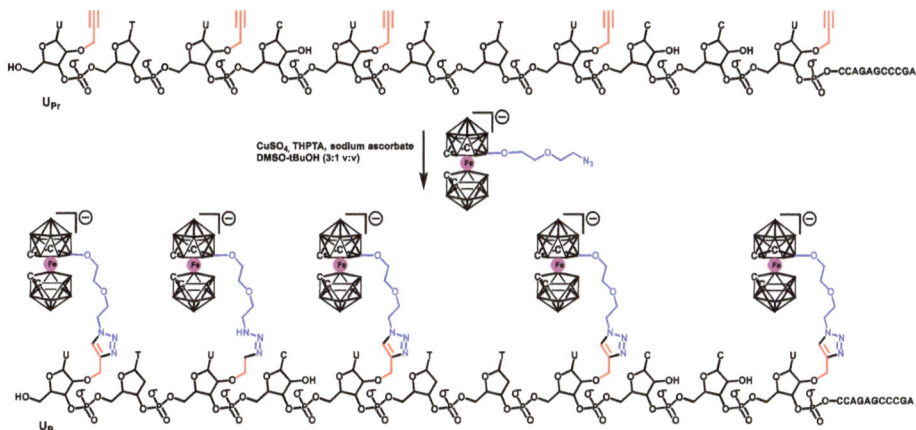

Scheme 8. Antisense UB-type DNA oligonucleotide targeted EGFR obtained via Cu(I)-assisted "click" conjugation, bearing five Fe (III)-metallacarborane units. Hydrogen atoms of C–H groups and B–H groups of the boron clusters, symbolized by the polygon vertex, were omitted for clarity.

Table 1. Binding affinity of the anti-epidermal growth factor receptor (EGFR) metallacarborane antisense oligonucleotides to the complementary RNA strand assessed by thermal dissociation assay (Tm), lipophilicity (Log*P*), RP-HPLC mobility (retention time R_T), and antisense activity (antisense oligodeoxyribonucleotide (ASO) activity) assessed by a dual fluorescence assay (DFA) against EGFR mRNA.

Name	Oligonucleotide Sequence [a]	Tm [b] [°C]	ASO Activity [c] [%]	Log*P* (R_T [d] [min])	Ref.
NM	5'-d(TTTCTTTTCCTCCAGAGCCCGA)-3'	69.5	65 ± 1	−1.17 ± 0.36 (11.6)	[35]
TB$_7$	5'-d(TTTCTTT**TB$_7$**TCCTCCAGAGCCCGA)-3'	62.5	46 ± 4	−0.60 ± 0.15	[35]
TB$_{11}$	5'-d(TTTCTTTTCC**TB$_{11}$**CCAGAGCCCGA)-3'	57.5	82 ± 2	nd	[35]
TB$_{7,11}$	5'-d(TTTCTTT**TB$_7$**TCC**TB$_{11}$**CCAGAGCCCGA)-3'	45.0	24 ± 2	nd	[35]
UB$_7$	5'-d(TTTCTTT**UB$_7$**TCCTCCAGAGCCCGA)-3'	65.5	34 ± 2	−0.71 ± 0.42 (17.5)	[35,37]
UB$_{11}$	5'-d(TTTCTTTTCC**UB$_{11}$**CCAGAGCCCGA)-3'	67.5	89 ± 2	nd	[35]
UB$_{7,11}$	5'-d(TTTCTTT**UB$_7$**TCC**UB$_{11}$**CCAGAGCCCGA)-3'	63.5	68 ± 1	nd (20.1)	[35,37]
UB$_{2,5,8,11}$	5'-d(T**UB$_2$**TC**UB$_5$**TT**UB$_8$**CC**UB$_{11}$**CCAGAGCCCGA)-3'	60.9	54 ± 2	(22.2)	[37]
UB$_{1,3,5,8,11}$	5'-d(**UB$_1$**T**UB$_3$**C**UB$_5$**TT**UB$_8$**CC**UB$_{11}$**CCAGAGCCCGA)-3'	60.4	62 ± 3	(23.9)	[37]

[a] The thymidine (TB) or uridine (UB) metallacarborane conjugates are inserted in the indicated positions of the oligonucleotide chains. [b] Melting temperatures (Tm) of ASO/RNA duplexes are given with ±0.5 °C accuracy. [c] Antisense activity of anti-EGFR ASO in HeLa cells at 200 nM concentration in DFA. [d] R_T [min]—RP HPLC retention time on a C18 reverse phase HPLC column, a gradient of acetonitrile in aqueous ammonium acetate solution (0.1 M CH$_3$COONH$_4$) from 0% to 70% of acetonitrile as an eluting solvent system was used.

5. Physicochemical and Biological Properties of Metallacarborane Modified Oligonucleotides of TB- and UB-Types

Binding affinity to complementary sequences. The affinity of the metallacarborane modified oligomers of TB- or UB-type (Table 1) to their complementary target RNA strand 3'-AAA GAA AAG GAG GUC UCG GGCU-5' was assessed by the thermal dissociation assay. The measurement of melting temperatures (Tm) has shown that metallacarborane modified ASO/RNA duplexes are thermodynamically less stable than the reference duplex containing the non-modified ASO strand (NM-ASO) complexed with complementary RNA (NM-ASO/RNA). However, the presence of the UB-type modifications decreased the duplex stability less than those of the TB-type. This is because the nucleotide of UB-type with the metallacarborane moiety present at the 2'-position adopts preferentially the C3'-*endo* conformation typical for the RNA units, and thus it well accommodates in the DNA/RNA helices, known for adopting the overall conformation of the A-type.

In contrast, in the TB unit, the metallacarborane group occupies the 3-*N* position of nucleobase, therefore, there is a disruption of the Watson–Crick base pairing with the complementary adenine nucleotide. Thus, one may conclude that ASOs modified with the UB-type units are advantageous for recognition of the target mRNA sequences and activation of the RNase H, a prerequisite for gene expression regulation by the antisense approach [38].

Lipophilicity of UB- and TB-type ASOs. The metallacarborane units lead to substantially increased lipophilicity of their oligonucleotide conjugates (listed in Table 1) compared with the non-modified reference precursor MN-ASO, as shown by the values of the partition coefficients in the water/1-octanol bi-phase system (Log*P*) [35] and by the changes of chromatographic mobility on a C18 reverse phase HPLC column (R_T are given in parentheses in Table 1) [35,37]. The increasing number of the UB units loaded leads to the linear increase of the lipophilicity of the UB-type ASOs [35,37].

Circular dichroism (CD) measurements of UB-type ASOs. CD is a powerful tool for monitoring structural changes of DNA and RNA duplexes resulting from the presence of modified units in their strands. As shown previously, the CD spectra of the UB-type ASO bound to the complementary RNA strand exhibit only minute deviations from the CD profile of the reference, non-modified DNA/RNA duplex [35,37]. This observation indicates that the UB-type units inserted in the oligonucleotide chain slightly alter the structure of the ASO/RNA duplexes. This feature is important for recognition of mRNA and for induction of the RNase H activity. In this respect, the UB-type ASO metallacarborane conjugates may be useful in the antisense or boron neutron capture therapy (BNCT) applications.

Infrared (IR) activity of metallacarborane labeled DNA. To date, detection of nucleic acids using infrared light sensitive labels and IR spectroscopy has not been used very often, mostly because of the lack of labels absorbing electromagnetic radiation within a 1900–2600 cm^{-1} window, where nucleic acids (and typical organic molecules) are transparent. Boron clusters attached to DNA or RNA oligonucleotides make a difference because IR spectra of carborane ($C_2B_{10}H_{12}$) or metallacarborane [3-Co-(1,2-C_2B_9H11)$_2$]$^-$ moieties contain a unique band attributed to the B–H vibration at ca. 2600 cm^{-1} [39,40]. The HMSC method allowed for further improvement of this labeling approach. Incrusting of an oligonucleotide probe with multiple metallacarborane labels, each containing 18 B–H bonds, resulted in significant enhancement of the diagnostic B–H signal and increased sensitivity of the detection [37]. For ASOs of the UB-type, containing four or five metallacarborane units at positions 2, 5, 8, 11 or 1, 3, 5, 8, 11 of the oligonucleotide chain, the B–H related bands occurred in the 2500–2650 cm^{-1} region, similar to the signals in free Fe(III)-metallacarborane [3-Fe-(1,2-C_2B_9H11)$_2$]$^-$, as well as the signals of the B–H bonds found in uridine nucleoside bearing Fe(III)-metallacarborane modification at 2'-position [37]. Thus, the metallacarborane moieties inserted into the oligonucleotide chain show the bands characteristic for the B–H vibration and can be used as an infrared light sensitive marker.

Biological evaluation of the UB-type ASOs. The silencing activity (inhibition of protein biosynthesis) of boron cluster-modified ASOs was tested against the mRNA of EGFR protein in a dual florescence assay (DFA), in which the fusion plasmid coding green fluorescent protein (GFP) and the target fragment of EGFR (pEGFP-EGFR) was exogenously expressed in HeLa cells, and was assessed by comparison of the level of fluorescence of EGFP and of the control red fluorescent protein (RFP) simultaneously expressed from pDsRed-N1 plasmid. Considerable silencing activity, depending upon the location of the modification within the oligonucleotide chain, was observed for the selected ASOs.

In the antisense approach, the ASO oligonucleotides operate by sequence-specific hybridization to the complementary RNA followed by recruitment of RNase H, which hydrolyses the target RNA. Although the requirements that the ASOs have to meet to inhibit a protein expression are not known precisely, modifications in a sugar moiety, for example, change of the sugar type or its orientation, are reported to affect the RNase H activation [41]. From the other side, borane clusters are known for binding to the cellular proteins [42] and modulation of their cellular function. Therefore, the observed differences in the silencing activity of the ASO-metallacarborane conjugates may be of different origin. In the reported case, single modified ASOs of the TB- and UB-type in position 11 of the ASO chain (as in TB$_{11}$ and UB$_{11}$) used in 200 nM concentration exhibit enhanced silencing potential compared with the reference ASO (by ca. 20%, Table 1). In contrast, location of the metallacarborane units at position 7 of ASOs (as in TB$_7$ and UB$_7$) decreased their silencing activity. For highly modified UB-type ASOs, the silencing effect is balanced by positive and negative features of the metallacarborane units, leading to ASOs of similar antisense activity as the reference ASO.

Redox properties of Fe (III)-metallacarborane ASO conjugates. The tested ASOs, when used at low concentrations (up to 50 μM), exhibited pro-oxidative activity by inducing reactive oxygen species (ROS) production and an increase in the mitochondrial activity of HeLa cells. In contrast, when used at higher concentrations, the ASOs exhibited antioxidative properties, lowering the level of ROS [37].

Inside cells, the highly metallacarborane-loaded ASOs containing redox-active Fe(III) ions generate ROS as a result of oxidative stress. At low concentrations, these compounds enhance the metabolic activity. Along with the increase of concentration, the amount of Fe(III) ions (free radicals scavenger) also increases, so the antioxidative properties of the ASOs are observed. When concentration of ASOs reaches ≈200 nM, the ROS levels are the same as in the control cells, suggesting that the rates of ROS formation and elimination are equal. It is suggested that the silencing of EGFR by the use of high concentrations of oligonucleotide–metallocarborane conjugates occurs without noticeable increase of ROS concentration. It should be emphasized that the delivery of unmodified antisense and nonsense oligonucleotides did not result in a significant increase of the ROS levels. Therefore, it is hypothesized that the high metallacarborane-loaded DNA oligomers may be used as potential BNCT and antisense oligonucleotide (dual-action) anticancer agents.

6. Summary and Conclusions

DNA modified with boron cluster-metal complexes, metallacarboranes [M(C$_2$B$_9$H$_{11}$)$_2$], represent a new class of metal containing nucleic acids. Metallacarboranes provide probably the most versatile platform for modification of nucleic acids with metals, though this immense potential is awaiting full exploration. Not all metallacarboranes are stable in a biological (water) environment, but a plethora of metallacarboranes suitable for use as metal carriers for biomolecules' modification can be selected. The methods described in this overview are modular and versatile, allowing practical access to innovative metal-containing DNA. These and new methods developed in future may open the way for more widespread use of metal-labeled nucleic acids for a variety of applications.

Author Contributions: Conceptualization, Z.J.L. and A.B.O. Methodology, Z.J.L., A.B.O. and B.N. Validation, Z.J.L., A.B.O. and B.N. Formal Analysis, Z.J.L., A.B.O. and B.N. Investigation, Z.J.L., A.B.O. and B.N. Resources, Z.J.L. Data Curation, Z.J.L., A.B.O. and B.N. Writing—Original Draft Preparation, Z.J.L. and A.B.O Writing—Review & Editing, Z.J.L., A.B.O. and B.N. Visualization, Z.J.L., A.B.O. and B.N. Supervision, Z.J.L. Project Administration, Z.J.L. Funding Acquisition, Z.J.L., B.N. and A.B.O.

Funding: This research was financially supported by The National Science Centre in Poland, Grant number 2015/16/W/ST5/00413 for years 2015–2021.

Conflicts of Interest: The authors declare no conflict of interest.

References

1. Seeman, N.C.; Sleiman, H.F. DNA nanotechnology. *Nat. Rev. Mater.* 2018. [CrossRef]
2. Chen, Y.-J.; Groves, B.; Muscat, R.A.; Seelig, G. DNA nanotechnology from the test tube to the cell. *Nat. Nano.* **2015**, *10*, 748–760. [CrossRef]
3. Chen, C.; Yang, Z.; Tang, X. Chemical modification of nucleic acid drugs and their delivery systems for gene-based therapy. *Med. Res. Rev.* **2018**, *38*, 829–869. [CrossRef]
4. Lesnikowski, Z.J. DNA as a platform for new biomaterials. Metal-containing nucleic acids. *Curr. Org. Chem.* **2007**, *11*, 355–381. [CrossRef]
5. Yang, H.; Metera, K.L.; Sleiman, H.F. DNA modified with metal complexes: Applications in the construction of higher order metal–DNA nanostructures. *Coord. Chem. Rev.* **2010**, *254*, 2403–2415. [CrossRef]

6. Northwestern University; Mirkin, C.A.; Hucp, J.T.; Farha, O.K.; Spokojny, A.M.; Mulfort, K.L. Metal-Organic Framework Materials Based on Icosahedral Boranes and Carboranes. US Patent 2009/0025556, 29 January 2009.

7. Center of Microbiology and Virology PAS; Lesnikowski, Z.J.; Olejniczak, A. Nucleoside Derivative, Modified Oligonucleotide, Method for Their Synthesis and Application Thereof. US Patent 2007/0009889, 11 January 2007.

8. Fuji Photo Film Co., Ltd.; Yoshihiko, M.; Yoshihiko, A.; Masashi, O.; Makoto, T.; Shigeo, T.; Kenichi, Y. Quantitative Detection of Nucleic Acids by Differential Hybridization Using Electrochemical Labeling of Samples. JP Patent 2002000299, 8 February 2002.

9. Yu, Z.; Cowan, J.A. Metal complexes promoting catalytic cleavage of nucleic acids—Biochemical tools and therapeutics. *Curr. Opin. Chem. Biol.* **2018**, *43*, 37–42. [CrossRef]

10. Zhang, Y.; Tu, J.; Wang, D.; Zhu, H.; Maity, S.K.; Qu, X.; Bogaert, B.; Pei, H.; Zhang, H. Programmable and multifunctional DNA-based materials for biomedical applications. *Adv. Mater.* **2018**, *30*. [CrossRef]

11. Grimes, R.N. *Carboranes*, 3rd ed.; Academic Press: London, UK, 2016; ISBN 978-0-12-801894-1.

12. Hey-Hawkins, E.; Vinas Teixidor, C. *Boron Based Compounds: Emerging Applications in Medicine*; John Wiley & Sons Inc.: Hoboken, NJ, USA, 2018.

13. Leśnikowski, Z.J. Challenges and opportunities for the application of boron clusters in drug design. *J. Med. Chem.* **2016**, *59*, 7738–7758. [CrossRef]

14. Grimes, R.N. Metallacarboranes in the new millennium. *Coord. Chem. Rev.* **2000**, *200–202*, 773–811. [CrossRef]

15. Farras, P.; Juarez-Perez, E.J.; Lepsik, M.; Luque, R.; Nunez, R.; Teixidor, F. Metallacarboranes and their interactions: theoretical insights and their applicability. *Chem. Soc. Rev.* **2012**, *41*, 3445–3463. [CrossRef]

16. Wojtczak, B.A.; Olejniczak, A.B.; Lesnikowski, Z.J. Nucleoside Modification with Boron Clusters and Their Metal Complexes. In *Current Protocols in Nucleic Acid Chemistry*; Beaucage, S.L., Ed.; John Wiley & Sons Inc.: Hoboken, NJ, USA, 2018; Chapter 4, Unit 4.37; pp. 1–26.

17. Lesnikowski, Z.J. Boron clusters—A new entity for DNA-oligonucleotide modification. *Eur. J. Org. Chem.* **2004**, *2003*, 4489–4500. [CrossRef]

18. Kwiatkowska, A.; Sobczak, M.; Mikolajczyk, B.; Janczak, S.; Olejniczak, A.B.; Sochacki, M.; Lesnikowski, Z.J.; Nawrot, B. siRNAs modified with boron cluster and their physicochemical and biological characterization. *Bioconjug. Chem.* **2013**, *24*, 1017–1026. [CrossRef] [PubMed]

19. Balintová, J.; Simonova, A.; Białek-Pietras, M.; Olejniczak, A.B.; Lesnikowski, Z.J.; Hocek, M. Carborane-linked 2'-deoxyuridine 5'-*O*-triphosphateas building block for polymerase synthesis of carborane-modified DNA. *Bioorg. Med. Chem. Lett.* **2017**, *27*, 4786–4788. [CrossRef]

20. Meldal, M.; Tornoe, C.W. Cu-catalyzed azide alkyne cycloaddition. *Chem. Rev.* **2008**, *108*, 2952–3015. [CrossRef] [PubMed]

21. Olejniczak, A.B. Metallacarboranes for the labelling of DNA-synthesis of oligonucleotides bearing a 3,3'-iron-1,2-1',2'-dicarbollide complex. *Can. J. Chem.* **2011**, *89*, 465–470. [CrossRef]

22. Olejniczak, A.B.; Mucha, P.; Grüner, B.; Lesnikowski, Z.J. DNA-dinucleotides bearing a 3',3'-cobalt- or 3',3'-iron-1,2,1',2'-dicarbollide complex. *Organometallics* **2007**, *26*, 3272. [CrossRef]

23. Ziółkowski, R.; Olejniczak, A.B.; Górski, Ł.; Janusik, J.; Leśnikowski, Z.J.; Malinowska, E. Electrochemical detection of DNA hybridization using metallacarborane unit. *Bioelectrochemistry* **2012**, *87*, 78–83.

24. Olejniczak, A.B.; Kierzek, R.; Wickstrom, E.; Lesnikowski, Z.J. Synthesis, physicochemical and biological studies of anti-IRS-1 oligonucleotides containing carborane and/or metallacarborane modification. *J. Organomet. Chem.* **2013**, *747*, 201–210. [CrossRef]

25. Atkinson, T.; Smith, M. Solid phase synthesis of oligodeoxyrybonucleotides by the phosphite-triester method. In *Oligonucleotide Synthesis: A Practical Approach*; Gait, M., Ed.; IRL: Oxford, UK, 1984; pp. 35–81.

26. Olejniczak, A.B.; Plesek, J.; Kriz, O.; Lesnikowski, Z.J. A nucleoside conjugate containing a metallacarborane group and its incorporation into a DNA oligonucleotide. *Angew. Chem.* **2003**, *115*, 5918–5921. [CrossRef]

27. Matuszewski, M.; Kiliszek, A.; Rypniewski, W.; Lesnikowski, Z.J.; Olejniczak, A.B. Nucleoside bearing boron clusters and their phosphoramidites—Building blocks for modified oligonucleotide synthesis. *New J. Chem.* **2015**, *39*, 1202–1221. [CrossRef]

28. Olejniczak, A.B.; Koziołkiewicz, M.; Leśnikowski, Z.J. Carboranyl oligonucleotides. 4. Synthesis, and physicochemical studies of oligonucleotides containing 2'-*O*-(o-carboran-1-yl)methyl group. *Antisense Nucl. Acid Drug Deve.* **2002**, *12*, 79–94. [CrossRef]

29. Jelen, F.; Olejniczak, A.B.; Kourilova, A.; Lesnikowski, Z.J.; Palecek, E. Electrochemical DNA Detection Based on Polyhedral Boron Cluster Label. *Anal. Chem.* **2009**, *81*, 840–844. [CrossRef]

30. Valliant, J.F.; Morel, P.; Schaffer, P.; Kaldis, J.H. Carboranes as ligands for the preparation of organometallic Tc and Re Radiopharmaceuticals. Synthesis of $[M(CO)_3(\eta^5\text{-}2,3\text{-}C_2B_9H_{11})]^-$ and rac-$[M(CO)_3(\eta^5\text{-}2\text{-}R\text{-}2,3\text{-}C_2B_9H_{10})]^-$ (M = Re, ^{99}Tc; R = $CH_2CH_2CO_2H$) from $[M(CO)_3Br_3]^{2-}$. *Inorg. Chem.* **2002**, *41*, 628–630. [CrossRef]

31. Olejniczak, A.B. Nucleoside-metallacarborane conjugtes: Synthessia of a uridine-bearing 3,3,3-(CO)3-closo-3,1,2-ReC2B9H10 complex. *Arkivoc* **2012**, *viii*, 90–97.

32. El-Sagheer, A.H.; Brown, T. Click chemistry with DNA. *Chem. Soc. Rev.* **2010**, *39*, 1388–1405. [CrossRef]

33. Prescher, J.A.; Bertozzi, C.R. Chemistry in living systems. *Nat. Chem. Biol.* **2005**, *1*, 13–21. [CrossRef]

34. Grabowska, I.; Stachyra, A.; Góra-Sochacka, A.; Sirko, A.; Olejniczak, A.B.; Leśnikowski, Z.J.; Radecki, J.; Radecka, H. DNA probe modification with 3-iron bis(dicarbollide) for electrochemical determination of DNA sequence of Avian Influenza Virus H5N1. *Biosens. Bioelectron.* **2014**, *51*, 170–176. [CrossRef]

35. Ebenryter-Olbińska, K.; Kaniowski, D.; Sobczak, M.; Wojtczak, B.A.; Janczak, S.; Wielgus, E.; Nawrot, B.; Leśnikowski, Z.J. Versatile method for the site specific modification of DNA with boron clusters—Anti-EGFR antisense oligonucleotide case. *Chem. Eur. J.* **2017**, *23*, 16535–16546. [CrossRef]

36. Amblard, F.; Cho, J.H.; Schinazi, R.F. The Cu(I)-catalyzed Huisgen azide-alkyne 1,3-dipolar cycloaddition reaction in nucleoside, nucleotide and oligonucleotide chemistry. *Chem. Rev.* **2009**, *109*, 4207–4220. [CrossRef]

37. Kaniowski, D.; Ebenryter-Olbinska, K.; Sobczak, M.; Wojtczak, B.; Janczak, S.; Lesnikowski, Z.J.; Nawrot, B. High Boron-loaded DNA-Oligomers as Potential Boron Neutron Capture Therapy and Antisense Oligonucleotide Dual-Action Anticancer Agents. *Molecules* **2017**, *22*, 1393. [CrossRef]

38. Crooke, S.T. Molecular Mechanisms of Antisense Oligonucleotides. *Nucleic Acid Ther.* **2017**, *27*, 70–77. [CrossRef]

39. Olejniczak, A.B.; Sut, A.; Wróblewski, A.E.; Leśnikowski, Z.J. Infrared spectroscopy of nucleoside and DNA-oligonucleotide conjugates labeled with carborane and metallacarborane. *Vib. Spectrosc.* **2005**, *39*, 177–185. [CrossRef]

40. Leites, L.A. Vibrational spectroscopy of carboranes and parent boranes and its capabilities in carborane chemistry. *Chem. Rev.* **1992**, *92*, 279–323.

41. Kher, G.; Trehan, S.; Misra, A. Antisense Oligonucleotides and RNA Interference. In *Challenges in Delivery of Therapeutic Genomics and Proteomics*, 1st ed.; Misra, A., Ed.; Elsevier: Burlington, MA, USA, 2011; pp. 325–386. ISBN 9780123849649.

42. Goszczyński, T.M.; Fink, K.; Kowalski, K.; Leśnikowski, Z.J.; Boratyński, J. Interactions of boron clusters and their derivatives with serum albumin. *Sci. Rep.* **2017**, *7*, 9800. [CrossRef]

International Journal of
Molecular Sciences

MDPI

Review

Modulating Chemosensitivity of Tumors to Platinum-Based Antitumor Drugs by Transcriptional Regulation of Copper Homeostasis

Yu-Hsuan Lai [1,2], Chin Kuo [1], Macus Tien Kuo [3] and Helen H. W. Chen [1,4,*]

[1]   Department of Radiation Oncology, National Cheng Kung University Hospital, College of Medicine,
       National Cheng Kung University, Tainan 70428, Taiwan; coscoscos.tw@hotmail.com (Y.-H.L.);
       tiffa663@gmail.com (C.K.)
[2]   Institute of Clinical Medicine, College of Medicine, National Cheng Kung University, Tainan 70428, Taiwan
[3]   Department of Translational Molecular Pathology, The University of Texas MD Anderson Cancer Center,
       Houston, TX 77054, USA; tienkuo@sbcglobal.net
[4]   Center of Applied Nanomedicine, National Cheng Kung University, Tainan 70101, Taiwan
*   Correspondence: helen@mail.ncku.edu.tw; Tel.: +886-6235-3535

Received: 23 April 2018; Accepted: 12 May 2018; Published: 16 May 2018

Abstract: Platinum (Pt)-based antitumor agents have been effective in treating many human malignancies. Drug importing, intracellular shuffling, and exporting—carried out by the high-affinity copper (Cu) transporter ($hCtr1$), Cu chaperone (Ato x1), and Cu exporters (ATP7A and ATP7B), respectively—cumulatively contribute to the chemosensitivity of Pt drugs including cisplatin and carboplatin, but not oxaliplatin. This entire system can also handle Pt drugs via interactions between Pt and the thiol-containing amino acid residues in these proteins; the interactions are strongly influenced by cellular redox regulators such as glutathione. $hCtr1$ expression is induced by acute Cu deprivation, and the induction is regulated by the transcription factor specific protein 1 (Sp1) which by itself is also regulated by Cu concentration variations. Copper displaces zinc (Zn) coordination at the zinc finger (ZF) domains of Sp1 and inactivates its DNA binding, whereas Cu deprivation enhances Sp1-DNA interactions and increases Sp1 expression, which in turn upregulates $hCtr1$. Because of the shared transport system, chemosensitivity of Pt drugs can be modulated by targeting Cu transporters. A Cu-lowering agent (trientine) in combination with a Pt drug (carboplatin) has been used in clinical studies for overcoming Pt-resistance. Future research should aim at further developing effective Pt drug retention strategies for improving the treatment efficacy.

Keywords: high-affinity copper transporter; $hCtr1$; Sp1; cisplatin; ovarian cancers; drug-resistance

1. Introduction

Platinum (Pt)-based drugs represent an extraordinary accomplishment in antitumor inorganic metal drug development [1]. They have been the mainstay of cancer chemotherapy in many different types of human malignancies for the last four decades since FDA approved *cis*-diamminedichloroplatinum(II) (cisplatin, cDDP) in 1978 [1,2]. These drugs are effective in treating advanced testicular cancers and ovarian cancers with cure rates of about 90% and 70%, respectively [3]. Together with carboplatin (Cbp) and oxaliplatin (Oxl), these Pt-based agents have shown a wide spectrum of antitumor activities including cancers of lung, bladder, breast, colon, and head and neck. It has been recognized that Pt drugs can attack many cellular targets including the plasma membrane, cellular organelles (mitochondria and endo-lysosomes), endoplasmic reticulum and cytoskeleton (see review in ref. [4] and references therein), but DNA damages are the principle cause of Pt drug-induced cell lethality. Another important factor that affects the treatment efficacy is the transport system which includes drug accumulation, intracellular drug shuffling and drug efflux (Figure 1A). This system cumulatively regulates the steady-state of intracellular drug contents which are directly correlated with the extent of DNA damage and cellular lethality [5,6]. Reduced cellular Pt content is an important hallmark of cDDP resistance in a wide variety of drug-resistant cell lines and tumors derived from Pt-refractory patients [7–9].

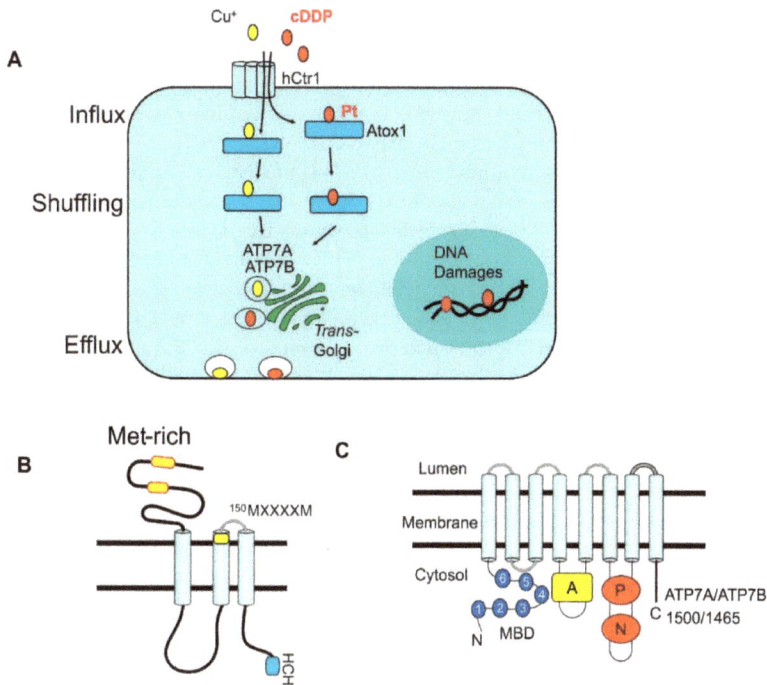

Figure 1. The similarity of transport systems between Cu(I) and cDDP. (**A**) routings of Cu(I) and cDDP from influx, intracellular trafficking (shuffling), to efflux; (**B**) schematic diagram depicting structure of *hCtr1*; (**C**) structure of ATP7A and ATP7B.

Previous works suggested that cDDP enters cells via a simple passive diffusion mechanism, because drug import is unsaturable and cannot be competed by cDDP analogues [6,10,11]. These findings suggest that cDDP influx may not require a transporter or carrier. However, because cDDP is a highly polar compound and its ability to across cellular membrane is limited, it is believed that the primary mechanism of Pt drug transport requires transporters. Indeed, studies have demonstrated that several copper (Cu) transporters are actively involved in the import, intracellular distribution, and export of Pt drugs [3]. These observations demonstrated that Cu ions and cDDP share similar transport mechanisms and mutually interfere with their cellular accumulations. In this review, we will first discuss the underlying mechanisms of how the Cu transporter system impedes Pt drugs from entering, following intracellular trafficking, and then exiting the cells. We will then focus on the transcriptional regulation of Cu transporter that bears clinical relevance to the treatment efficacy of Pt-based cancer chemotherapy.

2. Roles of the Copper Transporters in Pt Drug Transport Mechanism

2.1. Cisplatin Importers

It has been established that the high-affinity copper transporter 1 (Ctr1, also known as SLC31A1) is the primary transporter for cDDP [3,12]. Ctr1 also transports Cbp with reduced efficacy, but does not transport Oxl. cDDP-resistant cell lines show reduced accumulation of cDDP but not Oxl or satraplatin (JM216), and no cross-resistance to these analogues [13]. The major influx transporters of Oxl are organic cation/carnitine transporter (OCT1 and OCT2) [14,15]. Ctr1 is an evolutionarily conserved Cu(I) ion transporter shared from yeast to humans. The important roles of Ctr1 in cDDP uptake were initially demonstrated by genetic ablations in yeast cells and in mouse embryonic fibroblasts, which exhibited decreased cDDP uptake and increased cDDP resistance [16]. This was also confirmed by the finding that overexpression of human Ctr1 (*hCtr1*) in small cell lung cancer cells increases cDDP uptake [17]. However, these observations may depend on cell sources, because overexpressing ectopic *hCtr1* was seen in cDDP-resistant cells but not in its drug sensitive cells [18,19]. While the precise mechanism of this discrepancy is not known, it may be relevant to the cellular capacity of Cu homeostasis and regulation mechanisms, which will be discussed below.

The *hCtr1* is a 190-amino acid membrane protein that contains three transmembrane domains with the N-terminus extracellularly located and the C-terminus inside the cytoplasm. Site-directed mutagenic analyses have identified multiple conserved amino acids in *hCtr1* that are critical for transport of Cu(I) and cDDP transports. These include [40]MMMMxM in the N-terminal extracellular domain (where M denotes methionine and x denotes variable amino acids), [150]MxxxxM in the second transmembrane domains [20], [189]HCH in the cytoplasmic C-terminal domain, and others [21] (Figure 1B). However, mutations at the GG-4 motif ([167]GxxxG) reduces Cu(I) but not cDDP transport [20], suggesting the subtle differences in *hCtr1*-mediated transport between these substrates. It is suggested that cDDP (chemical formula: $cis[Pt(Cl_2(NH_3)_2)]$ interacts with the extracellularly exposed methionine (M)-motifs by forming the $[Pt(Met)Cl(NH_3)_2]$ intermediate, resulting in induced *hCtr1* conformational changes and stabilization of the trimeric formation as revealed by protein cross-linking agents [20]. cDDP and Cu(I) are thought to transverse through the axis of the trimeric *hCtr1* channel and move inward by an intermolecular sulfur-sulfur exchange mechanism [3]. Electron cryostallography of 2D protein crystals in a native phospholipid bilayer provides an estimate of ~9 angstroms for the pore size of the *hCtr1* homotrimer configuration [22]. cDDP has two each of chloride (Cl) and aminonia (NH_3) ligands coordinated to the central Pt in a square-planar structure. While the precise molecular dimension of cDDP is unclear, it is estimated to be at least three to four orders of magnitude (~1 nm) larger than the pore size [23], suggesting that conformational changes are involved in *hCtr1*-mediated cDDP passing.

Humans also have a low-affinity Cu transporter hCtr2 (SLC31A2) arisen by gene duplication from *hCtr1* [24]. hCtr2 shares substantial structural homology with *hCtr1*, but has only 143 amino acid residues because the majority of the N-terminal Met-rich sequence is missing. *mCtr2*-knockout mice show elevated Cu accumulation in several tissues with increased mCtr1 expression [25]. Expression of hCtr2 is mainly localized in intracellular vesicles. Recent studies demonstrated that mCtr2 interacts with mCtr1 and causes truncation of mCtr1 through cleavage of its ectodomain by a cathepsin protease, resulting in substantially reduced Cu(I) and cDDP transport capacities, and thus affecting the effectiveness of cDDP killing [26]. However, another recent study using CRISPR-Cas9 genomic editing strategy to knockout hCtr2 in two human tumor cell lines demonstrated only modest changes in cDDP sensitivity compared to the parental cell lines [27]. The discrepancies of these results are yet to be determined. In 40 ovarian cancers, it was reported that while high *hCtr1* expression is associated with chemosensitivity of Pt-based drugs, patients with low *hCtr1* and high hCtr2 in their tumors have poor treatment outcomes and shorter overall survival (OS) time [28]. In another study, it also reported that ovarian cancer patients with high hCtr2/*hCtr1* ratios in the tumor lesions are resistant to Pt-based chemotherapy [29]. These results demonstrated that hCtr2, in conjunction with *hCtr1*, may also involve in chemosensitivity of Pt drugs.

2.2. Cisplatin Chaperone

Once inside the cells, cDDP, like Cu(I), is carried away by different Cu chaperones for intracellular delivery to various compartments [30]. Antioxidant protein 1 (Atox1), one of the Cu chaperones, receives Cu(I) directly from the C-terminal end of *hCtr1* at the cytoplasmic membrane. The Cu-Atxo1 complex delivers Cu(I) to the Cu-efflux pumps, ATP7A and ATP7B, which are two P-type ATPases and located at trans-Golgi network (TGN) [31–33] (Figure 1A). The driving force for this directional trafficking is thought to be due to affinity gradients between protein partners along the route [34].

The human Atox1 contains 68 amino acids and has a $\beta_1\alpha_1\beta_2\beta_3\alpha_2\beta_4$ ferredox-like structure. The Cu binding motif ^{12}CXXC of Atox1 is located between the β_1 $\alpha v\delta$ α_1 loops [35]. It has been demonstrated that cDDP also binds this motif when delivering cDDP to ATP7A/ATP7B efflux pumps [36]. Deletion of Atox1 results in resistance to cDDP, indicating its important role in Pt drug sensitivity [37,38].

Atox1 also contains a nuclear targeting signal (^{38}KKTGK) between the α_2 $\alpha v\delta$ β_4 loop for Cu-dependent nuclear translocation of Atox1 [39]. Nuclear Atox1 functions as a transcriptional regulator for the mammalian cell proliferation gene *cyclin D1* by interacting with the 5′-GAAAGA sequence about 500 bp upstream of the transcription start site. Other Atox1-regulated genes include extracellular superoxide dismutase (*SOD3*) which encodes a secretory Cu-containing antioxidant enzyme [40], and *OCT4* which codes for a pluripotency factor in embryonic development [41]. The nuclear targeting property of Atox1 may deliver cDDP to elicit its lethal effect of DNA damage.

2.3. Role of ATP7A and ATP7B in cDDP Efflux

ATP7A and ATP7B, contain 1500 and 1465 amino acids, respectively, each have 8 transmembrane domains (TMDs). They share 67% amino acid sequence identity. Both ATPases contain several functionally conserved domains, i.e., six metal-binding domains (N-MBD) at the N-terminus, each with a CXXC motif; the nucleotide-binding domain (N-domain) for ATPase catalytic activity; the P-domain for phosphorylation at the ^{1207}Asp residue; and the A-domain for actuator/dephosphorylation (Figure 1C). There are also multiple Cu-binding sites located at transmembrane domain (TM) 4, TM5, and TM6 [42,43]. ATP7A is mainly expressed in the intestinal epithelium for Cu absorption from food. Mutations in ATP7A result in systemic Cu deficiency that causes the Menkes' disease. ATP7B is mainly located in the liver and brain. Mutations of ATP7B result in massive Cu buildup in these organs, resulting in Wilson's disease [44,45].

Detailed mechanisms of how Cu/Pt-Atox1 transfers the metal ions into ATP7A and ATP7B and how these metals are subsequently eliminated remain largely unknown. Current understanding suggests that similar protein folding of the CXXC metal binding motifs between Atox1 and the N-MBD of ATPases may facilitate rapid intermolecular metal transfer through electrostatics, hydrogen bonding, and hydrophobic interactions [46]. Another critical amino acid residue in Atox1 is ^{60}lysine which is essential for protein heterodimerization with ATPases for processing Cu(I) transfer [47,48].

While Cu-Atox1 can potentially interact with all the six N-MBDs of ATP7A and ATP7B, it preferentially interacts with the N-MBDs 1 to 4 [42]. Interactions of Cu-Atox1 with these N-MBDs induce conformational changes and activate ATPase catalytic activity by mobilizing N-MBD to cross-talk with the N-domain of ATP7B [49]. These interactions induce autophosphorylation of Asp^{1027} in the P domain by ATP, and phosphorylation of several serine residues in the TMDs by protein kinase D (PKD). However, it was found that deletion of the first five N-MBDs did not suppress the autophosphorylation of Asp^{1027}, whereas PKD-induced phosphorylation of the serine residues was downregulated [42,50]. Furthermore, recent results showed that native ATP7B forms dimeric configuration in the TGN, however, deleting the four MBDs from the N-terminus reshuffles the protein to the secretary vesicles, suggesting that the N-MBDs may also serve as a Golgi-retention signal [43].

These observations demonstrated the structural complexity involved in intramolecular transfer of metals by the mammalian ATPases.

Work on the bacterial system suggests that Cu chaperones can deliver Cu ions to TM-MBDs. Upon enzyme phosphorylation, Cu(I) is processed within the transmembrane region. Following the opening of the TM-MBD, Cu(I) is then released into the vesicles [51], which then traffic to the cellular membrane where Cu(I) is exported [52]. It was demonstrated that hydrolysis of one ATP molecule is sufficient to confer one molecule of Cu(I) translocation by the microbial ortholog CopA [51]. However, it is currently unclear whether this holds true for the mammalian Cu-ATPases.

Evidence has accumulated indicating that Pt(II) shares substantial similarity in coordination chemistry with Cu(I). The Cu-binding CXXC-motifs in Atox1 and ATP7A/ATP7B are also involved in cDDP binding [53]. Binding of cDDP to these sites may [54] or may not [55] displace Cu(I) binding. Mechanistic details of cDDP translocation in ATP7A/ATP7B is largely unknown because cellular levels of these ATPases are very low. Using microsomal fraction enriched in recombinant ATP7B (or ATP7A) absorbed onto a solid supported membrane, it was demonstrated that Pt drugs activate Cu-ATPases following a mechanism analogous to that of Cu(I) [54]. This in vitro study supports the roles of ATP7A and ATP7B in Pt drug transport. It has been reported that deletion of Atox1 resulted in an inability of cDDP delivery to ATP7A/ATP7B [37,38] and overexpression of ATP7A and ATP7B, resulting in impaired cDDP elimination, consequently resulting in cellular resistance to cDDP [56–59], Cbp [60], and Oxl [61].

3. Modulating cDDP Sensitivity through Redox Regulation of Cu Homeostasis

One of the important hallmarks in Pt-based chemotherapy is the induction of reactive oxygen species (ROS)-related oxidative stress that leads to cell killing [62]. It has been abundantly demonstrated that cDDP treatment induces expression of redox-regulating enzymes that are involved in the biosynthesis of glutathione (GSH), such as γ-glutamylcysteine synthetase (γGCS) and glutathione synthetase (GS) (see review in [62] and references therein). GSH is an abundant cellular redox regulator (in mM quantity) (see reviewers in [63,64]) (Figure 2). cDDP-resistant cell lines established by long-term exposure to cDDP display elevated expression of these enzymes, and depleting GSH using buthionine sulfoximine (BSO) reverses cDDP resistance [62,63,65]. Glutamate (Glu) and cysteine (Cys) are the upstream substrates of GSH biosynthesis, and their cellular transports are carried out by the xCT cysteine-glutamate antiporter (Figure 2). Pharmacological inhibition of xCT increases cDDP sensitivity [66]. cDDP-resistant cells also show increased glutamine (Gln) uptake, which is the precursor of Glu biosynthesis [67]. Moreover, Gln-fed rats reduced cDDP-induced nephrotoxicity in the proximal tubules [68]. These results, collectively, support the important roles of the GSH biosynthetic system in cDDP resistance.

However, we demonstrated that increased GSH by transfection with recombinant DNA encoding γGCS confers *sensitization*, but not resistance, to cDDP's cell killing [64,65]. γGCS is the rate-limiting enzyme for the biosynthesis of GSH. These findings demonstrate that elevated GSH does not per se confer cDDP resistance. We reason that the elevated expression of GSH and its biosynthetic enzymes observed in the cDDP-resistant cells under long-term cDDP exposure are likely due to CDDP-associated oxidative stress because these enzymes are redox-regulated. Instead, sensitization of the γGCS-transfected cells to cDDP is due to alternation of *hCtr1* expression. In another study, Franzini et al. [69] reported that overexpression of γ-glutamyltransferase (GGT) exhibited reduced GSH levels in the transfected cells which displayed increased resistance to cDDP. No description of the expression of *hCtr1* was mentioned in this work, because *hCtr1* as a cDDP transporter was reported in the subsequent year [70]. We will discuss this further in light of GSH regulation of Cu(I)/cDDP transport below.

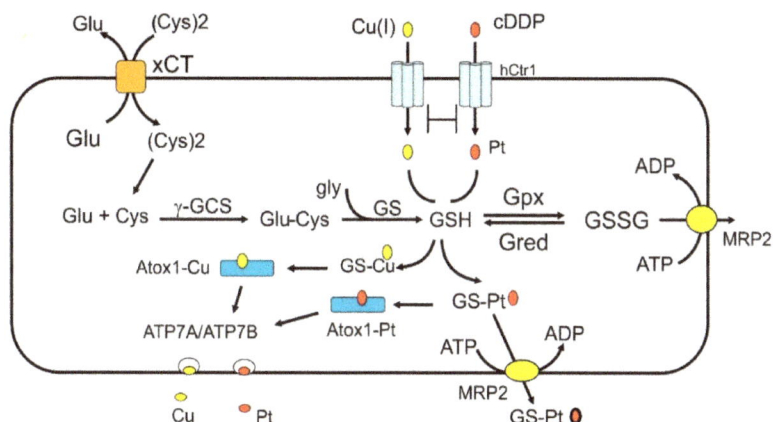

Figure 2. Regulation of Cu(I) and cDDP transports by the redox mechanism. The biosynthesis of GSH which plays important roles in redox regulation of Cu(I) and cDDP transport is shown here. The substrates of γGCS are glutamine and cysteine which are transported by xCT. GSH is oxidized to GSSG by Gpx and GSSG is reduced back to GSH by Gred. GSH can facilitate Cu(I) and Pt(II) delivery to Atox1 and ATP7A/ATP7B. GSH can upregulate *hCtr1* expression because of its chelation with Cu(I). GSSG and Pt-GS conjugate can be transported by the MRP2 efflux pump. Abbreviations: xCT, cysteine-glutamine anti-polar transporter; Gpx, glutathione peroxidase; GS, glutathione synthetase; γGCS, γ-glutamylcysteine synthetase; Gred, glutathione reductase; MRP2, multidrug resistance protein 2.

The redox system plays multiple roles in regulating Cu(I)/cDDP transport activities. First, extracellular Cu is normally present in an oxidized Cu(II) form. It is reduced to Cu(I) by cell membrane-associated reductases for the *hCtr1*, Atox1, and ATP7A/ATP7B transport cascade. Although Cu(I) can be converted into Cu(II) under oxidative stress conditions by the Fenton reaction, Cu(II) cannot be transported by these proteins. Second, Cu(I) and cDDP can bind to the major redox regulator, GSH and metallothionine (MT), resulting in reduction of bioavailability of Cu(I) that leads to increased *hCtr1* expression (see below). This can explain why GCS-transfection confers cDDP sensitivity [64,65]. Chelation of GSH by Cu(I) or Pt(II) induce GSH depletion, resulting in increased reactive oxygen species (ROS) production, which are generated from complex III in the mitochondria. The resultant ROS can modulate the expression of a whole spectrum of redox-regulating enzymes, including GSH peroxidase (Gpx) which converts GSH into GSSG, an oxidized form of GSH [63]. GSSG can be eliminated by the multidrug resistant protein transporter MRP2 (Figure 2). Third, the cytochrome c oxidase Cu chaperone (Cox17), which delivers Cu(I) to mitochondrial cytochrome c oxidase (CCO), can also deliver cDDP to mitochondria, the powerhouse of ROS production [71]. Platination of Cox17 affects the function of CCO, an important enzyme in the mitochondrial respiratory chain. Fourth, GSH can assist the conjugation of Cu(I) and Pt(II) with their chaperones and facilitate intracellular translocations of these metal ions. While apoAtox1 does not bind Cu-ATPases and is normally in monomeric configuration, Cu-Atox1 often exists as a dimer and its formation is strictly influenced by cellular redox conditions [72,73]. Under physiologic redox conditions, conjugation of Cu(I) and GSH exits as a polymer, which assists Cu(I) transferring to Atox1. GSH can also assist cDDP complex formation with Atox1. Moreover, Cu(I)-loaded Atox1 promotes the binding of cDDP and form the Atox1-Cu-Pt tertiary complex via sulfur-bridge linkages [74,75]. Likewise, GSH can facilitate the formation of Pt-Cox17 [71]. These findings demonstrate the important roles of GSH in Cu(I) and cDDP transport systems. Finally, Pt-(GSH)$_2$ conjugate is also a known substrate of MRP2 efflux pump [76,77] (Figure 2), and elevated expression of MRP2 is associated with cDDP resistance [78].

4. Modulating cDDP Sensitivity through Transcriptional Regulation of *hCtr1* Expression

4.1. Regulation of Ctr1 Internalization by Cu Bioavailability

The findings that cDDP highjacks the Cu-transport system for its own transport underscore the importance of Cu transporters in regulating cDDP sensitivity in cancer chemotherapy. Homeostatic regulation of Ctr1 expression is an evolutionarily conserved mechanism from yeast [79] to humans [3,80,81]. Levels of Ctr1 are induced under Cu starvation conditions but are downregulated by Cu overload. A previous study in the yeast *Saccharomyces cerevisiae* demonstrated that Cu-induced yCtr1 internalization is one mechanism of Cu(I) acquisition, and that the internalized yCtr1 is rapidly degraded [82]. Cu-induced *hCtr1* internalization has also been reported in human cells [83,84]. However, detailed mechanisms on how mobilization of Ctr1 movements in response to Cu availability remain to be investigated.

4.2. Transcriptional Regulation of Ctr1 Expression by Cu Bioavailability

Transcriptional regulation of Ctr1 involves a variety of transcription factors. These transcription regulators generally contain Cu-sensing domains which bear DNA binding activities and transactivation domains for transcriptional activation. Budding yeast Mac1 is the transcriptional regulator controlling the expression of *yCtr1* and *yCtr3* and reductase *Fre1*, together encoding the Cu transport system. The N-terminal 40 amino acids of Mac1 contains zinc finger (ZF)-like motifs which have DNA-binding activities. Two cysteine (C)-rich transactivation domains are located at its C-terminus. Under low Cu conditions, Mac1 activates the expression of Cu transport genes by binding to the promoters of these genes. At high Cu concentrations, four Cu(I) ions each bind the C-rich transactivation domains, inducing conformational changes of Mac1, resulting in its fall-off from the target genes and shutting down of transcription [85]. In the meantime, another transcriptional factor, Ace1, is activated through the formation of a tetracopper-thiolate cluster within the Cu regulatory domain [86,87], and binds the promoter of *CUP1* and *CRS5* encoding Cu-chelating metallothionein (MT) [88]. Thus, yeasts use Mac1 and Ace1 in coordinating their activation and inactivation in response to Cu depletion and repletion conditions, respectively, to regulate Cu homeostasis.

Zinc finger-like transcriptional factors are also involved in regulating *Ctr1* gene expression in other organisms. These include metal responsive transcription factor 1 (MTF-1) for the *dCtr1B* in *Drosophila* [89], CRR1 for the *Chlamydomonas* Cu transporters *Cyc6*, *CPX1*, and *CRD1* [90], and SQUAMOS promoter-binding protein like-7 (SPL-7) factor for three *Arabidopsis* Cu transporters, COPT1, COPT2 and COPT6 [91,92]. The schematic structure of transcription factors for *Ctr1* genes in different species is shown in Figure 3 [93].

In humans, we previously identified that specific protein (Sp1) is the transcription factor that regulates *hCtr1* expression [94,95]. Sp1 by itself is regulated by Cu homeostasis via transcriptional interactions with at least 10 Sp1-binding sites located at the Sp1 promoter. High Cu conditions downregulate Sp1 expression (Figure 4A), whereas reduced Cu conditions by Cu chelation upregulate Sp1 (Figure 4B). Sp1 in turn regulates the expression of *hCtr1* in response to Cu concentration variations accordingly (Figure 4A,B) by interacting with two Sp1-binding sites located 25 bp downstream from the *hCtr1* transcriptional start site. Systemic mutations of these Sp1-binding sites abolish the Cu responsiveness of Sp1 and *hCtr1* expression [95].

Another transcriptional unit with opposite direction, named *FKBP133* (also known *KIAA0674*) encodes an FK506-binding protein-like transcript, and is located −201 bp upstream of the *hCtr1* locus. FKBP133 is also regulated by Cu stressed conditions using the same Sp1 binding sites for the *hCtr1* regulation (our unpublished result). How this Sp1-mediated bidirectional transcription and the fine-turning of its targeted gene expression under Cu stressed conditions remains to be investigated.

Species	Cu-Transporter	Transcription Regulator	Structure	References
Yeast	*yCtr1, yCtr3*	Mac1	2x ZF — 417 aa	[85]
		Ace1	225	
Drosophila	*dCtr1B*	MTF-1	6x ZF — 675	[89]
Chlamydomonas	*CPX1, Cyc6, CRD1*	CRR1	2x ZF — 1232	[90]
Arabidopsis	*COPT1, COPT2, COPT6*	SPL-7	2x ZF — 801	[91,92]
Human	*hCtr1*	Sp1	3x ZF — 618	[94,95]

Figure 3. Schematic diagrams showing the structures of transcription factors for copper transporters from different species. Black boxes refer to ZF-like domains; yellow boxes, transactivation domains (see the text for details).

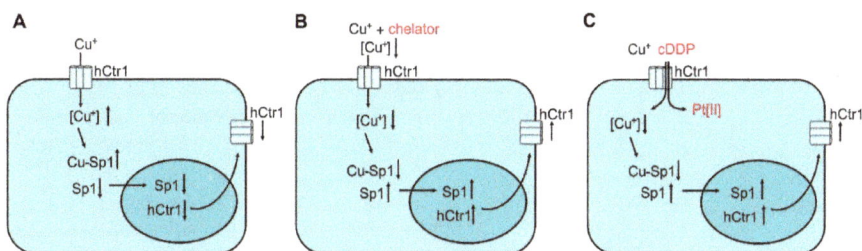

Figure 4. Mechanisms of transcriptional regulation of Sp1 and *hCtr1* in response to various challenges: (**A**) downregulation by Cu overload; (**B**) upregulation by Cu chelator; (**C**) upregulation by Cu(I) and cDDP combination.

4.3. The Sensing Mechanisms of Cu Bioavailability by Sp1

Sp1 is a ubiquitous transcription factor consisting of a DNA-binding domain at the C-terminus that contains three ZF and a transactivation domain that contains two serine/threonine-rich and two glutamine-rich (Q-rich 1 and Q-rich 2) subdomains (Figure 3). The ZF of Sp1 is constitutively bound by Zn(II) because apoSp1 is very unstable [96]. Each ZF consists of Cys2-His2 residues that is coordinated by one Zn(II) molecule. Elevated Cu ions displace Zn(II) binding of Sp1 [97]. Although this causes only minor structural alternations, the "Cu-finger" cannot interact with the *hCtr1* promoter [98]. Thus, Cu is a negative regulator of Sp1 by poisoning its ZF DNA binding domains.

Sp1 is a member of the Sp/KLF (Krüppel-like factor) transcription factor family sharing the general three copies of C2H2-type ZFs [99]. However, ZF domains in this family have a variety of structures and at least eight different topologies have been categorized; many of these ZF-binding proteins do not respond to Cu challenges [100]. Even Sp3, which is the closest member of Sp1, does not regulate *hCtr1* expression [94].

Interaction between cDDP and Sp1 is very weak at best, although it can interact with other ZF-containing proteins, i.e., the retroviral protein NCp7 [101] and DNA polymerase 1 [102]. However, it has been reported that the reactivity of Pt(II) with ZFs can be modulated by reducing agents such as tris(2-carboxyethyl)phosphine) (TCEP) [103] which is commonly used to maintain cysteine residues at the reduced state [102]. In contrast, the *trans*-platinum thiazole Pt complex [PtCl2(NH3)(thiazole)] is highly reactive towards Sp1-ZF2, but the resulting Pt-thiazole complex prevents nuclear trafficking of Sp1 [103]. Moreover, the reactivity of Sp1-ZF3 with the cDDP complex is increased when the NH3 ligands of cDDP are replaced by the chelating ethylenediamine (en) in [PtCl2(en)] [104]. These reactions cause conformational distortion of ZFs, rendering it unable to bind DNA. These results illustrate the capacity for interaction of Pt compounds with ZF-proteins. However, the pharmacological relevance of the interaction of Pt compounds with ZF proteins remains to be further studied.

We previously demonstrated that cDDP can transcriptionally induce Sp1 and *hCtr1* expression in time- and concentration-dependent manners [105]. Since cDDP does not directly act upon Sp1, these results can be interpreted by virtue that cDDP acts as a competitor for *hCtr1*-mediated Cu(I) transport, resulting in reduced cellular Cu levels that leads to upregulation of Sp1/*hCtr1* (Figure 4C). In this context, cDDP may be considered as a Cu-lowering agent, further supporting the integral role of cDDP in Cu homeostasis regulation.

4.4. The Capacity of hCtr1 Regulation and Cellular Cu Bioavailability

Copper is an essential micronutrient for growth but is toxic when in excess. This is evidenced by the findings that ablation of both murine *mCtr1* alleles results in embryonic lethality and the associated toxic effects in Menkes' and Wilson's diseases due to Cu anomalies in the intestine and liver, respectively, as mentioned above. It has been reported that almost all cellular Cu(I) are chelated by cellular constituents and only a small fraction of Cu(I) is bioavailable [106]. Thus, the bioavailable Cu pools have to be tightly regulated. While a yeast (*S. cerevisiae*) cell contains as high as 0.01–0.1 M of total Cu ions, but the bioavailable Cu(I) concentration is estimated to be only in the range of 8.9×10^{-17} to 5.1×10^{-23} M. This was estimated using the highly Cu(I)-selective and -sensitive transcription factors Mac1 and Ace1 for the lower and upper limits, respectively [107]. Estimation of bioavailable Cu(I) pools in human cells cannot be similarly carried out using Sp1 as a probe because of its abundant target genes.

It is conceivable that bioavailable Cu(I) pools are different from cell type to cell type and play a critical role in determining the capacity of Cu homeostasis regulation under Cu stressed conditions. Intestinal epithelium is the primary cell type acquiring Cu from food. Intestinal epithelial cell-specific knockout mice show drastic reduction of Cu in different organs, i.e., about 95% in the liver but only about 30% in the kidneys as compared respectively with those in normal mice [108], refracting differential capacities of Cu homeostatic regulation in different tissues.

A great variety of human malignancies have shown elevated Cu levels in the serum and tumors, and elevated Cu levels are often positively correlated with cancer progression (see review in [109] and references therein). Elevated Cu levels in tumors are correlated with high levels of *hCtr1* in lung cancers [110]. Likewise, Sp1 expression levels are elevated in many human tumors, including tumors of stomach, pancreas, breast, brain, and thyroid (see [111] and references therein). These findings suggest that capacity of Cu bioavailability regulation may differ between human cancers and their normal counterparts.

5. Modulation of hCtr1 Transcriptional Regulation for Overcoming cDDP Resistance in Cancer Chemotherapy

By analyzing a publicly available database containing 91 ovarian cancers, Ishida et al. [70] reported that patients with high *hCtr1* levels were associated with longer disease-free survival time after adjuvant chemotherapy with a Pt drug and taxane. We analyzed a database of 243 patients with endometrioid tumors of the ovary treated with first line Pt/taxane chemotherapy and found that patients with high *hCtr1* levels, but not ATP7A and ATP7B, have significantly longer progression-free survival (PFS) and OS than those with low *hCtr1* [20]. Recently, a more comprehensive study based on eight datasets containing 2149 patients from nine countries, Sun et al. [8] reported that high *hCtr1* expression is associated with favorable treatment outcomes in ovarian and lung cancer patients who underwent Pt-based chemotherapy. These results suggest that cancers with high *hCtr1* expression have a better response to Pt-based drugs, suggesting that transcriptional upregulation of *hCtr1* expression by Cu-lowering agents can be an attractive strategy for improving chemosensitivity to Pt drugs with the following considerations. Firstly, we demonstrated in a cultured cell study that *hCtr1* mRNA and protein levels can be induced by Cu chelators within one hour, but it takes several days for the upregulated *hCtr1* mRNA and protein to return to basal levels upon removal of the chelators. These results suggest that induced *hCtr1* by Cu chelation is considerably stable. Neither ATP7A nor ATP7B mRNA is altered under Cu chelation [112]. Secondly, as stated, while multiple mechanisms are involved cDDP resistance, reduced *hCtr1* expression is a common mechanism. Moreover, we found that cDDP-resistant cancer cells exhibit a greater magnitude of *hCtr1* upregulation by Cu-lowering agents as compared with their drug-sensitive counterparts. These results were demonstrated in three independent cDDP-resistant cell lines treated with three different Cu-lowering agents, i.e., trientine, D-pencillamide (D-pen), and tetrathiomolybdate (TM). The reduced levels of *hCtr1* in these cDDP-resistant cells were recovered to those comparable with the corresponding drug-sensitive counterparts. The effectiveness of these Cu-lowering agents have little cell line- and agent-specificities [112]. These findings demonstrate that Cu-lowering agents can overcome cDDP resistance [112]. Third, Cu-lowering agents by themselves have been in many clinical trials for treating various human malignancies because Cu is involved in tumor growth pathways, such as metastatic tumor angiogenesis [113] and the oncogenic transformation pathway [114,115].

Based on these pre-clinical observations, a phase I clinical trial using Cbp plus trientine in 55 patients with advanced malignancies, 45 of which had prior failure in Pt drug treatment, was conducted at MD Anderson Cancer. About 19% of patients ($n = 9$) who maintained low serum Cu levels after the treatments had significantly longer median PFS ($p = 0.001$) and OS ($p = 0.03$) as compared with those patients ($n = 38$) who did not [116,117]. However, while the response rate remains low—given the heterogeneity of patient population and multiple mechanisms of drug resistance that may be involved, and the intrinsic variation in the capacity of *hCtr1* induction by Cu-lowering agent as mentioned above—there are options for improvement using this strategy (see below).

6. Conclusions and Perspectives

Pt drugs have been effective in treating many human malignancies, however, treatment efficacy has been hampered by drug resistance. Mechanisms of cDDP resistance are complex, and defective drug transport is commonly associated with cDDP resistance [63,118,119]. Recent studies have established that the Pt drug transport system is an integral mechanism of Cu homeostasis regulation. Preclinical studies have established that interventions of expression of *hCtr1* transporter, Atox1 chaperone, and ATP7A/ATP7B efflux pumps can affect chemosensitivity of Pt drugs. However, targeting this transport system for overcoming cDDP resistance has not been translated into a therapeutic benefit. Our laboratories have discovered that *hCtr1*, but not ATP7A/ATP7B, expression can be transcriptionally modulated by Cu concentration variations via the transcriptional factor Sp1. The ZF domains of Sp1 are sensors of both Cu repletion and depletion conditions and transcriptionally down- and up-regulates Sp1 accordingly. Sp1 in turn regulates *hCtr1* expression, which is the major regulator of cellular Cu content. This constitutes the Cu-Sp1-*hCtr1* homeostatic self-regulatory loop [3,120].

Not only does *hCtr1* function as an important importer for cellular cDDP accumulation, its expression is also upregulated by cDDP, because cDDP functions as a potent competitor for *hCtr1*-mediated Cu transport, resulting in reduced intracellular Cu content. These observations have led to an exploratory clinical investigation using the Cu-lowering agent trientine to enhance *hCtr1* expression for overcoming Pt resistance in multiple cancer types [116]. Although the outcome of this first-in-human trial remains low, the established mechanistic basis has important potentials for further clinical studies in improving the treatment efficacy of Pt-based chemotherapy, with the following considerations.

Firstly, the ZF transcription regulators have been firmly assessed in Cu sensing to transcriptionally regulate Pt drug transporters in eukaryotic cells. There are 23,299 genes encoding ZF-containing proteins in the human genome [121]. This provides a vast wealth of opportunities for exploring other ZF regulators that may control the cellular concentration of Pt drugs. These ZF proteins may directly regulate the expression of Pt transporters, or indirectly function as intracellular metal chelators that regulate global Cu ion bioavailability which regulates the expression of the Pt transport system. Secondly, given the observations that Ctr1 expression is tissue-specific, and its responses to Cu depletion differs in different organs [108], it is important to elucidate tissue-specific mechanisms of *hCtr1* expression using the Cu-chelation strategy. This research may identify specific tumor types that are favorable for Cu chelation therapy. As an example, we found that expression of *hCtr1* mRNA and proteins are higher in 20 out of 20 ovarian tumor biopsies as compared with their adjacent naïve tissues in patients prior to chemotherapy (HHWC, unpublished data). These results, together with those published previously [8,70,112], may explain why ovarian cancers are preferentially sensitive to Pt-based chemotherapy [122,123], and why cDDP-refractory ovarian cancers may seem to have better response rates to the trientine plus Cbp combination therapy than other tumors in our exploratory clinical investigation [117]. Thirdly, previous clinical studies demonstrated that better treatment outcomes are associated in patients with reduced serum Cu levels after the trientine/Cbp combination therapy [116,117]. These results suggest that effective therapy may lie in the Cu-lowering ability of the treatment. Thus, identification of predictive biomarkers for Cu chelation in patients may be critical. Fourthly, the current clinical study used trientine; the effectiveness of other potent Cu-lowering agents such as desferal [124] and TM [125] still need to be explored. Fifthly, cytotoxic effects associated with combination therapy using a Pt drug and copper-lowering agent need to be critically evaluated. Finally, while we have learned much about *hCtr1* regulation, other transport components such as Atox1 and ATP7B/ATP7B remains a largely uncharted area of research. Exploitation of drug retention by targeting these transporters may be fruitful.

In summary, we have highlighted the importance of targeting drug transporters in this review. The described transcriptional regulation of *hCtr1* presented here may serve as the translational basis for future investigations for improving the treatment efficacy of cancer chemotherapy using Pt-based drugs.

Author Contributions: All authors have contributed in generating concepts, manuscript writing, and final approval of the manuscript.

Acknowledgments: This work was financially supported by the Center of Applied Nanomedicine, National Cheng Kung University from The Featured Areas Research Center Program within the framework of the Higher Education Sprout Project by the Ministry of Education (MOE) in Taiwan, and in part by grants from the Ministry of Science and Technology, Taiwan (MOST-105-2314-B-006-046-MY3 to Helen H. W. Chen, MOST-106-2314-B-006-018-MY2 to Yu-Hsuan Lai, National Cheng Kung University Hospital, Taiwan (NCKUH-10704014 to Yu-Hsuan Lai), and the National Cancer Institute, USA (CA149260 to Macus Tien Kuo).

Conflicts of Interest: The authors declare no conflict of interest.

Abbreviations

Atox1	antioxidant 1
BSO	buthionine sulfoximine
Cbp	carboplatin
Ctr1	the high-affinity copper transporter 1 (SLC31A1)
GCS	γ-glutamylcysteine synthetase
GGT	γ-glutamyltransferase
GS	glutathione synthetase
GSH	Glutathione

MTF-1	metal responsive transcription factor 1
D-pen	D-pencillamide
Oxl	oxaliplatin
ROS	reactive oxygen species
Sp1	specific protein
TM	tetrathiomolybdate
TGN	trans-Golgi network
xCT	cystine glutamate antipolar transporter

References

1. Muggia, F.M.; Bonetti, A.; Hoeschele, J.D.; Rozencweig, M.; Howell, S.B. Platinum Antitumor Complexes: 50 Years Since Barnett Rosenberg's Discovery. *J. Clin. Oncol.* **2015**, *33*, 4219–4226. [CrossRef] [PubMed]
2. Johnstone, T.C.; Suntharalingam, K.; Lippard, S.J. The Next Generation of Platinum Drugs: Targeted Pt(II) Agents, Nanoparticle Delivery, and Pt(IV) Prodrugs. *Chem. Rev.* **2016**, *116*, 3436–3486. [CrossRef] [PubMed]
3. Chen, H.H.; Chen, W.C.; Liang, Z.D.; Tsai, W.B.; Long, Y.; Aiba, I.; Fu, S.; Broaddus, R.; Liu, J.; Feun, L.G.; et al. Targeting drug transport mechanisms for improving platinum-based cancer chemotherapy. *Expert Opin. Ther. Targets* **2015**, *19*, 1307–1317. [CrossRef] [PubMed]
4. Gatti, L.; Cassinelli, G.; Zaffaroni, N.; Lanzi, C.; Perego, P. New mechanisms for old drugs: Insights into DNA-unrelated effects of platinum compounds and drug resistance determinants. *Drug Resist. Updates* **2015**, *20*, 1–11. [CrossRef] [PubMed]
5. Kim, E.S.; Lee, J.J.; He, G.; Chow, C.W.; Fujimoto, J.; Kalhor, N.; Swisher, S.G.; Wistuba, I.I.; Stewart, D.J.; Siddik, Z.H. Tissue platinum concentration and tumor response in non-small-cell lung cancer. *J. Clin. Oncol.* **2012**, *30*, 3345–3352. [CrossRef] [PubMed]
6. Hall, M.D.; Okabe, M.; Shen, D.W.; Liang, X.J.; Gottesman, M.M. The role of cellular accumulation in determining sensitivity to platinum-based chemotherapy. *Annu. Rev. Pharmacol. Toxicol.* **2008**, *48*, 495–535. [CrossRef] [PubMed]
7. Andrews, P.A.; Howell, S.B. Cellular pharmacology of cisplatin: Perspectives on mechanisms of acquired resistance. *Cancer Cells* **1990**, *2*, 35–43. [PubMed]
8. Sun, S.; Cai, J.; Yang, Q.; Zhao, S.; Wang, Z. The association between copper transporters and the prognosis of cancer patients undergoing chemotherapy: A meta-analysis of literatures and datasets. *Oncotarget* **2017**, *8*, 16036–16051. [CrossRef] [PubMed]
9. Long, Y.; Tsai, W.B.; Chang, J.T.; Estecio, M.; Wangpaichitr, M.; Savaraj, N.; Feun, L.G.; Chen, H.H.; Kuo, M.T. Cisplatin-induced synthetic lethality to arginine-starvation therapy by transcriptional suppression of ASS1 is regulated by DEC1, HIF-1alpha, and c-Myc transcription network and is independent of ASS1 promoter DNA methylation. *Oncotarget* **2016**, *7*, 82658–82670. [CrossRef] [PubMed]
10. Gately, D.P.; Howell, S.B. Cellular accumulation of the anticancer agent cisplatin: A review. *Br. J. Cancer* **1993**, *67*, 1171–1176. [CrossRef] [PubMed]
11. Ivy, K.D.; Kaplan, J.H. A re-evaluation of the role of *hCtr1*, the human high-affinity copper transporter, in platinum-drug entry into human cells. *Mol. Pharmacol.* **2013**, *83*, 1237–1246. [CrossRef] [PubMed]
12. Ohrvik, H.; Thiele, D.J. The role of Ctr1 and Ctr2 in mammalian copper homeostasis and platinum-based chemotherapy. *J. Trace Elem. Med. Biol.* **2015**, *31*, 178–182. [CrossRef] [PubMed]
13. Martelli, L.; Di Mario, F.; Ragazzi, E.; Apostoli, P.; Leone, R.; Perego, P.; Fumagalli, G. Different accumulation of cisplatin, oxaliplatin and JM216 in sensitive and cisplatin-resistant human cervical tumour cells. *Biochem. Pharmacol.* **2006**, *72*, 693–700. [CrossRef] [PubMed]
14. Buss, I.; Hamacher, A.; Sarin, N.; Kassack, M.U.; Kalayda, G.V. Relevance of copper transporter 1 and organic cation transporters 1–3 for oxaliplatin uptake and drug resistance in colorectal cancer cells. *Metall. Integr. Biomet. Sci.* **2018**, *10*, 414–425. [CrossRef] [PubMed]
15. Sprowl, J.A.; Ciarimboli, G.; Lancaster, C.S.; Giovinazzo, H.; Gibson, A.A.; Du, G.; Janke, L.J.; Cavaletti, G.; Shields, A.F.; Sparreboom, A. Oxaliplatin-induced neurotoxicity is dependent on the organic cation transporter OCT2. *Proc. Natl. Acad. Sci. USA* **2013**, *110*, 11199–11204. [CrossRef] [PubMed]

16. Ishida, S.; Lee, J.; Thiele, D.J.; Herskowitz, I. Uptake of the anticancer drug cisplatin mediated by the copper transporter Ctr1 in yeast and mammals. *Proc. Natl. Acad. Sci. USA* **2002**, *99*, 14298–14302. [CrossRef] [PubMed]

17. Song, I.S.; Savaraj, N.; Siddik, Z.H.; Liu, P.; Wei, Y.; Wu, C.J.; Kuo, M.T. Role of human copper transporter Ctr1 in the transport of platinum-based antitumor agents in cisplatin-sensitive and cisplatin-resistant cells. *Mol. Cancer Ther.* **2004**, *3*, 1543–1549. [PubMed]

18. Beretta, G.L.; Gatti, L.; Tinelli, S.; Corna, E.; Colangelo, D.; Zunino, F.; Perego, P. Cellular pharmacology of cisplatin in relation to the expression of human copper transporter CTR1 in different pairs of cisplatin-sensitive and -resistant cells. *Biochem. Pharmacol.* **2004**, *68*, 283–291. [CrossRef] [PubMed]

19. Rabik, C.A.; Maryon, E.B.; Kasza, K.; Shafer, J.T.; Bartnik, C.M.; Dolan, M.E. Role of copper transporters in resistance to platinating agents. *Cancer Chemother. Pharmacol.* **2009**, *64*, 133–142. [CrossRef] [PubMed]

20. Liang, Z.D.; Stockton, D.; Savaraj, N.; Tien Kuo, M. Mechanistic comparison of human high-affinity copper transporter 1-mediated transport between copper ion and cisplatin. *Mol. Pharmacol.* **2009**, *76*, 843–853. [CrossRef] [PubMed]

21. Du, X.; Wang, X.; Li, H.; Sun, H. Comparison between copper and cisplatin transport mediated by human copper transporter 1 (*hCtr1*). *Metall. Integr. Biomet. Sci.* **2012**, *4*, 679–685. [CrossRef] [PubMed]

22. Aller, S.G.; Unger, V.M. Projection structure of the human copper transporter CTR1 at 6-A resolution reveals a compact trimer with a novel channel-like architecture. *Proc. Natl. Acad. Sci. USA* **2006**, *103*, 3627–3632. [CrossRef] [PubMed]

23. Pottier, A.; Borghi, E.; Levy, L. New use of metals as nanosized radioenhancers. *Anticancer Res.* **2014**, *34*, 443–453. [PubMed]

24. Logeman, B.L.; Wood, L.K.; Lee, J.; Thiele, D.J. Gene duplication and neo-functionalization in the evolutionary and functional divergence of the metazoan copper transporters Ctr1 and Ctr2. *J. Biol. Chem.* **2017**, *292*, 11531–11546. [CrossRef] [PubMed]

25. Ohrvik, H.; Nose, Y.; Wood, L.K.; Kim, B.E.; Gleber, S.C.; Ralle, M.; Thiele, D.J. Ctr2 regulates biogenesis of a cleaved form of mammalian Ctr1 metal transporter lacking the copper- and cisplatin-binding ecto-domain. *Proc. Natl. Acad. Sci. USA* **2013**, *110*, E4279–E4288. [CrossRef] [PubMed]

26. Ohrvik, H.; Logeman, B.; Turk, B.; Reinheckel, T.; Thiele, D.J. Cathepsin Protease Controls Copper and Cisplatin Accumulation via Cleavage of the Ctr1 Metal-binding Ectodomain. *J. Biol. Chem.* **2016**, *291*, 13905–13916. [CrossRef] [PubMed]

27. Bompiani, K.M.; Tsai, C.Y.; Achatz, F.P.; Liebig, J.K.; Howell, S.B. Copper transporters and chaperones CTR1, CTR2, ATOX1, and CCS as determinants of cisplatin sensitivity. *Metall. Integr. Biomet. Sci.* **2016**, *8*, 951–962. [CrossRef] [PubMed]

28. Lee, Y.Y.; Choi, C.H.; Do, I.G.; Song, S.Y.; Lee, W.; Park, H.S.; Song, T.J.; Kim, M.K.; Kim, T.J.; Lee, J.W.; et al. Prognostic value of the copper transporters, CTR1 and CTR2, in patients with ovarian carcinoma receiving platinum-based chemotherapy. *Gynecol. Oncol.* **2011**, *122*, 361–365. [CrossRef] [PubMed]

29. Yoshida, H.; Teramae, M.; Yamauchi, M.; Fukuda, T.; Yasui, T.; Sumi, T.; Honda, K.; Ishiko, O. Association of copper transporter expression with platinum resistance in epithelial ovarian cancer. *Anticancer Res.* **2013**, *33*, 1409–1414. [PubMed]

30. Robinson, N.J.; Winge, D.R. Copper metallochaperones. *Annu. Rev. Biochem.* **2010**, *79*, 537–562. [CrossRef] [PubMed]

31. Flores, A.G.; Unger, V.M. Atox1 contains positive residues that mediate membrane association and aid subsequent copper loading. *J. Membr. Biol.* **2013**, *246*, 903–913. [CrossRef] [PubMed]

32. Kahra, D.; Kovermann, M.; Wittung-Stafshede, P. The C-Terminus of Human Copper Importer Ctr1 Acts as a Binding Site and Transfers Copper to Atox1. *Biophys. J.* **2016**, *110*, 95–102. [CrossRef] [PubMed]

33. Wu, X.; Yuan, S.; Wang, E.; Tong, Y.; Ma, G.; Wei, K.; Liu, Y. Platinum transfer from *hCtr1* to Atox1 is dependent on the type of platinum complex. *Metall. Integr. Biomet. Sci.* **2017**, *9*, 546–555. [CrossRef] [PubMed]

34. Banci, L.; Bertini, I.; Ciofi-Baffoni, S.; Kozyreva, T.; Zovo, K.; Palumaa, P. Affinity gradients drive copper to cellular destinations. *Nature* **2010**, *465*, 645–648. [CrossRef] [PubMed]

35. Wernimont, A.K.; Huffman, D.L.; Lamb, A.L.; O'Halloran, T.V.; Rosenzweig, A.C. Structural basis for copper transfer by the metallochaperone for the Menkes/Wilson disease proteins. *Nat. Struct. Biol.* **2000**, *7*, 766–771. [PubMed]

36. Boal, A.K.; Rosenzweig, A.C. Crystal structures of cisplatin bound to a human copper chaperone. *J. Am. Chem. Soc.* **2009**, *131*, 14196–14197. [CrossRef] [PubMed]

37. Hua, H.; Gunther, V.; Georgiev, O.; Schaffner, W. Distorted copper homeostasis with decreased sensitivity to cisplatin upon chaperone Atox1 deletion in Drosophila. *Biometals* **2011**, *24*, 445–453. [CrossRef] [PubMed]

38. Safaei, R.; Maktabi, M.H.; Blair, B.G.; Larson, C.A.; Howell, S.B. Effects of the loss of Atox1 on the cellular pharmacology of cisplatin. *J. Inorg. Biochem.* **2009**, *103*, 333–341. [CrossRef] [PubMed]

39. Itoh, S.; Kim, H.W.; Nakagawa, O.; Ozumi, K.; Lessner, S.M.; Aoki, H.; Akram, K.; McKinney, R.D.; Ushio-Fukai, M.; Fukai, T. Novel role of antioxidant-1 (Atox1) as a copper-dependent transcription factor involved in cell proliferation. *J. Biol. Chem.* **2008**, *283*, 9157–9167. [CrossRef] [PubMed]

40. Itoh, S.; Ozumi, K.; Kim, H.W.; Nakagawa, O.; McKinney, R.D.; Folz, R.J.; Zelko, I.N.; Ushio-Fukai, M.; Fukai, T. Novel mechanism for regulation of extracellular SOD transcription and activity by copper: Role of antioxidant-1. *Free Radic. Biol. Med.* **2009**, *46*, 95–104. [CrossRef] [PubMed]

41. Celauro, E.; Mukaj, A.; Fierro-Gonzalez, J.C.; Wittung-Stafshede, P. Copper chaperone ATOX1 regulates pluripotency factor OCT4 in preimplantation mouse embryos. *Biochem. Biophys. Res. Commun.* **2017**, *491*, 147–153. [CrossRef] [PubMed]

42. Inesi, G.; Pilankatta, R.; Tadini-Buoninsegni, F. Biochemical characterization of P-type copper ATPases. *Biochem. J.* **2014**, *463*, 167–176. [CrossRef] [PubMed]

43. Jayakanthan, S.; Braiterman, L.T.; Hasan, N.M.; Unger, V.M.; Lutsenko, S. Human Copper Transporter Atp7b (Wilson Disease Protein) Forms Stable Dimers in Vitro and in Cells. *J. Biol. Chem.* **2017**, *292*, 18760–18774. [CrossRef] [PubMed]

44. Cox, D.W.; Moore, S.D. Copper transporting P-type ATPases and human disease. *J. Bioenergy Biomembr.* **2002**, *34*, 333–338. [CrossRef]

45. Lutsenko, S.; Barnes, N.L.; Bartee, M.Y.; Dmitriev, O.Y. Function and regulation of human copper-transporting ATPases. *Physiol. Rev.* **2007**, *87*, 1011–1046. [CrossRef] [PubMed]

46. Arnesano, F.; Banci, L.; Bertini, I.; Thompsett, A.R. Solution structure of CopC: A cupredoxin-like protein involved in copper homeostasis. *Structure* **2002**, *10*, 1337–1347. [CrossRef]

47. Hussain, F.; Olson, J.S.; Wittung-Stafshede, P. Conserved residues modulate copper release in human copper chaperone Atox1. *Proc. Natl. Acad. Sci. USA* **2008**, *105*, 11158–11163. [CrossRef] [PubMed]

48. Xi, Z.; Shi, C.; Tian, C.; Liu, Y. Conserved residue modulates copper-binding properties through structural dynamics in human copper chaperone Atox1. *Metall. Integr. Biomet. Sci.* **2013**, *5*, 1566–1573. [CrossRef] [PubMed]

49. Yu, C.H.; Yang, N.; Bothe, J.; Tonelli, M.; Nokhrin, S.; Dolgova, N.V.; Braiterman, L.; Lutsenko, S.; Dmitriev, O.Y. The metal chaperone Atox1 regulates the activity of the human copper transporter ATP7B by modulating domain dynamics. *J. Biol. Chem.* **2017**, *292*, 18169–18177. [CrossRef] [PubMed]

50. Lewis, D.; Pilankatta, R.; Inesi, G.; Bartolommei, G.; Moncelli, M.R.; Tadini-Buoninsegni, F. Distinctive features of catalytic and transport mechanisms in mammalian sarco-endoplasmic reticulum Ca^{2+} ATPase (SERCA) and Cu^+ (ATP7A/B) ATPases. *J. Biol. Chem.* **2012**, *287*, 32717–32727. [CrossRef] [PubMed]

51. Gonzalez-Guerrero, M.; Arguello, J.M. Mechanism of Cu+-transporting ATPases: Soluble Cu^+ chaperones directly transfer Cu^+ to transmembrane transport sites. *Proc. Natl. Acad. Sci. USA* **2008**, *105*, 5992–5997. [CrossRef] [PubMed]

52. Pilankatta, R.; Lewis, D.; Inesi, G. Involvement of protein kinase D in expression and trafficking of ATP7B (copper ATPase). *J. Biol. Chem.* **2011**, *286*, 7389–7396. [CrossRef] [PubMed]

53. Safaei, R.; Adams, P.L.; Maktabi, M.H.; Mathews, R.A.; Howell, S.B. The CXXC motifs in the metal binding domains are required for ATP7B to mediate resistance to cisplatin. *J. Inorg. Biochem.* **2012**, *110*, 8–17. [CrossRef] [PubMed]

54. Tadini-Buoninsegni, F.; Bartolommei, G.; Moncelli, M.R.; Inesi, G.; Galliani, A.; Sinisi, M.; Losacco, M.; Natile, G.; Arnesano, F. Translocation of platinum anticancer drugs by human copper ATPases ATP7A and ATP7B. *Angew. Chem.* **2014**, *53*, 1297–1301. [CrossRef] [PubMed]

55. Palm, M.E.; Weise, C.F.; Lundin, C.; Wingsle, G.; Nygren, Y.; Bjorn, E.; Naredi, P.; Wolf-Watz, M.; Wittung-Stafshede, P. Cisplatin binds human copper chaperone Atox1 and promotes unfolding in vitro. *Proc. Natl. Acad. Sci. USA* **2011**, *108*, 6951–6956. [CrossRef] [PubMed]

56. Nakagawa, T.; Inoue, Y.; Kodama, H.; Yamazaki, H.; Kawai, K.; Suemizu, H.; Masuda, R.; Iwazaki, M.; Yamada, S.; Ueyama, Y.; et al. Expression of copper-transporting P-type adenosine triphosphatase (ATP7B) correlates with cisplatin resistance in human non-small cell lung cancer xenografts. *Oncol. Rep.* **2008**, *20*, 265–270. [PubMed]

57. Komatsu, M.; Sumizawa, T.; Mutoh, M.; Chen, Z.S.; Terada, K.; Furukawa, T.; Yang, X.L.; Gao, H.; Miura, N.; Sugiyama, T.; et al. Copper-transporting P-type adenosine triphosphatase (ATP7B) is associated with cisplatin resistance. *Cancer Res.* **2000**, *60*, 1312–1316. [PubMed]

58. Safaei, R.; Howell, S.B. Copper transporters regulate the cellular pharmacology and sensitivity to Pt drugs. *Crit. Rev. Oncol. Hematol.* **2005**, *53*, 13–23. [CrossRef] [PubMed]

59. Leonhardt, K.; Gebhardt, R.; Mossner, J.; Lutsenko, S.; Huster, D. Functional interactions of Cu-ATPase ATP7B with cisplatin and the role of ATP7B in the resistance of cells to the drug. *J. Biol. Chem.* **2009**, *284*, 7793–7802. [CrossRef] [PubMed]

60. Samimi, G.; Safaei, R.; Katano, K.; Holzer, A.K.; Rochdi, M.; Tomioka, M.; Goodman, M.; Howell, S.B. Increased expression of the copper efflux transporter ATP7A mediates resistance to cisplatin, carboplatin, and oxaliplatin in ovarian cancer cells. *Clin. Cancer Res.* **2004**, *10*, 4661–4669. [CrossRef] [PubMed]

61. Martinez-Balibrea, E.; Martinez-Cardus, A.; Musulen, E.; Gines, A.; Manzano, J.L.; Aranda, E.; Plasencia, C.; Neamati, N.; Abad, A. Increased levels of copper efflux transporter ATP7B are associated with poor outcome in colorectal cancer patients receiving oxaliplatin-based chemotherapy. *Int. J. Cancer* **2009**, *124*, 2905–2910. [CrossRef] [PubMed]

62. Brozovic, A.; Ambriovic-Ristov, A.; Osmak, M. The relationship between cisplatin-induced reactive oxygen species, glutathione, and BCL-2 and resistance to cisplatin. *Crit. Rev. Toxicol.* **2010**, *40*, 347–359. [CrossRef] [PubMed]

63. Stewart, D.J. Mechanisms of resistance to cisplatin and carboplatin. *Crit. Rev. Oncol. Hematol.* **2007**, *63*, 12–31. [CrossRef] [PubMed]

64. Chen, H.H.; Kuo, M.T. Role of glutathione in the regulation of Cisplatin resistance in cancer chemotherapy. *Met.-Based Drugs* **2010**, *2010*, 430939. [CrossRef] [PubMed]

65. Chen, H.H.; Song, I.S.; Hossain, A.; Choi, M.K.; Yamane, Y.; Liang, Z.D.; Lu, J.; Wu, L.Y.; Siddik, Z.H.; Klomp, L.W.; et al. Elevated glutathione levels confer cellular sensitization to cisplatin toxicity by up-regulation of copper transporter hCtr1. *Mol. Pharmacol.* **2008**, *74*, 697–704. [CrossRef] [PubMed]

66. Roh, J.L.; Kim, E.H.; Jang, H.; Shin, D. Aspirin plus sorafenib potentiates cisplatin cytotoxicity in resistant head and neck cancer cells through xCT inhibition. *Free Radic. Biol. Med.* **2017**, *104*, 1–9. [CrossRef] [PubMed]

67. Wangpaichitr, M.; Wu, C.; Li, Y.Y.; Nguyen, D.J.M.; Kandemir, H.; Shah, S.; Chen, S.; Feun, L.G.; Prince, J.S.; Kuo, M.T.; et al. Exploiting ROS and metabolic differences to kill cisplatin resistant lung cancer. *Oncotarget* **2017**, *8*, 49275–49292. [CrossRef] [PubMed]

68. Kim, H.J.; Park, D.J.; Kim, J.H.; Jeong, E.Y.; Jung, M.H.; Kim, T.H.; Yang, J.I.; Lee, G.W.; Chung, H.J.; Chang, S.H. Glutamine protects against cisplatin-induced nephrotoxicity by decreasing cisplatin accumulation. *J. Pharmacol. Sci.* **2015**, *127*, 117–126. [CrossRef] [PubMed]

69. Franzini, M.; Corti, A.; Lorenzini, E.; Paolicchi, A.; Pompella, A.; De Cesare, M.; Perego, P.; Gatti, L.; Leone, R.; Apostoli, P.; et al. Modulation of cell growth and cisplatin sensitivity by membrane gamma-glutamyltransferase in melanoma cells. *Eur. J. Cancer* **2006**, *42*, 2623–2630. [CrossRef] [PubMed]

70. Ishida, S.; McCormick, F.; Smith-McCune, K.; Hanahan, D. Enhancing tumor-specific uptake of the anticancer drug cisplatin with a copper chelator. *Cancer Cell* **2010**, *17*, 574–583. [CrossRef] [PubMed]

71. Zhao, L.; Cheng, Q.; Wang, Z.; Xi, Z.; Xu, D.; Liu, Y. Cisplatin binds to human copper chaperone Cox17: The mechanistic implication of drug delivery to mitochondria. *Chem. Commun.* **2014**, *50*, 2667–2669. [CrossRef] [PubMed]

72. Narindrasorasak, S.; Zhang, X.; Roberts, E.A.; Sarkar, B. Comparative analysis of metal binding characteristics of copper chaperone proteins, Atx1 and ATOX1. *Bioinorg. Chem. Appl.* **2004**, 105–123. [CrossRef] [PubMed]

73. Tanchou, V.; Gas, F.; Urvoas, A.; Cougouluegne, F.; Ruat, S.; Averseng, O.; Quemeneur, E. Copper-mediated homo-dimerisation for the HAH1 metallochaperone. *Biochem. Biophys. Res. Commun.* **2004**, *325*, 388–394. [CrossRef] [PubMed]

74. Dolgova, N.V.; Yu, C.; Cvitkovic, J.P.; Hodak, M.; Nienaber, K.H.; Summers, K.L.; Cotelesage, J.J.H.; Bernholc, J.; Kaminski, G.A.; Pickering, I.J.; et al. Binding of Copper and Cisplatin to Atox1 Is Mediated by Glutathione through the Formation of Metal-Sulfur Clusters. *Biochemistry* **2017**, *56*, 3129–3141. [CrossRef] [PubMed]

75. Xi, Z.; Guo, W.; Tian, C.; Wang, F.; Liu, Y. Copper binding promotes the interaction of cisplatin with human copper chaperone Atox1. *Chem. Commun.* **2013**, *49*, 11197–11199. [CrossRef] [PubMed]

76. Ishikawa, T. The ATP-dependent glutathione S-conjugate export pump. *Trends Biochem. Sci.* **1992**, *17*, 463–468. [CrossRef]

77. Ishikawa, T.; Ali-Osman, F. Glutathione-associated cis-diamminedichloroplatinum(II) metabolism and ATP-dependent efflux from leukemia cells. Molecular characterization of glutathione-platinum complex and its biological significance. *J. Biol. Chem.* **1993**, *268*, 20116–20125. [PubMed]

78. Yamasaki, M.; Makino, T.; Masuzawa, T.; Kurokawa, Y.; Miyata, H.; Takiguchi, S.; Nakajima, K.; Fujiwara, Y.; Matsuura, N.; Mori, M.; et al. Role of multidrug resistance protein 2 (MRP2) in chemoresistance and clinical outcome in oesophageal squamous cell carcinoma. *Br. J. Cancer* **2011**, *104*, 707–713. [CrossRef] [PubMed]

79. Dancis, A.; Yuan, D.S.; Haile, D.; Askwith, C.; Eide, D.; Moehle, C.; Kaplan, J.; Klausner, R.D. Molecular characterization of a copper transport protein in *S. cerevisiae*: An unexpected role for copper in iron transport. *Cell* **1994**, *76*, 393–402. [CrossRef]

80. Kuo, M.T.; Fu, S.; Savaraj, N.; Chen, H.H. Role of the human high-affinity copper transporter in copper homeostasis regulation and cisplatin sensitivity in cancer chemotherapy. *Cancer Res.* **2012**, *72*, 4616–4621. [CrossRef] [PubMed]

81. Howell, S.B.; Safaei, R.; Larson, C.A.; Sailor, M.J. Copper transporters and the cellular pharmacology of the platinum-containing cancer drugs. *Mol. Pharmacol.* **2010**, *77*, 887–894. [CrossRef] [PubMed]

82. Ooi, C.E.; Rabinovich, E.; Dancis, A.; Bonifacino, J.S.; Klausner, R.D. Copper-dependent degradation of the Saccharomyces cerevisiae plasma membrane copper transporter Ctr1p in the apparent absence of endocytosis. *EMBO J.* **1996**, *15*, 3515–3523. [PubMed]

83. Guo, Y.; Smith, K.; Lee, J.; Thiele, D.J.; Petris, M.J. Identification of methionine-rich clusters that regulate copper-stimulated endocytosis of the human Ctr1 copper transporter. *J. Biol. Chem.* **2004**, *279*, 17428–17433. [CrossRef] [PubMed]

84. Molloy, S.A.; Kaplan, J.H. Copper-dependent recycling of *hCtr1*, the human high affinity copper transporter. *J. Biol. Chem.* **2009**, *284*, 29704–29713. [CrossRef] [PubMed]

85. Jensen, L.T.; Posewitz, M.C.; Srinivasan, C.; Winge, D.R. Mapping of the DNA binding domain of the copper-responsive transcription factor Mac1 from Saccharomyces cerevisiae. *J. Biol. Chem.* **1998**, *273*, 23805–23811. [CrossRef] [PubMed]

86. Furst, P.; Hu, S.; Hackett, R.; Hamer, D. Copper activates metallothionein gene transcription by altering the conformation of a specific DNA binding protein. *Cell* **1988**, *55*, 705–717. [CrossRef]

87. Keller, G.; Bird, A.; Winge, D.R. Independent metalloregulation of Ace1 and Mac1 in Saccharomyces cerevisiae. *Eukaryot Cell* **2005**, *4*, 1863–1871. [CrossRef] [PubMed]

88. Thiele, D.J. ACE1 regulates expression of the Saccharomyces cerevisiae metallothionein gene. *Mol. Cell. Biol.* **1988**, *8*, 2745–2752. [CrossRef] [PubMed]

89. Selvaraj, A.; Balamurugan, K.; Yepiskoposyan, H.; Zhou, H.; Egli, D.; Georgiev, O.; Thiele, D.J.; Schaffner, W. Metal-responsive transcription factor (MTF-1) handles both extremes, copper load and copper starvation, by activating different genes. *Genes Dev.* **2005**, *19*, 891–896. [CrossRef] [PubMed]

90. Strenkert, D.; Schmollinger, S.; Sommer, F.; Schulz-Raffelt, M.; Schroda, M. Transcription factor-dependent chromatin remodeling at heat shock and copper-responsive promoters in Chlamydomonas reinhardtii. *Plant Cell* **2011**, *23*, 2285–2301. [CrossRef] [PubMed]

91. Garcia-Molina, A.; Xing, S.; Huijser, P. Functional characterisation of Arabidopsis SPL7 conserved protein domains suggests novel regulatory mechanisms in the Cu deficiency response. *BMC Plant Biol.* **2014**, *14*, 231. [CrossRef] [PubMed]

92. Yamasaki, H.; Hayashi, M.; Fukazawa, M.; Kobayashi, Y.; Shikanai, T. SQUAMOSA Promoter Binding Protein-Like7 Is a Central Regulator for Copper Homeostasis in Arabidopsis. *Plant Cell* **2009**, *21*, 347–361. [CrossRef] [PubMed]

93. Kuo, M.T.; Chen, H.H. Overcoming platinum drug resistance with copper-lowering agents. *Anticancer Res.* **2013**, *33*, 4157–4161.

94. Liang, Z.D.; Tsai, W.B.; Lee, M.Y.; Savaraj, N.; Kuo, M.T. Specificity protein 1 (sp1) oscillation is involved in copper homeostasis maintenance by regulating human high-affinity copper transporter 1 expression. *Mol. Pharmacol.* **2012**, *81*, 455–464. [CrossRef] [PubMed]

95. Song, I.S.; Chen, H.H.; Aiba, I.; Hossain, A.; Liang, Z.D.; Klomp, L.W.; Kuo, M.T. Transcription factor Sp1 plays an important role in the regulation of copper homeostasis in mammalian cells. *Mol. Pharmacol.* **2008**, *74*, 705–713. [CrossRef] [PubMed]

96. Bittel, D.C.; Smirnova, I.V.; Andrews, G.K. Functional heterogeneity in the zinc fingers of metalloregulatory protein metal response element-binding transcription factor-1. *J. Biol. Chem.* **2000**, *275*, 37194–37201. [CrossRef] [PubMed]

97. Yan, D.; Aiba, I.; Chen, H.H.; Kuo, M.T. Effects of Cu(II) and cisplatin on the stability of Specific protein 1 (Sp1)-DNA binding: Insights into the regulation of copper homeostasis and platinum drug transport. *J. Inorg. Biochem.* **2016**, *161*, 37–39. [CrossRef] [PubMed]

98. Yuan, S.; Chen, S.; Xi, Z.; Liu, Y. Copper-finger protein of Sp1: The molecular basis of copper sensing. *Metall. Integr. Biomet. Sci.* **2017**, *9*, 1169–1175. [CrossRef] [PubMed]

99. Wierstra, I. Sp1: Emerging roles—Beyond constitutive activation of TATA-less housekeeping genes. *Biochem. Biophys. Res. Commun.* **2008**, *372*, 1–13. [CrossRef] [PubMed]

100. Krishna, S.S.; Majumdar, I.; Grishin, N.V. Structural classification of zinc fingers: Survey and summary. *Nucleic Acids Res.* **2003**, *31*, 532–550. [CrossRef] [PubMed]

101. Anzellotti, A.I.; Liu, Q.; Bloemink, M.J.; Scarsdale, J.N.; Farrell, N. Targeting retroviral Zn finger-DNA interactions: A small-molecule approach using the electrophilic nature of trans-platinum-nucleobase compounds. *Chem. Biol.* **2006**, *13*, 539–548. [CrossRef] [PubMed]

102. Maurmann, L.; Bose, R.N. Unwinding of zinc finger domain of DNA polymerase I by cis-diamminedichloroplatinum(II). *Dalton Trans.* **2010**, *39*, 7968–7979. [CrossRef] [PubMed]

103. Chen, S.; Xu, D.; Jiang, H.; Xi, Z.; Zhu, P.; Liu, Y. Trans-platinum/thiazole complex interferes with Sp1 zinc-finger protein. *Angew. Chem.* **2012**, *51*, 12258–12262. [CrossRef] [PubMed]

104. Du, Z.; de Paiva, R.E.; Qu, Y.; Farrell, N. Tuning the reactivity of Sp1 zinc fingers with platinum complexes. *Dalton Trans.* **2016**, *45*, 8712–8716. [CrossRef] [PubMed]

105. Liang, Z.D.; Long, Y.; Chen, H.H.; Savaraj, N.; Kuo, M.T. Regulation of the high-affinity copper transporter (hCtr1) expression by cisplatin and heavy metals. *J. Biol. Inorg. Chem.* **2014**, *19*, 17–27. [CrossRef] [PubMed]

106. Rae, T.D.; Schmidt, P.J.; Pufahl, R.A.; Culotta, V.C.; O'Halloran, T.V. Undetectable intracellular free copper: The requirement of a copper chaperone for superoxide dismutase. *Science* **1999**, *284*, 805–808. [CrossRef] [PubMed]

107. Wegner, S.V.; Sun, F.; Hernandez, N.; He, C. The tightly regulated copper window in yeast. *Chem. Commun.* **2011**, *47*, 2571–2573. [CrossRef] [PubMed]

108. Nose, Y.; Kim, B.E.; Thiele, D.J. Ctr1 drives intestinal copper absorption and is essential for growth, iron metabolism, and neonatal cardiac function. *Cell Metab.* **2006**, *4*, 235–244. [CrossRef] [PubMed]

109. Gupte, A.; Mumper, R.J. Elevated copper and oxidative stress in cancer cells as a target for cancer treatment. *Cancer Treat. Rev.* **2009**, *35*, 32–46. [CrossRef] [PubMed]

110. Kim, E.S.; Tang, X.; Peterson, D.R.; Kilari, D.; Chow, C.W.; Fujimoto, J.; Kalhor, N.; Swisher, S.G.; Stewart, D.J.; Wistuba, I.I.; et al. Copper transporter CTR1 expression and tissue platinum concentration in non-small cell lung cancer. *Lung Cancer* **2014**, *85*, 88–93. [CrossRef] [PubMed]

111. Beishline, K.; Azizkhan-Clifford, J. Sp1 and the 'hallmarks of cancer'. *FEBS J.* **2015**, *282*, 224–258. [CrossRef] [PubMed]

112. Liang, Z.D.; Long, Y.; Tsai, W.B.; Fu, S.; Kurzrock, R.; Gagea-Iurascu, M.; Zhang, F.; Chen, H.H.; Hennessy, B.T.; Mills, G.B.; et al. Mechanistic basis for overcoming platinum resistance using copper chelating agents. *Mol. Cancer Ther.* **2012**, *11*, 2483–2494. [CrossRef] [PubMed]

113. Brem, S.; Grossman, S.A.; Carson, K.A.; New, P.; Phuphanich, S.; Alavi, J.B.; Mikkelsen, T.; Fisher, J.D.; New Approaches to Brain Tumor Therapy CNS Consortium. Phase 2 trial of copper depletion and penicillamine as antiangiogenesis therapy of glioblastoma. *Neuro-Oncology* **2005**, *7*, 246–253. [CrossRef] [PubMed]

114. Brady, D.C.; Crowe, M.S.; Greenberg, D.N.; Counter, C.M. Copper Chelation Inhibits BRAF(V600E)-Driven Melanomagenesis and Counters Resistance to BRAF(V600E) and MEK1/2 Inhibitors. *Cancer Res.* **2017**, *77*, 6240–6252. [CrossRef] [PubMed]

Int. J. Mol. Sci. **2018**, *19*, 1486

115. Garber, K.; BIOMEDICINE. Targeting copper to treat breast cancer. *Science* **2015**, *349*, 128–129. [CrossRef] [PubMed]

116. Fu, S.; Hou, M.M.; Wheler, J.; Hong, D.; Naing, A.; Tsimberidou, A.; Janku, F.; Zinner, R.; Piha-Paul, S.; Falchook, G.; et al. Exploratory study of carboplatin plus the copper-lowering agent trientine in patients with advanced malignancies. *Investig. New Drugs* **2014**, *32*, 465–472. [CrossRef] [PubMed]

117. Fu, S.; Naing, A.; Fu, C.; Kuo, M.T.; Kurzrock, R. Overcoming platinum resistance through the use of a copper-lowering agent. *Mol. Cancer Ther.* **2012**, *11*, 1221–1225. [CrossRef] [PubMed]

118. Siddik, Z.H. Cisplatin: Mode of cytotoxic action and molecular basis of resistance. *Oncogene* **2003**, *22*, 7265–7279. [CrossRef] [PubMed]

119. Cossa, G.; Gatti, L.; Zunino, F.; Perego, P. Strategies to improve the efficacy of platinum compounds. *Curr. Med. Chem.* **2009**, *16*, 2355–2365. [CrossRef] [PubMed]

120. Kuo, M.T.; Chen, H.H.; Song, I.S.; Savaraj, N.; Ishikawa, T. The roles of copper transporters in cisplatin resistance. *Cancer Metastasis Rev.* **2007**, *26*, 71–83. [CrossRef] [PubMed]

121. Klug, A. The discovery of zinc fingers and their applications in gene regulation and genome manipulation. *Annu. Rev. Biochem.* **2010**, *79*, 213–231. [CrossRef] [PubMed]

122. Kelland, L. The resurgence of platinum-based cancer chemotherapy. *Nat. Rev. Cancer* **2007**, *7*, 573–584. [CrossRef] [PubMed]

123. Jayson, G.C.; Kohn, E.C.; Kitchener, H.C.; Ledermann, J.A. Ovarian cancer. *Lancet* **2014**, *384*, 1376–1388. [CrossRef]

124. Chen, S.J.; Kuo, C.C.; Pan, H.Y.; Tsou, T.C.; Yeh, S.C.; Chang, J.Y. Desferal regulates *hCtr1* and transferrin receptor expression through Sp1 and exhibits synergistic cytotoxicity with platinum drugs in oxaliplatin-resistant human cervical cancer cells in vitro and in vivo. *Oncotarget* **2016**, *7*, 49310–49321. [CrossRef] [PubMed]

125. Brewer, G.J. The promise of copper lowering therapy with tetrathiomolybdate in the cure of cancer and in the treatment of inflammatory disease. *J. Trace Elem. Med. Biol.* **2014**, *28*, 372–378. [CrossRef] [PubMed]

International Journal of
Molecular Sciences

MDPI

Review

The Interaction of the Metallo-Glycopeptide Anti-Tumour Drug Bleomycin with DNA

Vincent Murray *, Jon K. Chen and Long H. Chung

School of Biotechnology and Biomolecular Sciences, University of New South Wales, Sydney,
NSW 2052, Australia; jonken.chen@gmail.com (J.K.C.); chunglonghoa@gmail.com (L.H.C.)
* Correspondence: v.murray@unsw.edu.au; Tel.: +61-2-9385-2028; Fax: +61-2-9385-1483

Received: 12 April 2018; Accepted: 24 April 2018; Published: 4 May 2018

Abstract: The cancer chemotherapeutic drug, bleomycin, is clinically used to treat several neoplasms including testicular and ovarian cancers. Bleomycin is a metallo-glycopeptide antibiotic that requires a transition metal ion, usually Fe(II), for activity. In this review, the properties of bleomycin are examined, especially the interaction of bleomycin with DNA. A Fe(II)-bleomycin complex is capable of DNA cleavage and this process is thought to be the major determinant for the cytotoxicity of bleomycin. The DNA sequence specificity of bleomycin cleavage is found to at 5′-GT* and 5′-GC* dinucleotides (where * indicates the cleaved nucleotide). Using next-generation DNA sequencing, over 200 million double-strand breaks were analysed, and an expanded bleomycin sequence specificity was found to be 5′-RTGT*AY (where R is G or A and Y is T or C) in cellular DNA and 5′-TGT*AT in purified DNA. The different environment of cellular DNA compared to purified DNA was proposed to be responsible for the difference. A number of bleomycin analogues have been examined and their interaction with DNA is also discussed. In particular, the production of bleomycin analogues via genetic manipulation of the modular non-ribosomal peptide synthetases and polyketide synthases in the bleomycin gene cluster is reviewed. The prospects for the synthesis of bleomycin analogues with increased effectiveness as cancer chemotherapeutic agents is also explored.

Keywords: bleomycin analogues; DNA interaction; double-strand breaks; genome-wide; next-generation sequencing; sequence specificity

1. Introduction

The bleomycins (Figure 1) are a group of structurally related metallo-glycopeptide antibiotics discovered by Umezama and colleagues [1] and are clinically used as cancer chemotherapeutic agents. The bleomycins are isolated from *Streptomyces verticillis* and the bacterium uses modular non-ribosomal peptide synthetases and polyketide synthases to synthesise bleomycin [2–6]. Bleomycin can be separated by chromatography into bleomycin A and B [1] and both A and B can be further separated into different components. The clinical product, Blenoxane, consists of 60% bleomycin A_2, 30% bleomycin B_2 and other minor components. The initial studies showed that bleomycin inhibited DNA synthesis in *E. coli* and HeLa cells and also suppressed the growth of cancer cells including Ehrlich carcinoma and mouse sarcoma [7]. In chemotherapy, bleomycin is currently used in combination with other drugs to treat malignant germ-cell tumours, Hodgkin's lymphoma and carcinomas of skin, head, and neck [8–12]. In particular, a treatment regimen of bleomycin, etoposide, and cisplatin is able to cure 90% of patients with testicular cancer [11]. Additionally, a combination of bleomycin, vinblastine, and cisplatin have been successfully used to treat metastatic ovarian cancer [13]. Bleomycin's cytotoxicity is attributed to its ability to cause double- and single-strand DNA breaks, which then lead to extended cell cycle arrest, apoptosis, and mitotic cell death [14,15]. Double-strand breaks are thought to be the most important for anti-tumour activity [16–19].

Int. J. Mol. Sci. **2018**, *19*, 1372'

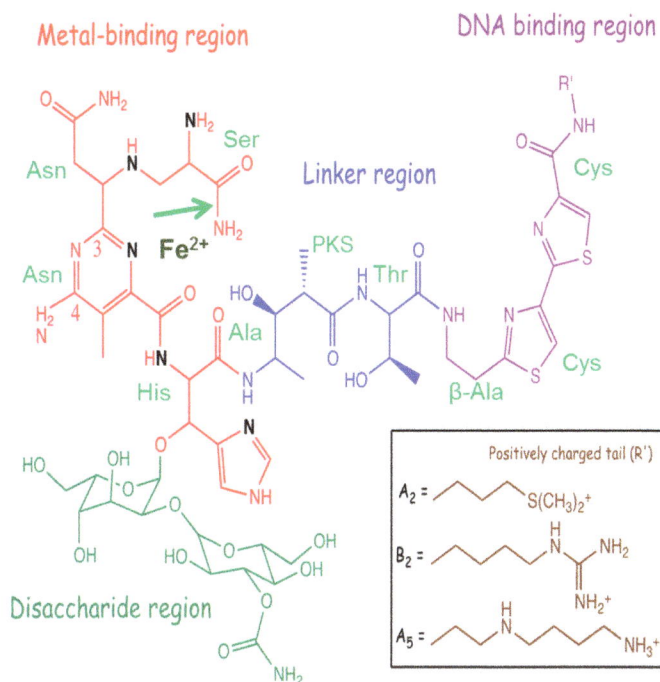

Figure 1. The structure of bleomycin with the four functional domains indicated. The precursor amino acids are shown as well as the propionate moiety derived from polyketide synthases (PKS). The positively charged tails (R') are shown for the various bleomycin congeners. The green arrow indicates the site of bleomycin hydrolase cleavage. The figure is adapted from [15] and reprinted by permission from Springer Nature.

2. Bleomycin Is Composed of Four Functional Domains

Bleomycin is a relatively large molecule and is composed of four structural domains (Figure 1): a metal-binding region; a linker region; a disaccharide; and a bithiazole tail [20]. Congeners of the bleomycin family are differentiated by the positively charged region on the bithiazole tail.

The metal binding region contains nitrogen atoms that are involved in coordination with transition metals [21–24]. Bleomycin is able to coordinate with a number of metals including: Fe, Co, Cu, Mn, V, and Zn. Fe ions are thought to be the biologically important divalent cations that are responsible for the cytotoxic activity of bleomycin. The metal-binding region is also thought to play an important role in the DNA sequence specificity of bleomycin. Structural studies of Co(III)-bleomycin bound to DNA show the interaction of the N_3 and C_4-NH_2 of the bleomycin pyrimidine moiety with the C_2-NH_2 and N_3 of guanine [20].

In conjunction with the metal-binding and linker region, the C-terminal bithiazole tail is important for DNA recognition and binding of bleomycin with DNA. The bithiazole tail and its C- terminal region is thought to contribute to DNA binding by intercalation or minor groove interaction [15,25,26]. With a positive charge on the R'-group, this tail facilitates the electrostatic interaction between bleomycin and DNA. Drugs with a planar fused ring system can interact with DNA via intercalation and hence it is expected that the coplanar bithiazole tail of bleomycin can intercalate between bases of the DNA to increase DNA affinity [23,27]. However, other studies pointed out that bleomycin might interact with DNA via minor grove binding [25,26,28]. More recent research using X-ray crystallography indicates that intercalative binding is the main form of DNA binding (see below) [15,20].

The function of the disaccharide region of bleomycin is not thoroughly understood because of the lack of analogues with varying sugar groups [29]. The disaccharide moiety consists of an L-gulose and 3-O-carbamoyl-D-mannose. A modified bleomycin complex, deglycobleomycin, where the disaccharide moiety is replaced with hydrogen, demonstrated a reduced efficiency of DNA cleavage [30]. Studies by the Hecht group have highlighted the role of the disaccharide moiety in the selectivity and uptake of bleomycin by tumour cells [31]. They demonstrated that the bleomycin disaccharide alone was sufficient for selective uptake by tumour cells [32–34]. The 3-O-carbamoyl-D-mannose moiety appears to be crucial for this function [35,36]. This is consistent with related work that provided evidence that this sugar region is involved in tumour cell uptake via glucose transport. Glycolysis is accelerated in tumour cells due to their high energy demands. Bleomycin may be effective in targeting cancer cells because the drug can be mistaken as glucose due to its disaccharide moiety and hence, freely enter into tumour cells via protein channels [33,34].

The linker region plays a role as a bridge that joins the metal binding and DNA binding domains [37]. The linker connects the metal binding region and bithiazole tail; and consists of a valerate and L-threonine subunit. Systematic modifications to the substituents of these subunits revealed their importance for the efficiency of bleomycin-mediated DNA cleavage [29,38–40]. Notably, these analogues exhibit significant reduction in DNA cleavage (compared to deglycobleomycin) but maintained a similar sequence specificity. On the other hand, analogues methylated at the valerate-threonine amide were reported to have diminished sequence specificity [41]. These studies suggest the importance of the linker in inducing an optimal conformation of bleomycin with respect to DNA binding. The X-ray crystal structure revealed that the valerate was hydrogen bonded to the minor groove of DNA [20].

3. Bleomycin DNA Cleavage Mechanism

When bleomycin, in its metal-free form, is administrated intravenously, it rapidly binds to Cu(II) in the blood serum to create a stable complex, bleomycin-Cu(II) [23]. When this complex is transferred inside a cell, bleomycin is reduced to bleomycin-Cu(I). The new complex can enter the nucleus and exchange with Fe(II) to form bleomycin-Fe(II).

The degradation of DNA by bleomycin is preceded by the formation of an activated form of bleomycin [15]. Activated bleomycin has been shown to form in the presence of O_2, when bleomycin-Fe(II) binds to O_2, forming bleomycin-Fe(II)-OO$^\bullet$, which then accepts an electron and H^+ to generate the "activated bleomycin" complex, bleomycin-Fe(III)-OOH. Regardless of the presence of DNA, the activated bleomycin complex decays rapidly, with a half-life of approximately 2 min at 4 °C and ultimately forms bleomycin-Fe(III) [42].

Activated bleomycin participates in the abstraction of the C4' hydrogen atom from the deoxyribose moiety of a pyrimidine nucleotide 3'- to a guanine (Figure 2) [23,28,43]. This results in the formation of a C4' radical intermediate that can proceed via two separate pathways, depending on the presence of O_2. In the absence of O_2, 98% of the C4' radical is converted to 4'-oxidised abasic sites [44]. On the other hand, in the presence of 1 atmosphere of O_2, approximately 70% undergoes strand scission and forms 3'-phosphoglycolate and 5'-phosphate ends, while the remainder forms 4'-oxidised abasic sites [45,46].

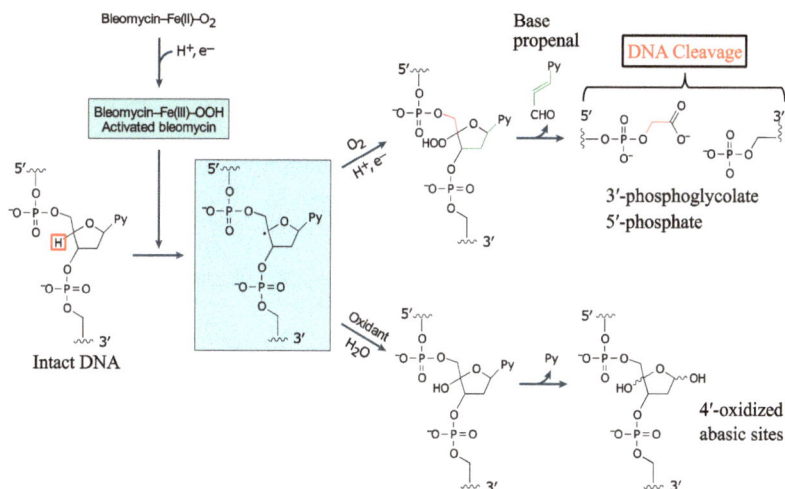

Figure 2. Mechanism of bleomycin-mediated DNA cleavage. The Bleomycin-Fe(III)-OOH activated form is generated in the presence of a one-electron reductant, Fe^{2+} and oxygen. The activated bleomycin then abstracts the hydrogen atom (red square) from C4′ of the deoxyribose moiety of DNA to form the intermediate 4′ radical. This intermediate can partition into two pathways. In the abundance of oxygen, the 4′ radical initiates a series of chemical transformations, leading to a direct strand break and producing 3′-phosphoglycolate and 5′-phosphate ends, and release of a base propenal. However, in the absence of oxygen, the intermediate reacts with an oxidant in the presence of water, generating 4′-oxidized abasic sites. The figure is adapted from [15] and reprinted by permission from Springer Nature.

In the oxygen-dependent pathway, the C4′ radical reacts with O_2 to form a peroxyl radical, which is then reduced to a 4′-hydroperoxide (Figure 2). The latter undergoes further chemical transformations that ultimately leads to DNA strand scission, releasing a base propenal. The resulting gap comprises 3′-phosphoglycolate and 5′-phosphate terminal ends. Alternatively, the C4′ radical can also initiate an oxygen-independent pathway where the C4′ radical undergoes oxidation to form a 4′-carbocation that is subsequently hydroxylated and finally generates a C4′-oxidised abasic site. The 4′-oxidised abasic site is unstable, with a half-life of 8–26 h at pH 7; it is also alkali labile and capable of undergoing β-elimination that results in DNA strand scission, forming a phosphate and 4′-ketodeoxyribose at the 5′- and 3′-ends, respectively [46,47].

Activated bleomycin has also been reported to catalyse the degradation of other cellular components including RNA [48], lipids [49], and proteins [50]. Cleavage of RNA by bleomycin was shown to be selective, depending on the secondary and tertiary structure of the substrate [51,52]. The significance of these bleomycin cellular targets is not fully understood since the cytotoxic nature of bleomycin is thought to mainly stem from its ability to mediate double-strand DNA cleavage [16–19].

4. Bleomycin Cleavage Specificity with Purified DNA

With the addition of Fe^{2+}, the cleavage of purified DNA by bleomycin can be observed. Using defined sequences of purified DNA, the DNA sequence specificity of bleomycin cleavage can be determined. Bleomycin preferentially cleaves DNA at particular dinucleotide sequences. DNA sequence specificity studies with purified plasmid sequences have indicated that the dinucleotides 5′-GT* and 5′-GC* are preferentially cleaved by bleomycin [53–68]. For ease of discussion in this review, the bleomycin cleavage site nucleotides are numbered with respect to the cleavage site at nucleotide position 0, and the positions discussed are from −3 to +2 (see Table 1). The * indicates the nucleotide at position 0 that is cleaved and destroyed during the reaction. Bleomycin has also been observed to cleave, to a lesser extent, at the dinucleotides GG*, GA*, AT*, AC*, and AA* [58,59].

Table 1. The preferred nucleotides at the bleomycin cleavage site.

Study	Type of Break	Preferred Individual Nucleotides						Consensus Sequence from the Individual Nucleotide Data	Consensus Sequence from Complete Sequence Data
Position		−3	−2	−1	0*	+1	+2		
Early ^{32}P-end-label experiments	SSB		T	G	T			5′-TGT*	
Random DNA sequence	SSB		T	G	T	A		5′-TGT*A	5′-TGT*A
Systematically altered RTGTAY clone	SSB	C > T	C = T	G	T	A	T = A	5′-YYGT*AW	
Purified DNA genome-wide preferred nucleotide (50k)	DSB	ns	T	G	T	A	T > A	5′-TGT*AW	5′-TGT*AT
Cellular DNA genome-wide preferred nucleotide (50k)	DSB	G	T	G	T	A	ns	5′-GTGT*A	5′-RTGT*AY

The preferred nucleotides at the bleomycin cleavage site are shown from the −3 to the +2 positions. Only the most highly preferred nucleotides are shown. The preferred nucleotides from the genome-wide data for cellular and purified human DNA are depicted for the top 50,000 cleavage sites. Note that the individual nucleotide analysis at the bleomycin cleavage site (second method), is shown in this Table except for the column on the right where the complete sequence (first method) data is shown. SSB—single-strand break; DSB—double-strand break; ns—not significant. The * indicates the nucleotide at position 0 that is cleaved and destroyed during the reaction.

Early studies noted that the relative intensity of cleavage between different sites with the same dinucleotide (e.g., 5′-GT*) can vary, suggesting that the DNA sequence determinant of the specificity of bleomycin cleavage is not confined to just two nucleotides [57,58]. Murray et al. found that the nucleotide immediately 5′ to the bleomycin dinucleotide cleavage site (position −2) affected the cleavage intensity; an adjacent thymine enhanced, whereas guanine and adenine reduced the cleavage intensity [58,59]. It was also observed that alternating purine-pyrimidine sequences were preferred cleavage sites for bleomycin [59].

Hecht's group utilised hairpin DNA containing a motif with a randomised sequence to identify variations that are strongly bound by bleomycin [69]. These hairpins consisted of single-stranded DNA of 64 nucleotides in length and when self-annealed, the hairpin DNA contained an 8 bp double-stranded motif with a randomised sequence at the centre. A mixture of these randomised hairpin DNAs was then incubated with resin-bound bleomycin and variations that strongly bound to bleomycin were isolated and sequenced. From this, they identified hairpin variations that were shown to strongly inhibit bleomycin cleavage of a 5′-GC-3′-containing hairpin in a competition assay [63,70]. It is interesting to note that among these hairpin variations, some did not contain any 5′-GT* or 5′-GC* cleavage sites and yet they exhibited strong competition against the assay hairpin. They also examined the sequence specificity of bleomycin cleavage with these hairpins. While the strongest cleavage was observed at the canonical 5′-GT* and 5′-GC* sites, they also reported efficient cleavage at non-conventional sites containing AT-rich dinucleotides [63,71]. Later studies reported that hairpins with a 5′-GCGT sequence bound strongly to bleomycin and produced high intensity double-strand breaks [66]. It is not clear whether the hairpin structures that were used in these experiments had unusual DNA microstructure that influenced the results.

5. Bleomycin Cleavage Specificity with Purified DNA Using Updated Technology

More recently, updated technology using capillary electrophoresis with laser-induced fluorescence (CE-LIF) and automated DNA sequencers, has permitted a more accurate and precise determination of the DNA sequence specificity of bleomycin DNA damage in longer purified DNA sequences [72–79]. In addition, the use of separate 5′- and 3′-end-labelling in a DNA sequence specificity study contributes to the precision of the process by greatly reducing end-label bias [78,79]. It should be noted that experiments with end-labelled DNA mainly detect single-strand breaks (see below).

Using these techniques, a more detailed bleomycin sequence specificity was elucidated using two random DNA sequence of 247 and 425 bp in length [78]. For a random DNA sequence to contain all the possible sequences, it must be sufficiently long. For a four bp recognition sequence, it must be at least 256 bp and for a five bp recognition sequence, it must be at least 1024 bp in length. Hence, for the 247 and 425 bp sequences, a 4 bp recognition sequence could be extracted. It was found that bleomycin preferentially cleaved at 5'-TGT*A DNA sequences [78].

Our genome-wide studies (see below) indicated that in human cells, the DNA sequence 5'-RTGT*AY (where R is G or A, and Y is T or C) was preferentially cleaved by bleomycin [80]. This 5'-RTGT*AY sequence was systematically altered by varying these nucleotides in a cloned DNA sequence. The bleomycin cleavage efficiency was then investigated in these variant nucleotide sequences. This study permitted the importance of flanking nucleotides around the 5'-GT, 5'-GT*A and 5'-TGT*A core sequences to be evaluated. It was observed that the preferred nucleotide sequence for high intensity bleomycin cleavage was 5'-YYGT*AW (where W is A or T) (Table 1). The DNA sequence that had the highest intensity of bleomycin cleavage was 5'-TCGT*AT and the seven highest intensity bleomycin cleavage sites conformed to the 5'-YYGT*AW consensus sequence [79]. This study permitted a precise evaluation of crucial neighbouring nucleotides that produced high intensity bleomycin DNA cleavage sites. This approach of systematically altering the nucleotides around bleomycin cleavage sites is a powerful method of analysis because every possible nucleotide is included in the analysis and no variant is omitted [79]. This is in contrast to an analysis of random DNA sequences where some nucleotide variants may not be present in the analysis. This study confirmed that 5'-GT, 5'-GT*A, and 5'-YGT*A were core sequences for high intensity bleomycin cleavage sites.

Human telomeres are composed of tandem repeats of the DNA sequence 5'-GGGTTA. Seventeen tandem repeats of this sequence were cloned into a plasmid DNA and the bleomycin sequence specificity was investigated [72,75]. It was found that bleomycin preferentially cleaved at 5'-GT dinucleotides in the telomeric sequence. Hence, telomeric DNA sequences are an important genomic site for bleomycin cleavage (Figure 3).

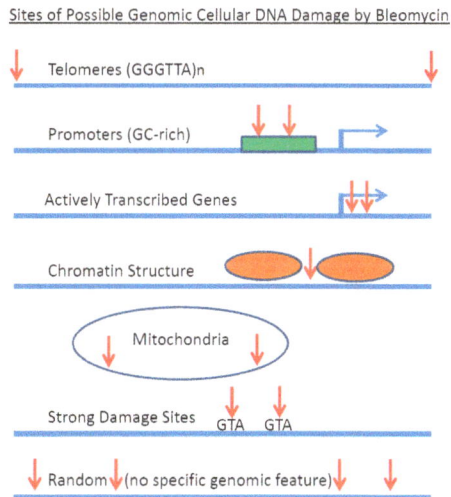

Figure 3. Sites of possible genomic cellular DNA damage by Bleomycin. Red arrows indicate bleomycin cleavage sites, while blue arrows are transcription start sites. The green rectangle is a promoter and the orange ovals are nucleosome cores.

6. Sequence Specificity at Bleomycin-Induced Abasic Sites

As well as cleaving DNA, bleomycin also produces abasic sites in DNA. The enzyme endonuclease IV is able to cleave DNA at abasic sites. Hence endonuclease IV can be used to examine the sequence specificity of bleomycin abasic site damage. It was found that bleomycin abasic DNA damage preferentially occurs at 5'-GC and 5'-GT sites compared with the 5'-GT preference observed for bleomycin-induced phosphodiester strand breaks [77].

Abasic DNA damage is enhanced in the absence of molecular oxygen and tumours are thought to be hypoxic [81,82]. Hence, the sequence specificity of bleomycin-induced abasic DNA damage is important for the effectiveness of anti-tumour agents and should be taken into account during the design of bleomycin analogues [2,29].

7. X-ray Crystal Structure of Bleomycin with DNA

An X-ray crystal structure of bleomycin bound to DNA has been determined [20]. In this structure, a Co(III)-bleomycin B_2 was bound to a DNA sequence containing a 5'-GT. It was shown that the bithiazole moiety intercalated into DNA. The metal binding and disaccharide domains were bound in the minor groove of DNA utilising hydrogen bonding. The Co(III) metal coordination involved the primary amine of β-aminoalanine as an axial ligand; and the secondary amine of β-aminoalanine, imidazole, histidine amide, and pyrimidine N1 as equatorial ligands. A modelled hydroperoxide ligand coordinated to Co(III) was perfectly positioned for abstraction of a hydrogen atom from C4'-H. A caveat about this X-ray crystal structure was that Co(III) was used instead of the physiologically important Fe(II).

An important feature of the X-ray crystal structure was that the bithiazole moiety was shown to be intercalated into DNA; whereas previously, minor groove binding of the bithiazole group was suggested [25,26,28]. The linker methylvalerate OH was found to be involved in hydrogen bonding to DNA. The NH_2 of 3-O-carbamoyl-D-mannose was also shown to hydrogen bond with DNA in the minor groove; whereas no function could be discerned for the L-gulose since it was not involved in DNA binding.

Concerning the sequence specific interaction of the Co(III)-bleomycin B_2 complex with DNA, a number of hydrogen bonds were made between the bleomycin complex and the minor groove of DNA. With reference to the 5'-TAGT*TAAC sequence used in the experiments, there were three hydrogen bonds with the crucial G nucleotide at position −1, one at position −2, and one at position −3. This explains the importance of the G nucleotide at position −1, but also shows that an extended bleomycin sequence specificity could be due to hydrogen bonding at positions upstream and downstream from the dinucleotide cleavage site. In addition, the hydroperoxide ligand was positioned close to the C4'-H of the T nucleotide at position 0 that would result in the cleavage of DNA at this T nucleotide.

The bleomycin complex was bound to a 5 bp section of DNA. The bithiazole group was intercalated between the 5'-T*T dinucleotide (positions 0 and +1); and the intercalation unwound the double helix 3' to the 5'-GT* dinucleotide (positions −1 and 0). These features have implications for the DNA sequence specificity of bleomycin (see below).

8. Mechanism of Bleomycin-Induced Double-Strand Break Formation

Double-strand breaks are primarily derived from single-strand breaks and the ratio of double-strand to single-strand breaks has been reported to be in the range 1:3 to 1:20 [16,27,83–86]. Both single-strand and double-strand break pathways share the common intermediate of a 4'-radical; however, the double-strand break process strictly occurs in the presence of oxygen [86–88]. A bleomycin double-strand cleavage event generates new DNA fragments with either blunt or 5'-staggered ends depending on the recognition sequences. A double-strand cleavage model proposed by Steighner and Povirk suggested that a single molecule of bleomycin carried out the cleavage events on both strands of DNA [84,86,87,89]. After cleaving at one site designated as the primary site, the molecule must be reactivated and swiftly relocate itself to the secondary site on the other strand to make another cut [84]. As a result, the ratio of double-strand to single-strand breaks is dependent on the likelihood of bleomycin reactivation after the primary cleavage, on the rate of molecular relocation and reorganisation, as well as the DNA sequences present. During the relocation, the DNA-binding domain (the bithiazole tail) plays a role in maintaining strong binding with the DNA so that the molecule is able to relocate itself to the secondary cleavage site without dissociating from the DNA. Therefore, a good primary cleavage site is more likely to give rise to the secondary cleavage site because in this case the bleomycin molecule has a strong interaction with DNA from its primary contact and its dissociation is low [89]. Through structural studies of bleomycin-Co(II)-OOH with 2D NMR, the mechanism of this rearrangement was proposed to involve the rotation of the partially intercalated bleomycin molecule after the primary cleavage event [23,88,90]. Acting together with the tail, the linker region acts as a tether to bring over the cleaving metal-binding domain to the secondary cleavage site. This might explain why modifying the linker region not only affects the efficiency of DNA cleavage but also reduces the double-strand to single-strand break ratio [91,92].

Studies with short sequences (500 bp) with a small number of sites (100 cleavage sites) have been used to detect and quantify double-strand breaks. It was proposed that a double-strand break primarily comes from a strong primary site which is 5'-GY* (5'-GT* or 5'-GC*) and the secondary cleavage site is determined by the nucleotide that is 3'- to the Y nucleotide. If this 3'-adjacent nucleotide is a pyrimidine, then the second cleavage will generate blunt-end fragments. In contrast, if it is a purine, then 5'-staggered-end fragments are produced [27,83,84,86,87,89]. Their studies also found that the palindromic site 5'-GT*AC was a hotspot for double-strand cleavage.

It was also observed that, subsequent to the cleavage of the primary site, the secondary site can partition into the formation of either a strand scission, containing the 3'-phosphoglycolate and 5'-phosphate ends; or a 4'-oxidised abasic site. In contrast, the primary cleavage site was always observed to form 3'-phosphoglycolates at the double-strand breaks [84]. Additionally, Absalon et al. reported that bleomycin-induced double-strand breaks do not form under oxygen-depleted conditions and this further supports the observation that the primary cleavage site must be a bleomycin-induced strand scission [87].

The single-strand to double-strand cleavage ratio of bleomycin was found to be conserved over a large range of bleomycin concentrations [83,86]. This observation is not consistent with the theory that the formation of double-strand breaks arises via the accumulation of random and independent single-strand cleavages, where one would expect the single-strand to double-strand break ratio to decrease when the concentration of bleomycin is increased. These findings have led to the proposal of a model which involves the reactivation of a single bleomycin to produce a double-strand break via cleavage on both strands.

9. The Sequence Specificity in Intact Human Cells

Bleomycin DNA damage has also been examined in intact human cells [93]. The bleomycin DNA sequence specificity was determined in human cells with repetitive centromeric alphoid DNA sequences [93], telomeric DNA sequences [74], globin [94–96], and retinoblastoma genes [97], with the DNA sequence specificity again found to be concentrated at the dinucleotides 5'-GT* and 5'-GC* in human cells. For the globin and retinoblastoma DNA sequences, bleomycin was able to footprint transcription factors and positioned nucleosomes in human cells. Bleomycin is known to cleave in the linker region of nucleosomes [93,98,99] and hence bleomycin is also a useful agent to probe chromatin structure in human cells (Figure 3).

Human telomeric sequences are composed of the tandemly repeated DNA sequence $(GGGTTA)_n$. Since the 5'-GT* dinucleotide is a main site for bleomycin cleavage, telomeric sequences are expected to be a major site for bleomycin cleavage (Figure 3). This was found to be the case where bleomycin preferentially cleaved at telomeric 5'-GT* dinucleotides in human cells [74]. Since telomeres are important in chromosome replication, it has been hypothesised that telomeres could be a crucial genomic site for the cytotoxicity of bleomycin [74,75].

10. Sequence Specificity of Bleomycin Double-Strand Breaks in the Entire Human Genome

In our laboratory, the sequence specificity of bleomycin-induced double-strand breaks in the entire human genome has been studied by use of massively parallel next-generation sequencing technology. Contrary to the approximately 100 sites investigated with the previous techniques, the next-generation sequencing technology has enabled more than 200 million bleomycin double-strand break sites in both cellular and purified DNA to be assessed in order to determine the genome-wide sequence specificity [80,99–101]. It should be noted that the genome-wide studies mainly detect double-strand breaks, whereas the experiments with purified end-labelled DNA sequences (discussed above) mainly detect single-strand breaks.

The genome-wide study calculated the frequency of occurrence of dinucleotide, trinucleotide, tetranucleotide, pentanucleotide, and hexanucleotide DNA sequences at bleomycin cleavage sites as well as the nucleotide frequency at each position for the ten nucleotides 5' to the cleavage site and eleven nucleotides 3' to the cleavage site. These analyses were performed for the 50,000 highest intensity cleavage sites to reveal the genome-wide DNA sequence specificity of bleomycin in human cells. This genome-wide method gave a longer preferred bleomycin cleavage site DNA sequence specificity than previous methods. For the 50,000 highest intensity cleavage sites, the preferred bleomycin cleavage sites were at 5'-GT* dinucleotide sequences, 5'-GT*A and 5'-TGT* trinucleotide sequences, 5'-TGTA tetranucleotide sequences and 5'-ATGT*A pentanucleotide sequences. For cellular DNA, the hexanucleotide DNA sequence 5'-RTGT*AY was the most highly cleaved DNA sequence. This finding strongly agreed with the observation that alternating purine-pyrimidine sequences are preferentially cleaved by bleomycin [78,80]. The core bleomycin cleavage site is probably the 5'-TGT*A tetranucleotide sequence since it was more frequently cleaved by about three-fold compared to the next-ranked tetranucleotide sequence.

There were differences between the genome-wide DNA sequence specificity of bleomycin cleavage in purified DNA compared with cellular DNA. These differences mainly occurred at sequences flanking the core 5'-TGT*A tetranucleotide cleavage sequence. With cellular DNA, the most highly cleaved sequence was the hexanucleotide 5'-RTGT*AY; whereas it was the pentanucleotide 5'-TGT*AT for purified DNA (Table 1).

There are two methods for analysing the genome-wide DNA sequence specificity of bleomycin cleavage. In the first method, the entire sequence present at the bleomycin cleavage site is examined; this is the method of analysis discussed in the previous paragraph and involves dinucleotides, trinucleotides, tetranucleotides, pentanucleotides, and hexanucleotides. In the second method, the individual nucleotides present at the bleomycin cleavage site are examined by frequency calculations. Using the second method, the statistically preferred nucleotides for cellular DNA were GTGT*A; whereas it was TGT*AW for purified human DNA. Again, the core sequence was the same, 5'-TGT*A, but there were variations in the flanking sequences at positions −3 and +2 (Table 1).

However, the first method of analysis (the complete sequence) gives probably the most accurate representation of the sequence specificity of bleomycin DNA cleavage since the second method suffers from the drawback that the nucleotides at the bleomycin cleavage site may not exist together at the actual cleavage sequence [80].

11. Comparison of the Bleomycin Genome-Wide DNA Sequence Specificity with Purified Plasmid DNA Sequences

The genome-wide bleomycin DNA sequence specificity was compared to results obtained from end-labelled purified plasmid DNA sequences. By examining the bleomycin sequence specificity in a cloned section of human DNA, a comparison could be made between a DNA sequence in purified DNA and the identical sequence in the genome-wide data. Human mitochondrial DNA sequences were utilised for this task after cloning into plasmids. Two sections of human mitochondrial DNA were investigated, and it was found that at individual bleomycin cleavage sites, there was a very low level of correlation in the intensity of bleomycin cleavage in the two environments [78]. However, at an overall level the bleomycin sequence specificity, 5′-TGT*A, was similar in the two environments.

As described above, the sequence in the human genome that was preferentially cleaved by bleomycin was found to be 5′-RTGT*AY [80]. This sequence along with systematically altered nucleotide variations, was placed into a plasmid construct called the RTGT*AY plasmid. The consensus DNA sequence derived from the most highly cleaved bleomycin cleavage sites with this plasmid was 5′-YYGT*AW [79]; while from the genome-wide data, it was 5′-TGT*AW for purified genomic DNA from the individual nucleotide data; and 5′-TGT*AT for purified genomic DNA from the complete sequence data (Table 1). There was a consistent core of 5′-GT*A in these consensus sequences. At the −3 position, it was C > T for the plasmid, and no statistically significant nucleotide preference for the purified genome-wide (Table 1). At the −2 position, it was C = T for the plasmid, and T for the purified genome-wide. At the +2 position, it was T = A for the plasmid, and T > A for the purified genome-wide.

The differences between the plasmid and the purified genome-wide can probably be attributed to the techniques used to obtain the data; the end-labelled plasmid data detected single-strand breaks, while purified genome-wide detected double-strand breaks. A similar process is thought to occur in the production of these breaks, but a double-strand break is thought to be a more extreme event [15,84].

During the genome-wide procedure, double-strand breaks are ligated to linkers before being added to the Illumina flowcell. This ligation procedure may not be sequence independent and may introduce sequence bias into the results; whereas the CE-LIF end-labelling procedure is simpler and has fewer steps.

The main drawback of the end-labelling procedure is that only a small number of DNA sequence sites can be examined; whereas hundreds of millions of double-strand breaks can be examined in the genome-wide procedure. In addition, the sequences in the genome-wide experiments are essentially random; whereas the sequence composition is constrained in plasmid constructs.

12. Comparison of the Bleomycin Genome-Wide Sequence Specificity in Cellular DNA Compared with Purified Genomic DNA Sequences

There were differences between the genome-wide DNA sequence specificity of bleomycin in purified DNA compared with cellular DNA. Both sets of data were derived from the Illumina system and therefore double-strand breaks were detected for both types of DNA environments.

The consensus DNA sequence derived from the most highly cleaved bleomycin cleavage sites was 5′-TGT*AW for purified genomic DNA and 5′-GTGT*A for cellular DNA from the individual nucleotide data; and 5′-TGT*AT for purified genomic DNA and 5′-RTGT*AY for cellular DNA from the complete sequence data (Table 1). The main differences were at the −3 and +2 positions since the core 5′-TGT*A was the same for the two environments. At the −3 position, no statistically significant nucleotide preference was found for the purified genome-wide, and G or R for the cellular genome-wide (Table 1). At the +2 position, it was T > A for the purified genome-wide and no nucleotide preference or Y for the cellular genome-wide.

The environment of cellular DNA is different to purified DNA in a number of ways. Cellular DNA is complexed with proteins and DNA-bound cellular proteins have a large influence on the interaction of bleomycin with cellular DNA. Bleomycin is a relatively large molecule (1500 daltons) and has difficulty accessing DNA bound to proteins. Cellular DNA is mainly found complexed with histones in the form of nucleosomes. Nucleosome cores are known to protect DNA from bleomycin cleavage [98,102]. DNA binding proteins, for example transcription factors, have also been observed to protect DNA from bleomycin cleavage [94–97,102]. Proteins bound to DNA can distort the structure of DNA and this can lead to an alteration in the DNA sequence specificity of bleomycin cleavage in cellular DNA compared to purified genomic DNA. In addition, cellular DNA is supercoiled and hence has a distorted DNA structure that could also lead to changes in bleomycin sequence specificity.

DNA is also a dynamic molecule in cells where DNA replication and transcription produce transient single-stranded regions that will result in an altered bleomycin interaction. The cellular environment contains many different chemical constituents and the cation bound to bleomycin can vary inside cells where not all of the Cu(I)- or Cu(II)-bleomycin may be exchanged with Fe(II); whereas in the purified experiment, the metal cation bound to bleomycin can be completely controlled.

13. Conformation of DNA and the DNA Sequence Specificity of Bleomycin

The DNA microstructure is likely to be very important for the intensity of bleomycin cleavage at each lesion site. The conformation of DNA is derived from the DNA sequence where the order of bases has a major influence on the DNA microstructure [103–106]. Hence, the consensus DNA sequence that is found from the DNA sequence specificity experiments will give rise to a particular DNA structure that is highly conducive to bleomycin cleavage; while other DNA sequences will have a different DNA structure that is less productive for bleomycin cleavage. The features of DNA structure that are likely to be important for bleomycin binding and cleavage are: the intercalation of the bithiazole group into the DNA helix; the productive binding in the minor groove by the metal binding, linker, and disaccharide domains; and the positioning of the complex to abstract a hydrogen atom from C4'-H. If these major features are optimally present, then a high intensity bleomycin cleavage site is likely to occur.

The other important parameter is the interactions of the bleomycin molecule with specific nucleotides in DNA. These interactions were revealed by the X-ray crystal structure of bleomycin with DNA and indicated that hydrogen bonding between the bleomycin complex and particular nucleotides, especially the G nucleotide at position −1, in the minor groove of DNA were important. Hence, a combination of specific DNA sequence interactions and the microstructure of DNA are likely to be the main determinant of the DNA sequence specificity of bleomycin. Of course, the DNA sequence informs the microstructure of DNA and the two are interrelated.

The X-ray crystal structure found that the bleomycin complex was bound to a 5 bp section of DNA [20]. This is consistent with the pentanucleotide and hexanucleotide consensus sequences presented above for the DNA sequence specificity of bleomycin. The bithiazole group was intercalated between the dinucleotide at positions 0 and +1.

The binding of proteins to DNA and other cellular parameters will also alter the DNA conformation and lead to differences in bleomycin DNA cleavage in cells compared to purified DNA.

14. Chromatin Structure Affects the Interaction of Bleomycin with Cellular DNA

As mentioned above, chromatin structure affects the interaction of bleomycin with cellular DNA where nucleosome cores and other DNA binding proteins prevent bleomycin from cleaving DNA [102]. It has been demonstrated that bleomycin targets the linker region of nucleosomes rather than the core DNA and it can be used to footprint chromatin structure in human cells [98,107,108].

Our genome-wide Illumina next-generation sequencing studies revealed an enhanced cleavage pattern for bleomycin at transcription start sites (Figure 4). The peaks of the enhanced bleomycin cleavage were approximately 200 bp apart. This implies that positioned nucleosomes are present at transcription start sites and that bleomycin preferentially cleaves in the linker region of the nucleosome and the nucleosome core protects DNA from bleomycin cleavage [109–111]. Hence, bleomycin can be used to detect chromatin structure at actively transcribed genes [99].

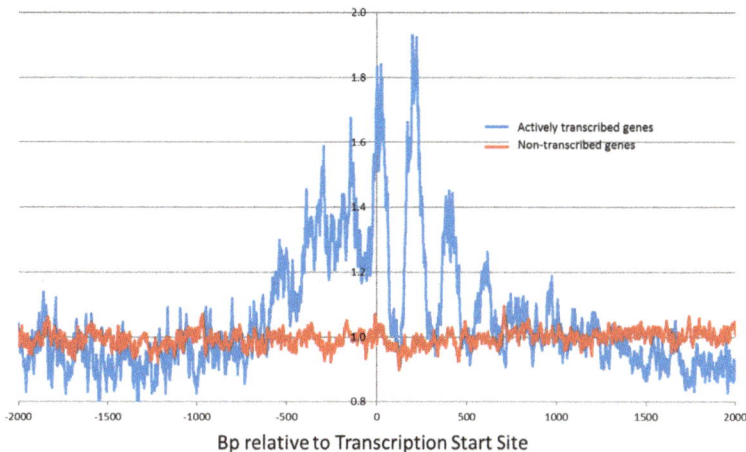

Figure 4. Bleomycin damage at Transcription Start Sites (TSSs) for actively transcribed genes (**blue**) and non-transcribed genes (**red**) in human HeLa cells. The cellular/purified DNA ratio is on the *y*-axis and this ratio removed any bias due to the DNA sequence preference of bleomycin cleavage. These data were compiled from the 24,402 most highly-expressed genes in HeLa cells and 82,596 non-transcribed genes in HeLa cells [99]. Note that the spacing between peaks is approximately 200 bps and probably corresponds to phased nucleosomes in these transcribed regions. The *x*-axis is the position of DNA cleavage at each nucleotide (bp) relative to the TSS. Reprinted by permission from Springer Nature [99].

Bleomycin preferentially damaged actively transcribed genes (Figure 4) and the degree of bleomycin cleavage correlated with the level of transcription [99]. This enhanced bleomycin damage occurred within 1000 bp of transcription start sites (TSSs). The 143,600 identified human TSSs were split into non-transcribed genes (82,596) and transcribed genes (61,004) for HeLa cells. The bleomycin cleavage pattern at highly transcribed gene TSSs was greatly enhanced compared with non-transcribed gene TSSs. Genes that are actively being transcribed have a more open chromatin structure compared with non-transcribed genes; hence bleomycin will be able to cleave at these active genes compared with the more closed chromatin structure of non-transcribed genes. There were also differences that depended on whether the sense or antisense strand were analysed.

For individual genes, the degree of bleomycin cleavage at TSSs was also determined and a ratio of cellular/purified DNA cleavage was calculated. As expected, it was found that highly transcribed genes had a higher cellular/purified ratio than non-expressed genes. In particular the following genes had a high cellular/purified ratio: SERF2, TM4SF1, and TUBA1B (tubulin). Targeting of actively transcribed genes in conjunction with bleomycin could enhance the cancer chemotherapeutic efficacy of bleomycin [112,113]. In combination with nucleic acid-based techniques that target these crucial genes, bleomycin cytotoxicity could be increased by focusing on these important genes.

15. Cancer Signal Transduction Pathways Affected by Bleomycin

It has been estimated for bleomycin that 500 double-strand breaks or 150,000 single-strand breaks were required to induce apoptosis in Chinese hamster fibroblasts [19]. Bleomycin appeared to mediate two types of cell death, depending on the intracellular accumulation of the drug. By using electropermeabilisation, Tounekti et al. could control the amount of bleomycin that is introduced into DC-3F Chinese hamster fibroblasts and the human head and neck carcinoma cell line A-253 [114]. At low cellular concentrations, cells were observed to enter cell cycle arrest at the G2-M phase that eventually led to mitotic death. On the other hand, at high concentrations, bleomycin was proposed to act as a micronuclease since it was observed to rapidly cause significant fragmentation to the DNA and induced an apoptosis-like pathway.

Treatment of cells with DNA-damaging agents, such as bleomycin, is associated with the activation of p53 [115,116]. Subsequently, the activated p53 can mediate apoptosis via the regulation of the Bcl-2 protein family and the release of cytochrome c from the mitochondria. For example, bleomycin treatment of HEp-2 cells resulted in apoptosis via the activation of p53 and a conformational change of Bax, coupled with the release of cytochrome c and the apoptosis-inducing factor [116].

Bleomycin has been observed to induce p53- and Bcl-2-independent apoptosis in squamous carcinoma cells [117]. Furthermore, p53-deficient HL-60 cells were able to induce mitochondria-mediated apoptosis [118]. Alternatively, the apoptosis-inducing factor promotes apoptosis via a caspase-independent cell death. The release of the apoptosis-inducing factor and the DNase endonuclease G from the mitochondria, and subsequent translocation into the nucleus, causes large-scale DNA fragmentation [119,120].

Bleomycin has a major effect on a number of signal transduction pathways. With regard to the signal transduction pathways involved in cancer, we utilised several tools to visualise the effect of bleomycin on these pathways. Using the Comparative Toxicogenomics Database and the Database for Annotation, Visualization, and Integrated Discovery tools, the bleomycin-affected genes were mapped onto the Kyoto Encyclopedia of Genes and Genomes (KEGG) Pathways in Cancer (Figure 5) [121]. The red stars in Figure 5 indicate the proteins/genes that were affected by bleomycin and it reveals that bleomycin impacts a number of signal transduction pathways involved in cancer. These include the PI3K-Akt, MAPK, p53, PPAR, and other pathways. This results in changes to caspases in apoptotic pathways; changes in proliferation, repair, DNA damage response, cell cycle, angiogenesis, evading apoptosis, differentiation, and insensitivity to anti-growth signals.

Note that the signal transduction pathways in cancer are highly complex and not fully understood; but the important point revealed in Figure 5 is that bleomycin acts at a number of these pathways and this could provide an explanation for the effectiveness of bleomycin as a cancer chemotherapeutic agent. Instead of affecting a single section of one pathway or a single crucial modification site, bleomycin impacts on a number of signal transduction pathways. Hence, it is more difficult for a tumour cell to develop resistance to these multiple blocks in signal transduction pathways compared with a single blockage that could be overcome by a single mutation/modification. In order to develop resistance to the multiple blocks caused by bleomycin, a number of resistance mechanisms must be generated simultaneously which is much harder than for a single block [121–124].

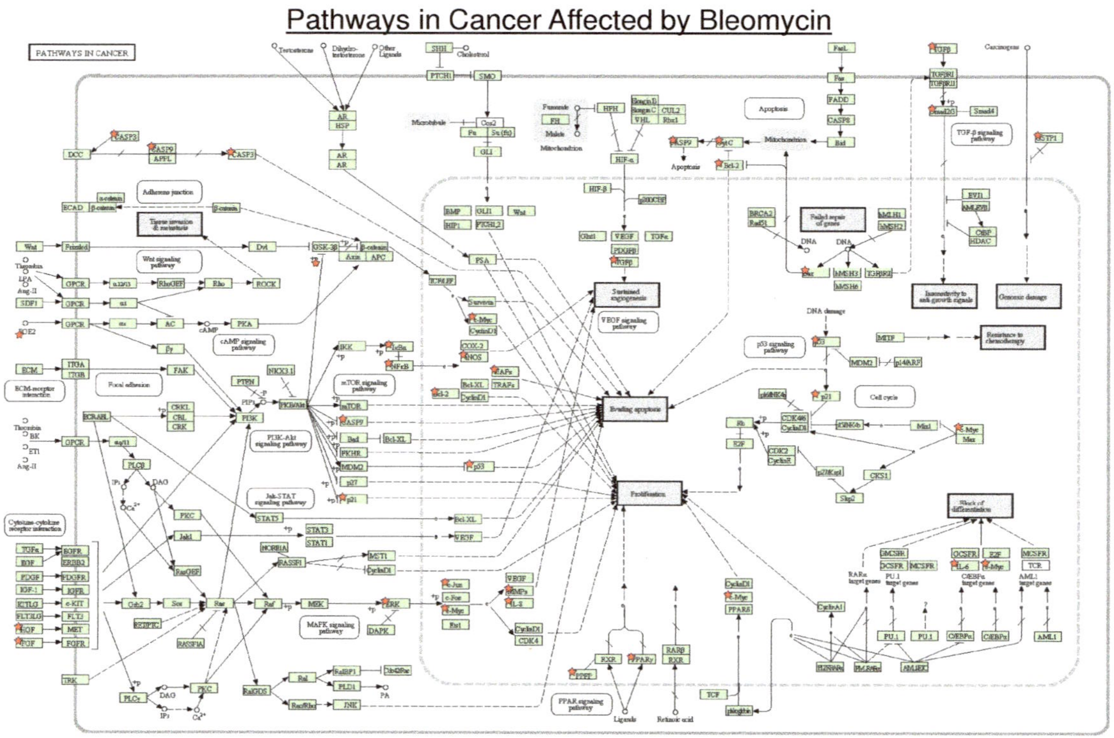

Figure 5. Pathways in Cancer affected by Bleomycin. In order to determine the reactions, genes and gene products that were affected by bleomycin, the Comparative Toxicogenomics Database (Available online: http://ctdbase.org/) was utilised with a requirement of a minimum of three hits. The Database for Annotation, Visualization, and Integrated Discovery (DAVID) tool (Available online: http://david.abcc.ncifcrf.gov/home.jsp) was then used to map these genes onto the Kyoto Encyclopedia of Genes and Genomes (KEGG) Pathways in Cancer map [121]. A step in the Pathways in Cancer KEGG map that was affected by bleomycin is indicated by a red star.

16. Repair of Bleomycin-Induced DNA Damage

16.1. Processing the 3'-Phosphoglycolate Termini

The direct scission of phosphodiester linkages in DNA by bleomycin results in the formation of 3'-phosphoglycolate termini at the site of cleavage. The repair of these lesions requires the removal of the 3'-phosphoglycolate ends and an array of enzymes has been discovered that possess this 3'-phosphodiesterase activity. These include apurinic/apyrimidinic endonuclease I (APE1) and APE2; tyrosyl-DNA phosphodiesterase 1 (TDP1); human nuclease Artemis; and Aprataxin [125,126]. It is generally thought that these enzymes work together to remove specific subsets of the bleomycin-induced 3'-phosphoglycolates, although the precise details are not fully understood.

Izumi et al. showed that the APE1 was a limiting factor in the repair of bleomycin-induced DNA damage [127]. The repair of bleomycin-damaged DNA by a human cellular extract was shown to be enhanced when supplemented with exogenous APE1. However, while APE1 is highly efficient at processing 3'-phosphoglycolates at gapped DNA, the enzyme was observed to be less efficient at removing phosphoglycolates from a 3'-terminus that is located at a recessed or blunt end; and highly inefficient at 3'-overhangs [128].

APE2 was shown to be able to remove phosphoglycolates from 3'-recessed ends [129,130], although it is not known how it compares with APE1.

The endonuclease activity of Artemis can trim long overhangs that contain 3'-phosphoglycolate ends, however, it appears to be less efficient at processing short overhangs [131–133]. Fibroblasts from patients who are deficient in Artemis are more sensitive to bleomycin [133]. Transfection of wild-type Artemis cDNA into the deficient fibroblasts restored its resistance to bleomycin [134].

TDP1 was reported to preferentially remove 3'-phosphoglycolates on single-strand DNA or at double-strand breaks [135,136]. Additionally, processing of phosphoglycolates at 3'-overhangs was observed to be three times more efficient, compared to blunt ends; and 10 times more efficient, compared to recessed ends [137]. This enzyme converts the phosphoglycolate terminus into phosphates, which is then likely to be a substrate for the polynucleotide kinase/phosphatase. Cellular extracts from patients deficient in TDP1 were unable to process phosphoglycolates at single-stranded DNA or 3'-overhangs [138]. $Tdp1^{-/-}$ mice and $Tdp1^{-/-}$ chicken DT10 cells were reported to be hypersensitive to bleomycin [139,140].

In summary, it is likely that TDP1 is the main enzyme responsible for removing 3'-phosphoglycolate termini from bleomycin cleaved DNA although other enzymes may also play a role.

16.2. Genome-Wide Bleomycin Repair

In our genome-wide experiments with bleomycin in HeLa cells, we examined the repair of 3'-phosphoglycolate termini at gene transcription start sites (TSSs) [100]. We found that repair of bleomycin DNA damage preferentially occurred at actively transcribed genes. The most actively transcribed genes had the highest level of repair, while the least actively transcribed genes had the lowest level of repair that was close to non-transcribed genes. There were also differences in repair that depended on whether the transcribed or non-transcribed strand was analysed.

16.3. Single-Strand Break Damage Repair

Bleomycin-generated single-strand breaks and 4'-oxidised abasic sites can be repaired by the base excision repair pathway [15,141]. As described above, repair of single-strand breaks is preceded by the removal of the blocking 3'-phosphoglycolate terminus, probably by TDP1 in human cells. Bleomycin-induced 4'-oxidised abasic sites are efficiently processed by the apurinic/apyrimidinic endonuclease activity of APE1, which incises the deoxyribose backbone 5' to the abasic site. This generates a nick in the DNA and produces a terminal 5'-phosphate at the 4'-oxidised abasic site. APE1 also recruits DNA polymerase β, which possesses a lyase activity that subsequently removes the 5'-phosphate 4'-oxidised abasic site, generating a single nucleotide gap in the DNA [142,143].

The single nucleotide gap is then filled and ligated by DNA polymerase β and the XRCC1/DNA ligase 3 complex, respectively [144]. Alternatively, the gap-filling process in base excision repair can also be undertaken via the long patch repair pathway.

16.4. Repair of Double-Strand Breaks

Double-strand breaks can be repaired via homologous recombination or the non-homologous end joining pathway. Both pathways involve an array of protein factors and are still actively being studied (reviewed in [145,146]). Similar to the single-strand damage repair, repair of the bleomycin-induced double-strand breaks requires the removal of the 3′-phosphoglycolate ends.

16.5. Repair and Bleomycin Resistance

It is expected that the DNA repair machineries can contribute to the resistance of DNA damaging agents such as bleomycin [102]. Suppression of DNA repair proteins can confer sensitivity of cells to bleomycin and other DNA damaging agents. Conversely, increased expression of proteins known to be involved in DNA repair have also been shown to increase the resistance of cells. For example, overexpression of APE1 in germ cell tumours was shown to increase resistance of these cells to bleomycin [147].

17. Cellular Transport of Bleomycin

Bleomycin is administered intravenously as a mixture (Blenoxane), containing predominantly the A_2 and B_2 congeners in metal-free form [47]. Once administered, bleomycin has a terminal half-life in the plasma of approximately 90 min, and 65% of it is excreted in the urine within 24 h [12,148,149]. Cu(II) from blood plasma binds to the intravenous bleomycin, forming bleomycin-Cu(II), which is believed to be the complex that is transported into the cells [15,150]. Once inside the cell, it is proposed that the bleomycin-Cu(II) complex is reduced to bleomycin-Cu(I) by cysteine and glutathione [46,151,152]. The Cu(I) ligand can be displaced by Fe(II) and this results in the active bleomycin-Fe(II) complex [153,154].

Bleomycin is a large molecule and hydrophilic, which makes it difficult to cross the cell membrane. However, studies with cobalt-bound bleomycin have shown that the drug binds to a receptor protein on the plasma membrane and then enters the cell via vesicles or receptor-mediated endocytosis. A 250 kDa protein was identified in Chinese hamster fibroblasts as well as in a human head and neck carcinoma cell line [155,156]. However, this protein has not yet been characterised. The subsequent release of bleomycin from endocytotic vesicles is also not fully understood.

A different transport mechanism was reported from studies using *S. cerevisiae* and fluorescently labelled bleomycin. It was found that the L-carnitine transporter Agp2 was able to carry out the uptake of bleomycin. Also, these studies indicated that bleomycin might share a common transport pathway with spermidine or polyamine [157,158]. The pathway requires two kinases Ptk2 and Sky1 to upregulate the transport. Yeast mutants that lack the carnitine transporter Agp2 as well as kinases Ptk2 and Sky1 were shown to have their bleomycin uptake level significantly decreased. The fate of bleomycin entering cells via the energy-dependent transport pathway hinges on whether it is sequestered into cytoplasmic vacuoles for cellular detoxification. The cytotoxicity of bleomycin is only present when it is not degraded by the vacuoles and can diffuse into the cytosol and ultimately the nucleus for DNA cleavage. The human analogue of the L-carnitine transporter, hCT2, was also identified and demonstrated to be involved in bleomycin uptake. The hCT2 protein is highly expressed in human testicular cells, which might explain the effectiveness of bleomycin in the treatment of testicular cancer. In contrast, resistance to the drug seen in human colon and breast cancer might be attributed to the fact that hCT2 is poorly expressed in those tissues [159].

The ability of cells to accumulate intracellular bleomycin is a factor that may contribute to bleomycin resistance or sensitivity. Certain cell membrane proteins, notably the human high affinity L-carnitine transporter, can modulate the accumulation of and subsequently the sensitivity of tumour cells to bleomycin. It was found that overexpression of the yeast polyamine transporter TPO1 increased resistance to bleomycin [160]. The yeast TPO1 transporter was previously found to be involved in the efflux of polyamines from the cell [161,162].

18. Bleomycin Hydrolase

Pulmonary toxicity is the major dose-limiting side effect of cancer treatment with bleomycin. Up to 46% of patients encounter pneumonitis or lung inflammation. Long-term usage can lead to lung fibrosis and up to 3% of patients face fatal consequences from pulmonary toxicity [12,15,163].

Bleomycin-induced lung inflammation is thought to result from the low level of bleomycin hydrolase in lung tissue since it is 5–15 fold lower than other tissues [164]. Bleomycin hydrolase, an aminopeptidase, is an enzyme that biochemically deactivates bleomycin. This enzyme catalyses the hydrolysis of the carboxamide group of the β-aminoalanine moiety and forms deamido bleomycin [165]. Deamido bleomycin exhibited DNA cleavage activity, with similar sequence specificity as bleomycin, but was significantly reduced in its ability to mediate double-strand breaks [85,166]. Additionally, studies have shown that the level of bleomycin hydrolase in the lungs correlated with the susceptibility to pulmonary toxicity [164,167]. It is also known that human skin and lungs have low levels of expression of this enzyme, which correlates with the major side effects of the bleomycin treatment [12]. Expression of the yeast bleomycin hydrolase homologue (known to metabolise bleomycin) in mouse NIH3T3 cells increased its resistance to bleomycin—an effect that was reversed when the cells were treated with the cysteine proteinase inhibitor E-64 [168]. In another study, bleomycin hydrolase was knocked-down by RNAi and HeLa cells became 3.4-fold more sensitive to bleomycin [169]. In addition, the bleomycin hydrolase gene has been identified as a methylated tumour suppressor gene in a hepatocellular carcinoma [170].

However, the significance of bleomycin hydrolase in mediating resistance against bleomycin is controversial since other studies have reported that bleomycin hydrolase did not appear to confer resistance in yeast cells [171].

19. Bleomycin Analogues

There are a number of naturally occurring bleomycin analogues. Studies have reported differences in the cleavage profiles of bleomycin and its naturally occurring analogues, such as talisomycin and phleomycin [56,57]. Notably, the cleavage profile of talisomycin included enhancement for 5'-GA*-3' sites, which were rarely cleaved by bleomycin [57].

Bleomycin is a complex molecule that makes it difficult to conduct structure-activity studies and there have been relatively few bleomycin analogues produced and tested compared with other clinically-used anti-cancer agents.

Several bleomycin analogues have been produced by altering the His, Ala, Thr, and Cys (bithiazole) amino acids (Figure 1) [14,172]. Alterations to the His and Thr residues resulted in bleomycin analogues with decreased DNA cleavage. However, a phenyl or isopropyl replacement of the methyl group in Ala and a chlorinated bithiazole, produced bleomycin analogues with enhanced DNA cleavage.

The same research group was also able to alter the carbohydrate region of bleomycin A5 and produced three analogues with differences in the carbohydrate region [173]. They found that changes in the carbohydrate region led to major changes in the DNA cleaving ability of the bleomycin analogue. Shen et al. have also found that alterations in the carbohydrate region gave rise to large changes in the DNA cleaving ability of a bleomycin analogue [29].

The successful replacement of a methyl group with a bulky phenyl or isopropyl group at the Ala residue would indicate that other hydrophobic bulky substituents may produce more effective bleomycin analogues. There are also grounds to suspect that bulky group additions to the linker Thr may also lead to more effective analogues [172]. Chlorination of the bithiazole produced interesting bleomycin analogues and hence manipulations of this part of the molecule may also prove productive.

In addition, alterations to the disaccharide region resulted in analogues with altered properties and hence this region of the molecule would be a fruitful area to investigate. In particular the replacement of a hydroxyl group with a hydrogen atom resulted in a large change in the DNA cleaving ability of the bleomycin analogue [29].

Bleomycin is natively produced in *S. verticillis* from a large 120 kb gene cluster. However, *S. verticillus* is refractory to transformation, and DNA manipulation and recombinant DNA techniques have proved to be extremely difficult in this bacterium [5]. To avoid this problem, Shen and co-workers have utilised the closely-related bacterium *Streptomyces flavoviridis* to manipulate and express the bleomycin gene cluster [29]. Bleomycin is synthesised using non-ribosomal peptide synthetases and polyketide synthetases from a large 120 kb gene cluster [2–6]. On expressing this 120 kb bleomycin gene cluster in *S. flavoviridis*, three bleomycin analogues, zorbamycin (ZBM), BLM Z, and 6'-deoxy-BLM Z were produced and purified [29].

The structures of bleomycin and the three bleomycin analogues are shown in Figure 6 and they differ at a small number of positions shown in red. On expressing the bleomycin gene cluster in *S. flavoviridis*, the C-terminal tail is derived from the *S. flavoviridis* biosynthetic apparatus. Thus, the BLM Z, and 6'-deoxy-BLM Z C-terminal tails are similar to ZBM [29]. The structure of bleomycin and BLM Z are exactly the same apart from the C terminal tail (Figure 6).

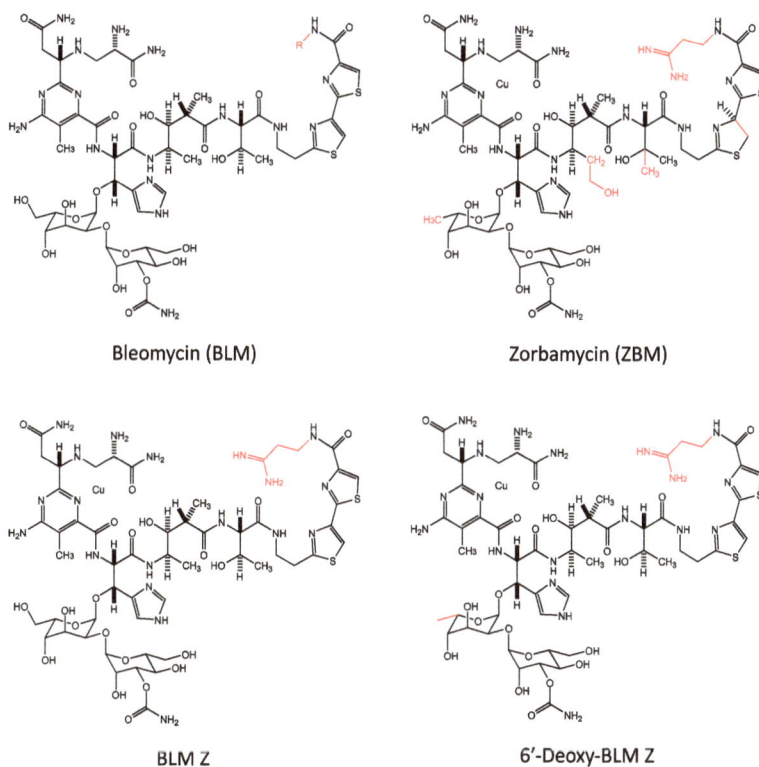

Figure 6. The chemical structures of bleomycin, 6'-deoxy-BLM Z, BLM Z, and ZBM. Differences in chemical structures are shown in red.

These three bleomycin analogues that were produced in *S. flavoviridis* were tested for their ability to cleave purified plasmid DNA and it was found that 6'-deoxy-BLM Z was the most efficient at DNA cleavage, followed by ZBM, BLM Z, and bleomycin [29].

The DNA sequence specificity of these three bleomycin analogues was also investigated and it was observed that bleomycin, BLM Z, and 6'-deoxy-BLM Z were very similar, but in comparison, ZBM had a different sequence specificity profile [174]. Bleomycin, BLM Z, and 6'-deoxy-BLM Z were found to mainly cleave at 5'-TGT*A sequences; while the cleavage preference of ZBM was 5'-TGT*G and 5'-TGT*A [174].

Using human HeLa cells, the cytotoxicity was examined and the IC_{50} was 2.9 μM for 6'-deoxy-BLM Z, 3.2 μM for BLM Z, 4.4 μM for bleomycin, and 7.9 μM for ZBM [101].

The genome-wide DNA sequence specificity of 6'-deoxy-BLM Z and ZBM was determined in human HeLa cells and compared with bleomycin [175]. More than 200 million double-strand breaks were analysed for each analogue. For 6'-deoxy-BLM Z, the individual nucleotide consensus sequence was 5'-GTGY*MC (where M is A or C); it was 5'-GTGY*MCA for ZBM; and 5'-GTGT*AC for bleomycin. The most highly ranked tetranucleotides were 5'-TGC*C and 5'-TGT*A for 6'-deoxy-BLM Z; 5'-TGC*C, 5'-TGT*A, and 5'-TGC*A for ZBM; and 5'-TGT*A for bleomycin. Hence, 6'-deoxy-BLM Z and ZBM had a preference for 5'-GC* and 5'-GT* dinucleotides, while it was 5'-GT* for bleomycin in human cellular DNA.

In experiments with purified human genomic DNA, the individual nucleotide consensus sequence was 5'-TGT*A for 6'-deoxy-BLM, 5'-RTGY*AYR for ZBM, and 5'-TGT*A for bleomycin. Thus, the purified genome-wide DNA sequence specificity was similar for bleomycin and 6'-deoxy-BLM, but was different for ZBM. In addition, the cellular DNA sequence specificities for the analogues, were different in cellular DNA compared with purified DNA. As mentioned above, there are many differences in the cellular environment compared with purified DNA, for example, chromatin structure. The differences in sequence specificity between the two environments are greater for the two analogues compared with bleomycin. Hence, the analogues must be more sensitive to these differences than bleomycin. It also shows that caution should be applied when extrapolating results with purified DNA to cellular DNA.

We also examined the effect of chromatin structure on the cellular DNA cleavage of the two analogues in comparison with bleomycin [101]. As for bleomycin, it was found that 6'-deoxy-BLM Z and ZBM preferentially cleaved at the transcription start sites (TSSs) of actively transcribed genes in human cells. The extent of preferential cleavage at the TSSs was quantified and it was observed to correlate with the cytotoxicity of the bleomycin analogues. This preferential cleavage at the TSSs is consistent with the concept that DNA double-strand breaks are the crucial lesion for the cytotoxicity of bleomycin.

As found for bleomycin, 6'-deoxy-BLM Z and ZBM cleaved in the linker region of the nucleosome [99–101]. These analogues were also able to detect positioned nucleosomes at the TSSs in human cells [99,101].

20. Production of Novel Bleomycin Analogues That Are Resistant to Cleavage by Bleomycin Hydrolase

The modular structure of non-ribosomal peptide synthetases and polyketide synthetases on the 120 kb gene cluster enables facile modification and manipulation of this cluster to produce novel bleomycin analogues [2–6]. Hence, the selective engineering of specific modules in the bleomycin synthetic pathway via combinatorial biosynthesis [29] opens the door for efficient production of novel bleomycin analogues that could have beneficial cancer chemotherapeutic properties.

As mentioned above, a major concern in the administration of bleomycin lies in its dose-limiting side effect, pulmonary toxicity, and up to 46% of patients encounter this side effect [12]. This toxicity is the major limitation for therapy and hence, it has been of interest to develop more effective analogues that can overcome this limitation. As mentioned above, bleomycin is inactivated by the endogenous enzyme bleomycin hydrolase and this enzyme is found at low levels in lung tissue. The production of bleomycin analogues that are not cleavable by human bleomycin hydrolase would result in lung tissue being as equally susceptible to bleomycin activity as other tissues. Hence this bleomycin analogue would be more effective as an anti-cancer agent because the selective lung toxicity has been eliminated.

Resistance to bleomycin in other tumour cell types via the over-expression of bleomycin hydrolase would also be bypassed.

The bleomycin hydrolase cleaves the amide bond in the β-aminoalanine moiety (green arrow in Figure 1). Manipulation of this part of the molecule could achieve the desired bleomycin analogue; namely an analogue with DNA cleaving ability but refractory to cleavage by bleomycin hydrolase. Other strategies, for example, the attachment of a large lipophilic group [176], may also prove viable.

There have been a number of studies where bleomycin analogues have been chemically synthesised [172]. However, none of these studies examined whether bleomycin hydrolase could cleave the bleomycin analogues.

21. Summary and Future Prospects

Bleomycin is a complex molecule and unlike other smaller anti-cancer agents, has not been extensively modified to conduct structure-activity studies.

There are several possible approaches to establish a more effective bleomycin analogue.

1. One approach is to produce bleomycin analogues that are resistant to cleavage by bleomycin hydrolase. The anti-tumour activity of bleomycin is limited by lung toxicity. The production of bleomycin analogues that are not cleaved by human bleomycin hydrolase will result in bleomycin analogues that are more effective as an anti-cancer agent because the lung toxicity would be eliminated.

2. The engineering of an analogue that has improved uptake into cells, or even better, preferential uptake into tumour cells would produce a more effective cancer chemotherapeutic agent. Alterations to the disaccharide region could be prime areas for modification to achieve this aim.

3. The production of analogues with faster/greater binding to DNA would target the analogue to the biological target DNA and eliminate side reaction with other molecules—the DNA targeting hypothesis [177,178]. The bithiazole tail could be modified to produce greater DNA binding.

4. The creation of an analogue that is more efficient in producing the "activated" intermediate could have beneficial properties.

5. More complicated and problematic would be the engineering of an analogue that is more effective at producing double-strand breaks compared with single-strand breaks, since double-strand breaks are thought to be the crucial lesion for the cytotoxicity of bleomycin.

6. Further investigations with genome-wide studies will determine the crucial genes that are preferentially cleaved by bleomycin. In combination with nucleic acid-based techniques that target these crucial genes, bleomycin cytotoxicity could be enhanced by focusing on these important genes.

7. Synergies could also be found with other nucleic acid and antibody-based novel therapies to enhance the action of these recently introduced therapeutics.

The clinical importance of bleomycin as a cancer chemotherapeutic agent suggests that improved analogues based on bleomycin can be developed. Experiments that provide a deeper molecular understanding of the crucial constituents of the bleomycin molecule could lead to directed synthesis of highly effective bleomycin analogues. Tools exist that provide a quantitative and precise platform whereby bleomycin analogues can be tested and compared. This information, along with knowledge of their structure and their efficacy, could provide an informed basis for the development of new more efficient anti-cancer analogues based on bleomycin.

Author Contributions: All authors contributed to the writing of this review paper.

Acknowledgments: Support of this work by the University of New South Wales, Science Faculty Research Grant Scheme is gratefully acknowledged. L.H.C. was supported by an Australian Postgraduate Award.

Conflicts of Interest: The authors declare no conflict of interest.

Abbreviations

APE1 and APE2	Apurinic/apyrimidinic endonuclease 1 and 2
CE-LIF	Capillary electrophoresis with laser-induced fluorescence detection
TSS	Transcription start site
TDP1	Tyrosyl-DNA phosphodiesterase 1
ZBM	zorbamycin
M	A or C nucleotides
R	G or A nucleotides
W	A or T nucleotides
Y	T or C nucleotides

References

1. Umezawa, H.; Maeda, K.; Takeuchi, T.; Okami, Y. New antibiotics, bleomycin A and B. *J. Antibiot.* **1966**, *19*, 200–209. [PubMed]
2. Du, L.; Sanchez, C.; Chen, M.; Edwards, D.J.; Shen, B. The biosynthetic gene cluster for the antitumor drug bleomycin from Streptomyces verticillus ATCC15003 supporting functional interactions between nonribosomal peptide synthetases and a polyketide synthase. *Chem. Biol.* **2000**, *7*, 623–642. [CrossRef]
3. Shen, B.; Du, L.; Sanchez, C.; Edwards, D.J.; Chen, M.; Murrell, J.M. Cloning and characterization of the bleomycin biosynthetic gene cluster from Streptomyces verticillus ATCC15003. *J. Nat. Prod.* **2002**, *65*, 422–431. [CrossRef] [PubMed]
4. Shen, B.; Du, L.; Sanchez, C.; Edwards, D.J.; Chen, M.; Murrell, J.M. The biosynthetic gene cluster for the anticancer drug bleomycin from Streptomyces verticillus ATCC15003 as a model for hybrid peptide-polyketide natural product biosynthesis. *J. Ind. Microbiol. Biotechnol.* **2001**, *27*, 378–385. [CrossRef] [PubMed]
5. Galm, U.; Wang, L.; Wendt-Pienkowski, E.; Yang, R.; Liu, W.; Tao, M.; Coughlin, J.M.; Shen, B. In vivo manipulation of the bleomycin biosynthetic gene cluster in Streptomyces verticillus ATCC15003 revealing new insights into its biosynthetic pathway. *J. Biol. Chem.* **2008**, *283*, 28236–28245. [CrossRef] [PubMed]
6. Galm, U.; Wendt-Pienkowski, E.; Wang, L.; Huang, S.X.; Unsin, C.; Tao, M.; Coughlin, J.M.; Shen, B. Comparative analysis of the biosynthetic gene clusters and pathways for three structurally related antitumor antibiotics: Bleomycin, tallysomycin, and zorbamycin. *J. Nat. Prod.* **2011**, *74*, 526–536. [CrossRef] [PubMed]
7. Umezawa, H. Studies on bleomycin. *J. Formos Med. Assoc.* **1969**, *68*, 569.
8. Williams, S.D.; Birch, R.; Einhorn, L.H.; Irwin, L.; Greco, F.A.; Loehrer, P.J. Treatment of disseminated germ-cell tumors with cisplatin, bleomycin, and either vinblastine or etoposide. *N. Engl. J. Med.* **1987**, *316*, 1435–1440. [CrossRef] [PubMed]
9. Stoter, G.; Kaye, S.B.; de Mulder, P.H.; Levi, J.; Raghavan, D. The importance of bleomycin in combination chemotherapy for good-prognosis germ cell carcinoma. *J. Clin. Oncol.* **1994**, *12*, 644–645. [CrossRef] [PubMed]
10. Neese, F.; Zaleski, J.M.; Loeb Zaleski, K.; Solomon, E.I. Electronic Structure of Activated Bleomycin: Oxygen Intermediates in Heme versus Non-Heme Iron. *J. Am. Chem. Soc.* **2000**, *122*, 11703–11724. [CrossRef]
11. Einhorn, L.H. Curing metastatic testicular cancer. *Proc. Natl. Acad. Sci. USA* **2002**, *99*, 4592–4595. [PubMed]
12. Froudarakis, M.; Hatzimichael, E.; Kyriazopoulou, L.; Lagos, K.; Pappas, P.; Tzakos, A.G.; Karavasilis, V.; Daliani, D.; Papandreou, C.; Briasoulis, E. Revisiting bleomycin from pathophysiology to safe clinical use. *Crit. Rev. Oncol. Hematol.* **2013**, *87*, 90–100. [CrossRef] [PubMed]
13. Carlson, R.W.; Sikic, B.I.; Turbow, M.M.; Ballon, S. Combination cisplatin, vinblastine, and bleomycin chemotherapy (PVB) for malignant germ-cell tumors of the ovary. *J. Clin. Oncol.* **1983**, *1*, 645–651. [CrossRef] [PubMed]
14. Leitheiser, C.J.; Smith, K.L.; Rishel, M.J.; Hashimoto, S.; Konishi, K.; Thomas, C.J.; Li, C.; McCormick, M.M.; Hecht, S.M. Solid-Phase Synthesis of Bleomycin Group Antibiotics. Construction of a 108-Member Deglycobleomycin Library. *J. Am. Chem. Soc.* **2003**, *125*, 8218–8227. [CrossRef] [PubMed]
15. Chen, J.; Stubbe, J. Bleomycins: Towards better therapeutics. *Nat. Rev. Cancer* **2005**, *5*, 102–112. [CrossRef] [PubMed]

16. Mirabelli, C.K.; Huang, C.H.; Fenwick, R.G.; Crooke, S.T. Quantitative measurement of single- and double-strand breakage of DNA in Escherichia coli by the antitumor antibiotics bleomycin and talisomycin. *Antimicrob. Agents Chemother.* **1985**, *27*, 460–467. [CrossRef] [PubMed]

17. Sikic, B.I. Biochemical and cellular determinants of bleomycin cytotoxicity. *Cancer Surv.* **1986**, *5*, 81–91. [PubMed]

18. Povirk, L.F. DNA damage and mutagenesis by radiomimetic DNA-cleaving agents: Bleomycin, neocarzinostatin and other enediynes. *Mutat. Res. Fundam. Mol. Mech. Mutagen.* **1996**, *355*, 71–89. [CrossRef]

19. Tounekti, O.; Kenani, A.; Foray, N.; Orlowski, S.; Mir, L. The ratio of single-to double-strand DNA breaks and their absolute values determine cell death pathway. *Br. J. Cancer* **2001**, *84*, 1272. [CrossRef] [PubMed]

20. Goodwin, K.D.; Lewis, M.A.; Long, E.C.; Georgiadis, M.M. Crystal structure of DNA-bound Co(III) bleomycin B2: Insights on intercalation and minor groove binding. *Proc. Natl. Acad. Sci. USA* **2008**, *105*, 5052–5056. [CrossRef] [PubMed]

21. Stubbe, J.; Kozarich, J.W.; Wu, W.; Vanderwall, D.E. Bleomycins: A structural model for specificity, binding, and double strand cleavage. *Acc. Chem. Res.* **1996**, *29*, 322–330. [CrossRef]

22. Lehmann, T.; Topchiy, E. Contributions of NMR to the Understanding of the Coordination Chemistry and DNA Interactions of Metallo-Bleomycins. *Molecules* **2013**, *18*, 9253–9277. [CrossRef] [PubMed]

23. Wu, W.; Vanderwall, D.E.; Stubbe, J.; Kozarich, J.W.; Turner, C.J. Interaction of Co.cntdot.Bleomycin A2 (Green) with d(CCAGGCCTGG)2: Evidence for Intercalation Using 2D NMR. *J. Am. Chem. Soc.* **1994**, *116*, 10843–10844. [CrossRef]

24. Deng, J.-Z.; Newman, D.J.; Hecht, S.M. Use of COMPARE analysis to discover functional analogues of bleomycin. *J. Nat. Prod.* **2000**, *63*, 1269–1272. [CrossRef] [PubMed]

25. Kuwahara, J.; Sugiura, Y. Sequence-specific recognition and cleavage of DNA by metallobleomycin: Minor groove binding and possible interaction mode. *Proc. Natl. Acad. Sci. USA* **1988**, *85*, 2459–2463. [CrossRef] [PubMed]

26. Manderville, R.A.; Ellena, J.F.; Hecht, S.M. Solution Structure of a Zn(II).Bleomycin A5-d(CGCTAGCG)2 Complex. *J. Am. Chem. Soc.* **1994**, *116*, 10851–10852. [CrossRef]

27. Povirk, L.F.; Hogan, M.; Dattagupta, N. Binding of bleomycin to DNA: Intercalation of the bithiazole rings. *Biochemistry* **1979**, *18*, 96–101. [CrossRef] [PubMed]

28. Abraham, A.T.; Zhou, X.; Hecht, S.M. Metallobleomycin-Mediated Cleavage of DNA Not Involving a Threading-Intercalation Mechanism. *J. Am. Chem. Soc.* **2001**, *123*, 5167–5175. [CrossRef] [PubMed]

29. Huang, S.X.; Feng, Z.; Wang, L.; Galm, U.; Wendt-Pienkowski, E.; Yang, D.; Tao, M.; Coughlin, J.M.; Duan, Y.; Shen, B. A designer bleomycin with significantly improved DNA cleavage activity. *J. Am. Chem. Soc.* **2012**, *134*, 13501–13509. [CrossRef] [PubMed]

30. Oppenheimer, N.J.; Chang, C.; Chang, L.H.; Ehrenfeld, G.; Rodriguez, L.O.; Hecht, S.M. Deglyco-bleomycin. Degradation of DNA and formation of a structurally unique Fe(II).CO complex. *J. Biol. Chem.* **1982**, *257*, 1606–1609. [PubMed]

31. Chapuis, J.-C.; Schmaltz, R.M.; Tsosie, K.S.; Belohlavek, M.; Hecht, S.M. Carbohydrate Dependent Targeting of Cancer Cells by Bleomycin−Microbubble Conjugates. *J. Am. Chem. Soc.* **2009**, *131*, 2438–2439. [CrossRef] [PubMed]

32. Yu, Z.; Paul, R.; Bhattacharya, C.; Bozeman, T.C.; Rishel, M.J.; Hecht, S.M. Structural Features Facilitating Tumor Cell Targeting and Internalization by Bleomycin and Its Disaccharide. *Biochemistry* **2015**, *54*, 3100–3109. [CrossRef] [PubMed]

33. Yu, Z.; Schmaltz, R.M.; Bozeman, T.C.; Paul, R.; Rishel, M.J.; Tsosie, K.S.; Hecht, S.M. Selective tumor cell targeting by the disaccharide moiety of bleomycin. *J. Am. Chem. Soc.* **2013**, *135*, 2883–2886. [CrossRef] [PubMed]

34. Schroeder, B.R.; Ghare, M.I.; Bhattacharya, C.; Paul, R.; Yu, Z.; Zaleski, P.A.; Bozeman, T.C.; Rishel, M.J.; Hecht, S.M. The disaccharide moiety of bleomycin facilitates uptake by cancer cells. *J. Am. Chem. Soc.* **2014**, *136*, 13641–13656. [CrossRef] [PubMed]

35. Bhattacharya, C.; Yu, Z.; Rishel, M.J.; Hecht, S.M. The Carbamoylmannose Moiety of Bleomycin Mediates Selective Tumor Cell Targeting. *Biochemistry* **2014**, *53*, 3264–3266. [CrossRef] [PubMed]

36. Madathil, M.M.; Bhattacharya, C.; Yu, Z.; Paul, R.; Rishel, M.J.; Hecht, S.M. Modified Bleomycin Disaccharides Exhibiting Improved Tumor Cell Targeting. *Biochemistry* **2014**, *53*, 6800–6810. [CrossRef] [PubMed]

37. Dabrowiak, J.C. The coordination chemistry of bleomycin: A review. *J. Biol. Inorg. Biochem.* **1980**, *13*, 317–337. [CrossRef]

38. Boger, D.L.; Ramsey, T.M.; Cai, H.; Hoehn, S.T.; Stubbe, J. Definition of the Effect and Role of the Bleomycin A 2 Valerate Substituents: Preorganization of a Rigid, Compact Conformation Implicated in Sequence-Selective DNA Cleavage. *J. Am. Chem. Soc.* **1998**, *120*, 9149–9158. [CrossRef]

39. Boger, D.L.; Ramsey, T.M.; Cai, H.; Hoehn, S.T.; Stubbe, J. A Systematic Evaluation of the Bleomycin A 2 L -Threonine Side Chain: Its Role in Preorganization of a Compact Conformation Implicated in Sequence-Selective DNA Cleavage. *J. Am. Chem. Soc.* **1998**, *120*, 9139–9148. [CrossRef]

40. Rishel, M.J.; Hecht, S.M. Analogues of bleomycin: Synthesis of conformationally rigid methylvalerates. *Org. Lett.* **2001**, *3*, 2867–2869. [CrossRef] [PubMed]

41. Boger, D.L.; Teramoto, S.; Cai, H. N-methyl threonine analogues of deglycobleomycin A2: Synthesis and evaluation. *Bioorg. Med. Chem.* **1997**, *5*, 1577–1589. [CrossRef]

42. Burger, R.M.; Peisach, J.; Horwitz, S.B. Activated bleomycin. A transient complex of drug, iron, and oxygen that degrades DNA. *J. Biol. Chem.* **1981**, *256*, 11636–11644. [PubMed]

43. McLean, M.J.; Dar, A.; Waring, M.J. Differences between sites of binding to DNA and strand cleavage for complexes of bleomycin with iron of cobalt. *J. Mol. Recognit.* **1989**, *1*, 184–192. [CrossRef] [PubMed]

44. Stubbe, J.; Kozarich, J.W. Mechanisms of bleomycin-induced DNA degradation. *Chem. Rev.* **1987**, *87*, 1107–1136. [CrossRef]

45. Chen, B.; Zhou, X.; Taghizadeh, K.; Chen, J.; Stubbe, J.; Dedon, P.C. GC/MS Methods To Quantify the 2-Deoxypentos-4-ulose and 3′-Phosphoglycolate Pathways of 4′ Oxidation of 2-Deoxyribose in DNA: Application to DNA Damage Produced by γ Radiation and Bleomycin. *Chem. Res. Toxicol.* **2007**, *20*, 1701–1708. [CrossRef] [PubMed]

46. Pitié, M.; Pratviel, G.V. Activation of DNA Carbon−Hydrogen Bonds by Metal Complexes. *Chem. Rev.* **2010**, *110*, 1018–1059. [CrossRef] [PubMed]

47. Burger, R.M. Cleavage of Nucleic Acids by Bleomycin. *Chem. Rev.* **1998**, *98*, 1153–1170. [CrossRef] [PubMed]

48. Magliozzo, R.S.; Peisach, J.; Ciriolo, M.R. Transfer RNA is cleaved by activated bleomycin. *Mol. Pharmacol.* **1989**, *35*, 428–432. [PubMed]

49. Ekimoto, H.; Takahashi, K.; Matsuda, A.; Takita, T.; Umezawa, H. Lipid peroxidation by bleomycin-iron complexes in vitro. *J. Antibiot.* **1985**, *38*, 1077–1082. [CrossRef] [PubMed]

50. Rana, T.M.; Meares, C.F. Transfer of oxygen from an artificial protease to peptide carbon during proteolysis. *Proc. Natl. Acad. Sci. USA* **1991**, *88*, 10578–10582. [CrossRef] [PubMed]

51. Carter, B.J.; Murty, V.S.; Reddy, K.S.; Wang, S.N.; Hecht, S.M. A role for the metal binding domain in determining the DNA sequence selectivity of Fe-bleomycin. *J. Biol. Chem.* **1990**, *265*, 4193–4196. [PubMed]

52. Morgan, M.A.; Hecht, S.M. Iron(II) Bleomycin-Mediated Degradation of a DNA-RNA Heteroduplex. *Biochemistry* **1994**, *33*, 10286–10293. [CrossRef] [PubMed]

53. Takeshita, M.; Grollman, A.P.; Ohtsubo, E.; Ohtsubo, H. Interaction of bleomycin with DNA. *Proc. Natl. Acad. Sci. USA* **1978**, *75*, 5983–5987. [CrossRef] [PubMed]

54. Takeshita, M.; Kappen, L.S.; Grollman, A.P.; Eisenberg, M.; Goldberg, I.H. Strand scission of deoxyribonucleic acid by neocarzinostatin, auromomycin, and bleomycin: Studies on base release and nucleotide sequence specificity. *Biochemistry* **1981**, *20*, 7599–7606. [CrossRef] [PubMed]

55. D'Andrea, A.D.; Haseltine, W.A. Sequence specific cleavage of DNA by the antitumor antibiotics neocarzinostatin and bleomycin. *Proc. Natl. Acad. Sci. USA* **1978**, *75*, 3608–3612. [CrossRef] [PubMed]

56. Kross, J.; Henner, W.D.; Hecht, S.M.; Haseltine, W.A. Specificity of deoxyribonucleic acid cleavage by bleomycin, phleomycin, and tallysomycin. *Biochemistry* **1982**, *21*, 4310–4318. [CrossRef] [PubMed]

57. Mirabelli, C.K.; Ting, A.; Huang, C.H.; Mong, S.; Crooke, S.T. Bleomycin and talisomycin sequence-specific strand scission of DNA: A mechanism of double-strand cleavage. *Cancer Res.* **1982**, *42*, 2779–2785. [PubMed]

58. Murray, V.; Martin, R.F. Comparison of the sequence specificity of bleomycin cleavage in two slightly different DNA sequences. *Nucleic Acids Res.* **1985**, *13*, 1467–1481. [CrossRef] [PubMed]

59. Murray, V.; Tan, L.; Matthews, J.; Martin, R.F. The sequence specificity of bleomycin damage in three cloned DNA sequences that differ by a small number of base substitutions. *J. Biol. Chem.* **1988**, *263*, 12854–12859. [PubMed]

60. Nightingale, K.P.; Fox, K.R. DNA structure influences sequence specific cleavage by bleomycin. *Nucleic Acids Res.* **1993**, *21*, 2549–2555. [CrossRef] [PubMed]

61. Murray, V. A survey of the sequence-specific interaction of damaging agents with DNA: Emphasis on anti-tumour agents. *Prog. Nucleic Acid Res. Mol. Biol.* **2000**, *63*, 367–415.

62. Lewis, M.A.; Long, E.C. Fluorescent intercalator displacement analyses of DNA binding by the peptide-derived natural products netropsin, actinomycin, and bleomycin. *Bioorg. Med. Chem.* **2006**, *14*, 3481–3490. [CrossRef] [PubMed]

63. Ma, Q.; Akiyama, Y.; Xu, Z.; Konishi, K.; Hecht, S.M. Identification and cleavage site analysis of DNA sequences bound strongly by bleomycin. *J. Am. Chem. Soc.* **2009**, *131*, 2013–2022. [CrossRef] [PubMed]

64. Bozeman, T.C.; Nanjunda, R.; Tang, C.; Liu, Y.; Segerman, Z.J.; Zaleski, P.A.; Wilson, W.D.; Hecht, S.M. Dynamics of bleomycin interaction with a strongly bound hairpin DNA substrate, and implications for cleavage of the bound DNA. *J. Am. Chem. Soc.* **2012**, *134*, 17842–17845. [CrossRef] [PubMed]

65. Segerman, Z.J.; Roy, B.; Hecht, S.M. Characterization of bleomycin-mediated cleavage of a hairpin DNA library. *Biochemistry* **2013**, *52*, 5315–5327. [CrossRef] [PubMed]

66. Tang, C.; Paul, A.; Alam, M.P.; Roy, B.; Wilson, W.D.; Hecht, S.M. A short DNA sequence confers strong bleomycin binding to hairpin DNAs. *J. Am. Chem. Soc.* **2014**, *136*, 13715–13726. [CrossRef] [PubMed]

67. Roy, B.; Tang, C.; Alam, M.P.; Hecht, S.M. DNA methylation reduces binding and cleavage by bleomycin. *Biochemistry* **2014**, *53*, 6103–6112. [CrossRef] [PubMed]

68. Roy, B.; Hecht, S.M. Hairpin DNA sequences bound strongly by bleomycin exhibit enhanced double-strand cleavage. *J. Am. Chem. Soc.* **2014**, *136*, 4382–4393. [CrossRef] [PubMed]

69. Akiyama, Y.; Ma, Q.; Edgar, E.; Laikhter, A.; Hecht, S.M. Identification of Strong DNA Binding Motifs for Bleomycin. *J. Am. Chem. Soc.* **2008**, *130*, 9650–9651. [CrossRef] [PubMed]

70. Akiyama, Y.; Ma, Q.; Edgar, E.; Laikhter, A.; Hecht, S.M. A Novel DNA Hairpin Substrate for Bleomycin. *Org. Lett.* **2008**, *10*, 2127–2130. [CrossRef] [PubMed]

71. Giroux, R.A.; Hecht, S.M. Characterization of bleomycin cleavage sites in strongly bound hairpin DNAs. *J. Am. Chem. Soc.* **2010**, *132*, 16987–16996. [CrossRef] [PubMed]

72. Murray, V.; Nguyen, T.V.; Chen, J.K. The Use of Automated Sequencing Techniques to Investigate the Sequence Selectivity of DNA Damaging Agents. *Chem. Biol. Drug Des.* **2012**, *80*, 1–8. [CrossRef] [PubMed]

73. Paul, M.; Murray, V. Use of an Automated Capillary DNA Sequencer to Investigate the Interaction of Cisplatin with Telomeric DNA Sequences. *Biomed. Chromatog.* **2012**, *26*, 350–354. [CrossRef] [PubMed]

74. Nguyen, H.T.Q.; Murray, V. The DNA sequence specificity of bleomycin cleavage in telomeric sequences in human cells. *J. Biol. Inorg. Chem.* **2012**, *17*, 1209–1215. [CrossRef] [PubMed]

75. Nguyen, T.V.; Murray, V. Human telomeric DNA sequences are a major target for the anti-tumour drug, bleomycin. *J. Biol. Inorg. Chem.* **2012**, *17*, 1–9. [CrossRef] [PubMed]

76. Nguyen, T.V.; Chen, J.K.; Murray, V. Bleomycin DNA damage: Anomalous mobility of 3′-phosphoglycolate termini in an automated capillary DNA sequencer. *J. Chromatogr. B* **2013**, *913*, 113–122. [CrossRef] [PubMed]

77. Chen, J.K.; Murray, V. The determination of the DNA sequence specificity of bleomycin-induced abasic sites. *J. Biol. Inorg. Chem.* **2016**, *21*, 395–406. [CrossRef] [PubMed]

78. Chung, L.H.; Murray, V. The mitochondrial DNA sequence specificity of the anti-tumour drug bleomycin using end-labeled DNA and capillary electrophoresis and a comparison with genome-wide DNA sequencing. *J. Chromatogr. B* **2016**, *1008*, 87–97. [CrossRef] [PubMed]

79. Gautam, S.D.; Chen, J.K.; Murray, V. The DNA sequence specificity of bleomycin cleavage in a systematically altered DNA sequence. *J. Biol. Inorg. Chem.* **2017**, *22*, 881–892. [CrossRef] [PubMed]

80. Murray, V.; Chen, J.K.; Tanaka, M.M. The genome-wide DNA sequence specificity of the anti-tumour drug bleomycin in human cells. *Mol. Biol. Rep.* **2016**, *43*, 639–651. [CrossRef] [PubMed]

81. Harris, A.L. Hypoxia—A key regulatory factor in tumour growth. *Nat. Rev. Cancer* **2002**, *2*, 38–47. [CrossRef] [PubMed]

82. Brown, J.M.; Wilson, W.R. Exploiting tumour hypoxia in cancer treatment. *Nat. Rev. Cancer* **2004**, *4*, 437–447. [CrossRef] [PubMed]

83. Povirk, L.F.; Wübker, W.; Köhnlein, W.; Hutchinson, F. DNA double-strand breaks and alkali-labile bonds produced by bleomycin. *Nucleic Acids Res.* **1977**, *4*, 3573–3580. [CrossRef] [PubMed]

84. Povirk, L.F.; Han, Y.H.; Steighner, R.J. Structure of bleomycin-induced DNA double-strand breaks: Predominance of blunt ends and single-base 5′ extensions. *Biochemistry* **1989**, *28*, 5808–5814. [CrossRef] [PubMed]

85. Huang, C.-H.; Mirabelli, C.K.; Jan, Y.; Crooke, S.T. Single-strand and double-strand deoxyribonucleic acid breaks produced by several bleomycin analogs. *Biochemistry* **1981**, *20*, 233–238. [CrossRef] [PubMed]

86. Absalon, M.J.; Kozarich, J.W.; Stubbe, J. Sequence Specific Double-Strand Cleavage of DNA by Fe-Bleomycin. 1. The Detection of Sequence-Specific Double-Strand Breaks Using Hairpin Oligonucleotides. *Biochemistry* **1995**, *34*, 2065–2075. [CrossRef] [PubMed]

87. Absalon, M.J.; Wu, W.; Kozarich, J.W.; Stubbe, J. Sequence-Specific Double-Strand Cleavage of DNA by Fe-Bleomycin. 2. Mechanism and Dynamics. *Biochemistry* **1995**, *34*, 2076–2086. [CrossRef] [PubMed]

88. Chen, J.; Ghorai, M.K.; Kenney, G.; Stubbe, J. Mechanistic studies on bleomycin-mediated DNA damage: Multiple binding modes can result in double-stranded DNA cleavage. *Nucleic Acids Res.* **2008**, *36*, 3781–3790. [CrossRef] [PubMed]

89. Steighner, R.J.; Povirk, L.F. Effect of in vitro cleavage of apurinic/apyrimidinic sites on bleomycin-induced mutagenesis of repackaged lambda phage. *Mutat. Res. Genet. Toxicol.* **1990**, *240*, 93–100. [CrossRef]

90. Vanderwall, D.E.; Lui, S.M.; Wu, W.; Turner, C.J.; Kozarich, J.W.; Stubbe, J. A model of the structure of HOO-Co·bleomycin bound to d(CCAGTACTGG): Recognition at the d(GpT) site and implications for double-stranded DNA cleavage. *Chem. Biol.* **1997**, *4*, 373–387. [CrossRef]

91. Boger, D.L.; Honda, T.; Dang, Q. Total synthesis of bleomycin A2 and related agents. 2. Synthesis of (−)-pyrimidoblamic acid, epi-(+)-pyrimidoblamic acid,(+)-desacetamidopyrimidoblamic acid, and (−)-descarboxamidopyrimidoblamic acid. *J. Am. Chem. Soc.* **1994**, *116*, 5619–5630. [CrossRef]

92. Boger, D.L.; Colletti, S.L.; Teramoto, S.; Ramsey, T.M.; Zhou, J. Synthesis of key analogs of bleomycin A2 that permit a systematic evaluation of the linker region: Identification of an exceptionally prominent role for the L-threonine substituent. *Bioorg. Med. Chem.* **1995**, *3*, 1281–1295. [CrossRef]

93. Murray, V.; Martin, R.F. The sequence specificity of bleomycin-induced DNA damage in intact cells. *J. Biol. Chem.* **1985**, *260*, 10389–10391. [PubMed]

94. Cairns, M.J.; Murray, V. Influence of chromatin structure on bleomycin-DNA interactions at base pair resolution in the human beta-globin gene cluster. *Biochemistry* **1996**, *35*, 8753–8760. [CrossRef] [PubMed]

95. Kim, A.; Murray, V. A large "footprint" at the boundary of the human beta-globin locus control region hypersensitive site-2. *Int. J. Biochem. Cell Biol.* **2000**, *32*, 695–702. [CrossRef]

96. Kim, A.; Murray, V. Chromatin structure at the 3'-boundary of the human beta-globin locus control region hypersensitive site-2. *Int. J. Biochem. Cell Biol.* **2001**, *33*, 1183–1192. [CrossRef]

97. Temple, M.D.; Murray, V. Footprinting the 'essential regulatory region' of the retinoblatoma gene promoter in intact human cells. *Int. J. Biochem. Cell Biol.* **2005**, *37*, 665–678. [CrossRef] [PubMed]

98. Galea, A.M.; Murray, V. The influence of chromatin structure on DNA damage induced by nitrogen mustards and cisplatin analogues. *Chem. Biol. Drug Des.* **2010**, *75*, 578–589. [CrossRef] [PubMed]

99. Murray, V.; Chen, J.K.; Galea, A.M. The anti-tumour drug, bleomycin, preferentially cleaves at the transcription start sites of actively transcribed genes in human cells. *Cell. Mol. Life Sci.* **2014**, *71*, 1505–1512. [CrossRef] [PubMed]

100. Murray, V.; Chen, J.K.; Galea, A.M. Enhanced repair of bleomycin DNA damage at the transcription start sites of actively transcribed genes in human cells. *Mutat. Res. Fundam. Mol. Mech. Mutagen.* **2014**, *769*, 93–99. [CrossRef] [PubMed]

101. Chen, J.K.; Yang, D.; Shen, B.; Murray, V. Bleomycin analogues preferentially cleave at the transcription start sites of actively transcribed genes in human cells. *Int. J. Biochem. Cell Biol.* **2017**, *85*, 56–65. [CrossRef] [PubMed]

102. Bolzán, A.D.; Bianchi, M.S. DNA and chromosome damage induced by bleomycin in mammalian cells: An update. *Mutat. Res. Rev. Mutat. Res.* **2018**, *775*, 51–62. [CrossRef] [PubMed]

103. Calladine, C.R. Mechanics of sequence-dependent stacking of bases in B-DNA. *J. Mol. Biol.* **1982**, *161*, 343–352. [CrossRef]

104. Zgarbova, M.; Jurecka, P.; Lankas, F.; Cheatham, T.E., 3rd; Sponer, J.; Otyepka, M. Influence of BII Backbone Substates on DNA Twist: A Unified View and Comparison of Simulation and Experiment for All 136 Distinct Tetranucleotide Sequences. *J. Chem. Inf. Model.* **2017**, *57*, 275–287. [CrossRef] [PubMed]

105. Travers, A.A. The structural basis of DNA flexibility. *Philos. Trans. A Math. Phys. Eng. Sci.* **2004**, *362*, 1423–1438. [CrossRef] [PubMed]

106. Geggier, S.; Vologodskii, A. Sequence dependence of DNA bending rigidity. *Proc. Natl. Acad. Sci. USA* **2010**, *107*, 15421–15426. [CrossRef] [PubMed]

107. Kuo, M.T.; Hsu, T.C. Bleomycin causes release of nucleosomes from chromatin and chromosomes. *Nature* **1978**, *271*, 83–84. [CrossRef] [PubMed]

108. Kuo, M.T. Preferential damage of active chromatin by bleomycin. *Cancer Res.* **1981**, *41*, 2439–2443. [PubMed]

109. Jiang, C.; Pugh, B.F. A compiled and systematic reference map of nucleosome positions across the Saccharomyces cerevisiae genome. *Genome Biol.* **2009**, *10*, R109. [CrossRef] [PubMed]

110. Jiang, C.; Pugh, B.F. Nucleosome positioning and gene regulation: Advances through genomics. *Nat. Rev. Genet.* **2009**, *10*, 161–172. [CrossRef] [PubMed]

111. Schones, D.E.; Cui, K.; Cuddapah, S.; Roh, T.Y.; Barski, A.; Wang, Z.; Wei, G.; Zhao, K. Dynamic regulation of nucleosome positioning in the human genome. *Cell* **2008**, *132*, 887–898. [CrossRef] [PubMed]

112. Reed, S.D.; Fulmer, A.; Buckholz, J.; Zhang, B.; Cutrera, J.; Shiomitsu, K.; Li, S. Bleomycin/interleukin-12 electrochemogenetherapy for treating naturally occurring spontaneous neoplasms in dogs. *Cancer Gene Ther.* **2010**, *17*, 571–578. [CrossRef] [PubMed]

113. Liu, S.; Guo, Y.; Huang, R.; Li, J.; Huang, S.; Kuang, Y.; Han, L.; Jiang, C. Gene and doxorubicin co-delivery system for targeting therapy of glioma. *Biomaterials* **2012**, *33*, 4907–4916. [CrossRef] [PubMed]

114. Tounekti, O.; Pron, G.; Belehradek, J.; Mir, L.M. Bleomycin, an apoptosis-mimetic drug that induces two types of cell death depending on the number of molecules internalized. *Cancer Res.* **1993**, *53*, 5462–5469. [PubMed]

115. Nelson, W.G.; Kastan, M.B. DNA strand breaks: The DNA template alterations that trigger p53-dependent DNA damage response pathways. *Mol. Cell. Biol.* **1994**, *14*, 1815–1823. [CrossRef] [PubMed]

116. Brahim, S.; Aroui, S.; Abid, K.; Kenani, A. Involvement of C-jun NH2-terminal kinase and apoptosis induced factor in apoptosis induced by deglycosylated bleomycin in laryngeal carcinoma cells. *Cell Biol. Int.* **2009**, *33*, 964–970. [CrossRef] [PubMed]

117. Patel, V.; Ensley, J.F.; Gutkind, J.S.; Yeudall, W.A. Induction of apoptosis in head-and-neck squamous carcinoma cells by gamma-irradiation and bleomycin is p53-independent. *Int. J. Cancer* **2000**, *88*, 737–743. [CrossRef]

118. Gimonet, D.; Landais, E.; Bobichon, H.; Coninx, P.; Liautaud-Roger, F. Induction of apoptosis by bleomycin in p53-null HL-60 leukemia cells. *Int. J. Oncol.* **2004**, *24*, 313–319. [CrossRef] [PubMed]

119. Niikura, Y.; Dixit, A.; Scott, R.; Perkins, G.; Kitagawa, K. BUB1 mediation of caspase-independent mitotic death determines cell fate. *J. Cell Biol.* **2007**, *178*, 283–296. [CrossRef] [PubMed]

120. Lorenzo, H.K.; Susin, S.A. Therapeutic potential of AIF-mediated caspase-independent programmed cell death. *Drug Resist. Updates* **2007**, *10*, 235–255. [CrossRef] [PubMed]

121. Hardie, M.E.; Murray, V. The sequence preference of DNA cleavage by T4 Endonuclease VII. *Biochimie* **2018**, *146*, 1–13. [CrossRef] [PubMed]

122. Coulson, E.J. Does the p75 neurotrophin receptor mediate Abeta-induced toxicity in Alzheimer's disease? *J. Neurochem.* **2006**, *98*, 654–660. [CrossRef] [PubMed]

123. De Velasco, M.A.; Tanaka, M.; Yamamoto, Y.; Hatanaka, Y.; Koike, H.; Nishio, K.; Yoshikawa, K.; Uemura, H. Androgen deprivation induces phenotypic plasticity and promotes resistance to molecular targeted therapy in a PTEN-deficient mouse model of prostate cancer. *Carcinogenesis* **2014**, *35*, 2142–2153. [CrossRef] [PubMed]

124. Boridy, S.; Le, P.U.; Petrecca, K.; Maysinger, D. Celastrol targets proteostasis and acts synergistically with a heat-shock protein 90 inhibitor to kill human glioblastoma cells. *Cell Death Dis.* **2014**, *5*, e1216. [CrossRef] [PubMed]

125. Wilson, D.M. Processing of nonconventional DNA strand break ends. *Environ. Mol. Mutagen.* **2007**, *48*, 772–782. [CrossRef] [PubMed]

126. Povirk, L.F. Processing of Damaged DNA Ends for Double-Strand Break Repair in Mammalian Cells. *ISRN Mol. Biol.* **2012**, *2012*. [CrossRef] [PubMed]

127. Izumi, T.; Hazra, T.K.; Boldogh, I.; Tomkinson, A.E.; Park, M.S.; Ikeda, S.; Mitra, S. Requirement for human AP endonuclease 1 for repair of 3'-blocking damage at DNA single-strand breaks induced by reactive oxygen species. *Carcinogenesis* **2000**, *21*, 1329–1334. [CrossRef] [PubMed]

128. Suh, D.; Wilson, D.M., 3rd; Povirk, L.F. 3'-phosphodiesterase activity of human apurinic/apyrimidinic endonuclease at DNA double-strand break ends. *Nucleic Acids Res.* **1997**, *25*, 2495–2500. [CrossRef] [PubMed]

129. Burkovics, P.; Hajdú, I.; Szukacsov, V.; Unk, I.; Haracska, L. Role of PCNA-dependent stimulation of 3'-phosphodiesterase and 3'-5' exonuclease activities of human Ape2 in repair of oxidative DNA damage. *Nucleic Acids Res.* **2009**, *37*, 4247–4255. [CrossRef] [PubMed]

130. Burkovics, P.; Szukacsov, V.; Unk, I.; Haracska, L. Human Ape2 protein has a 3′-5′ exonuclease activity that acts preferentially on mismatched base pairs. *Nucleic Acids Res.* **2006**, *34*, 2508–2515. [CrossRef] [PubMed]

131. Moshous, D.; Callebaut, I.; de Chasseval, R.; Corneo, B.; Cavazzana-Calvo, M.; Le Deist, F.; Tezcan, I.; Sanal, O.; Bertrand, Y.; Philippe, N.; et al. Artemis, a Novel DNA Double-Strand Break Repair/V(D)J Recombination Protein, Is Mutated in Human Severe Combined Immune Deficiency. *Cell* **2001**, *105*, 177–186. [CrossRef]

132. Ma, Y.; Pannicke, U.; Schwarz, K.; Lieber, M.R. Hairpin Opening and Overhang Processing by an Artemis/DNA-Dependent Protein Kinase Complex in Nonhomologous End Joining and V(D)J Recombination. *Cell* **2002**, *108*, 781–794. [CrossRef]

133. Povirk, L.F.; Zhou, T.; Zhou, R.; Cowan, M.J.; Yannone, S.M. Processing of 3′-phosphoglycolate-terminated DNA double strand breaks by Artemis nuclease. *J. Biol. Chem.* **2007**, *282*, 3547–3558. [CrossRef] [PubMed]

134. Mohapatra, S.; Kawahara, M.; Khan, I.S.; Yannone, S.M.; Povirk, L.F. Restoration of G1 chemo/radioresistance and double-strand-break repair proficiency by wild-type but not endonuclease-deficient Artemis. *Nucleic Acids Res.* **2011**, *39*, 6500–6510. [CrossRef] [PubMed]

135. Inamdar, K.V.; Pouliot, J.J.; Zhou, T.; Lees-Miller, S.P.; Rasouli-Nia, A.; Povirk, L.F. Conversion of Phosphoglycolate to Phosphate Termini on 3′ Overhangs of DNA Double Strand Breaks by the Human Tyrosyl-DNA Phosphodiesterase hTdp1. *J. Biol. Chem.* **2002**, *277*, 27162–27168. [CrossRef] [PubMed]

136. Raymond, A.C.; Staker, B.L.; Burgin, A.B. Substrate Specificity of Tyrosyl-DNA Phosphodiesterase I (Tdp1). *J. Biol. Chem.* **2005**, *280*, 22029–22035. [CrossRef] [PubMed]

137. Zhou, T.; Akopiants, K.; Mohapatra, S.; Lin, P.-S.; Valerie, K.; Ramsden, D.A.; Lees-Miller, S.P.; Povirk, L.F. Tyrosyl-DNA phosphodiesterase and the repair of 3′-phosphoglycolate-terminated DNA double-strand breaks. *DNA Repair* **2009**, *8*, 901–911. [CrossRef] [PubMed]

138. Zhou, T.; Lee, J.W.; Tatavarthi, H.; Lupski, J.R.; Valerie, K.; Povirk, L.F. Deficiency in 3′-phosphoglycolate processing in human cells with a hereditary mutation in tyrosyl-DNA phosphodiesterase (TDP1). *Nucleic Acids Res.* **2005**, *33*, 289–297. [CrossRef] [PubMed]

139. Hirano, R.; Interthal, H.; Huang, C.; Nakamura, T.; Deguchi, K.; Choi, K.; Bhattacharjee, M.B.; Arimura, K.; Umehara, F.; Izumo, S.; et al. Spinocerebellar ataxia with axonal neuropathy: Consequence of a Tdp1 recessive neomorphic mutation? *EMBO J.* **2007**, *26*, 4732–4743. [CrossRef] [PubMed]

140. Murai, J.; Huang, S.Y.; Das, B.B.; Dexheimer, T.S.; Takeda, S.; Pommier, Y. Tyrosyl-DNA phosphodiesterase 1 (TDP1) repairs DNA damage induced by topoisomerases I and II and base alkylation in vertebrate cells. *J. Biol. Chem.* **2012**, *287*, 12848–12857. [CrossRef] [PubMed]

141. Caldecott, K.W. DNA single-strand break repair. *Exp. Cell Res.* **2014**, *329*, 2–8. [CrossRef] [PubMed]

142. Xu, Y.J.; Kim, E.Y.; Demple, B. Excision of C-4′-oxidized deoxyribose lesions from double-stranded DNA by human apurinic/apyrimidinic endonuclease (Ape1 protein) and DNA polymerase beta. *J. Biol. Chem.* **1998**, *273*, 28837–28844. [CrossRef] [PubMed]

143. Jacobs, A.C.; Kreller, C.R.; Greenberg, M.M. Long Patch Base Excision Repair Compensates for DNA Polymerase β Inactivation by the C4′-Oxidized Abasic Site. *Biochemistry* **2011**, *50*, 136–143. [CrossRef] [PubMed]

144. Caldecott, K.W. XRCC1 and DNA strand break repair. *DNA Repair* **2003**, *2*, 955–969. [CrossRef]

145. Jasin, M.; Rothstein, R. Repair of Strand Breaks by Homologous Recombination. *Cold Spring Harb. Perspect. Biol.* **2013**, *5*, a012740. [CrossRef] [PubMed]

146. Davis, A.J.; Chen, D.J. DNA double strand break repair via non-homologous end-joining. *Transl. Cancer Res.* **2013**, *2*, 130–143. [PubMed]

147. Robertson, K.A.; Bullock, H.A.; Xu, Y.; Tritt, R.; Zimmerman, E.; Ulbright, T.M.; Foster, R.S.; Einhorn, L.H.; Kelley, M.R. Altered expression of Ape1/ref-1 in germ cell tumors and overexpression in NT2 cells confers resistance to bleomycin and radiation. *Cancer Res.* **2001**, *61*, 2220–2225. [PubMed]

148. Alberts, D.S.; Chen, H.S.G.; Liu, R.; Himmelstein, K.J.; Mayersohn, M.; Perrier, D.; Gross, J.; Moon, T.; Broughton, A.; Salmon, S.E. Bleomycin pharmacokinetics in man. *Cancer Chemother. Pharmacol.* **1978**, *1*, 177–181. [CrossRef] [PubMed]

149. Dorr, R.T. Bleomycin pharmacology: Mechanism of action and resistance, and clinical pharmacokinetics. *Semin. Oncol.* **1992**, *19*, 3–8. [PubMed]

150. Kanao, M.; Tomita, S.; Ishihara, S.; Murakami, A.; Okada, H. Chelation of bleomycin with copper in vivo. *Chemotherapy* **1973**, *21*, 1305–1310.

151. Petering, D.H.; Byrnes, R.W.; Antholine, W.E. The role of redox-active metals in the mechanism of action of bleomycin. *Chem. Biol. Interact.* **1990**, *73*, 133–182. [CrossRef]

152. Ehrenfeld, G.M.; Shipley, J.B.; Heimbrook, D.C.; Sugiyama, H.; Long, E.C.; van Boom, J.H.; van der Marel, G.A.; Oppenheimer, N.J.; Hecht, S.M. Copper-dependent cleavage of DNA by bleomycin. *Biochemistry* **1987**, *26*, 931–942. [CrossRef] [PubMed]

153. Sugiura, Y.; Ishizu, K.; Miyoshi, K. Studies of metallobleomycins by electronic spectroscopy, electron spin resonance spectroscopy, and potentiometric titration. *J. Antibiot.* **1979**, *32*, 453–461. [CrossRef] [PubMed]

154. Ehrenfeld, G.M.; Rodriguez, L.O.; Hecht, S.M.; Chang, C.; Basus, V.J.; Oppenheimer, N.J. Copper(I)-bleomycin: Structurally unique complex that mediates oxidative DNA strand scission. *Biochemistry* **1985**, *24*, 81–92. [CrossRef] [PubMed]

155. Pron, G.; Belehradek, J.; Mir, L.M. Identification of a Plasma Membrane Protein That Specifically Binds Bleomycin. *Biochem. Biophys. Res. Commun.* **1993**, *194*, 333–337. [CrossRef] [PubMed]

156. Pron, G.; Mahrour, N.; Orlowski, S.; Tounekti, O.; Poddevin, B.; Belehradek, J.; Mir, L.M. Internalisation of the bleomycin molecules responsible for bleomycin toxicity: A receptor-mediated endocytosis mechanism. *Biochem. Pharmacol.* **1999**, *57*, 45–56. [CrossRef]

157. Aouida, M. A Genome-Wide Screen in Saccharomyces cerevisiae Reveals Altered Transport as a Mechanism of Resistance to the Anticancer Drug Bleomycin. *Cancer Res.* **2004**, *64*, 1102–1109. [CrossRef] [PubMed]

158. Aouida, M.; Leduc, A.; Poulin, R.; Ramotar, D. AGP2 Encodes the Major Permease for High Affinity Polyamine Import in Saccharomyces cerevisiae. *J. Biol. Chem.* **2005**, *280*, 24267–24276. [CrossRef] [PubMed]

159. Aouida, M.; Poulin, R.; Ramotar, D. The Human Carnitine Transporter SLC22A16 Mediates High Affinity Uptake of the Anticancer Polyamine Analogue Bleomycin-A5. *J. Biol. Chem.* **2010**, *285*, 6275–6284. [CrossRef] [PubMed]

160. Berra, S.; Ayachi, S.; Ramotar, D. Upregulation of the Saccharomyces cerevisiae efflux pump Tpo1 rescues an Imp2 transcription factor-deficient mutant from bleomycin toxicity. *Environ. Mol. Mutagen.* **2014**, *55*, 518–524. [CrossRef] [PubMed]

161. Uemura, T.; Tachihara, K.; Tomitori, H.; Kashiwagi, K.; Igarashi, K. Characteristics of the polyamine transporter TPO1 and regulation of its activity and cellular localization by phosphorylation. *J. Biol. Chem.* **2005**, *280*, 9646–9652. [CrossRef] [PubMed]

162. Igarashi, K.; Kashiwagi, K. Characteristics of cellular polyamine transport in prokaryotes and eukaryotes. *Plant. Physiol. Biochem.* **2010**, *48*, 506–512. [CrossRef] [PubMed]

163. Della Latta, V.; Cecchettini, A.; Del Ry, S.; Morales, M. Bleomycin in the setting of lung fibrosis induction: From biological mechanisms to counteractions. *Pharmacol. Res.* **2015**, *97*, 122–130. [CrossRef] [PubMed]

164. Lazo, J.S.; Humphreys, C.J. Lack of metabolism as the biochemical basis of bleomycin-induced pulmonary toxicity. *Proc. Natl. Acad. Sci. USA* **1983**, *80*, 3064–3068. [CrossRef] [PubMed]

165. Brömme, D.; Rossi, A.B.; Smeekens, S.P.; Anderson, D.C.; Payan, D.G. Human Bleomycin Hydrolase: Molecular Cloning, Sequencing, Functional Expression, and Enzymatic Characterization. *Biochemistry* **1996**, *35*, 6706–6714. [CrossRef] [PubMed]

166. Zou, Y.; Fahmi, N.E.; Vialas, C.; Miller, G.M.; Hecht, S.M. Total Synthesis of Deamido Bleomycin A$_2$, the Major Catabolite of the Antitumor Agent Bleomycin. *J. Am. Chem. Soc.* **2002**, *124*, 9476–9488. [CrossRef] [PubMed]

167. Schwartz, D.R.; Homanics, G.E.; Hoyt, D.G.; Klein, E.; Abernethy, J.; Lazo, J.S. The neutral cysteine protease bleomycin hydrolase is essential for epidermal integrity and bleomycin resistance. *Proc. Natl. Acad. Sci. USA* **1999**, *96*, 4680–4685. [CrossRef] [PubMed]

168. Pei, Z.; Calmels, T.P.; Creutz, C.E.; Sebti, S.M. Yeast cysteine proteinase gene ycp1 induces resistance to bleomycin in mammalian cells. *Mol. Pharmacol.* **1995**, *48*, 676–681. [PubMed]

169. Chen, J.; Chen, Y.; He, Q. Action of bleomycin is affected by bleomycin hydrolase but not by caveolin-1. *Int. J. Oncol.* **2012**, *41*, 2245–2252. [CrossRef] [PubMed]

170. Okamura, Y.; Nomoto, S.; Hayashi, M.; Hishida, M.; Nishikawa, Y.; Yamada, S.; Fujii, T.; Sugimoto, H.; Takeda, S.; Kodera, Y. Identification of the bleomycin hydrolase gene as a methylated tumor suppressor gene in hepatocellular carcinoma using a novel triple-combination array method. *Cancer Lett.* **2011**, *312*, 150–157. [CrossRef] [PubMed]

171. Wang, H.; Ramotar, D. Cellular resistance to bleomycin in Saccharomyces cerevisiae is not affected by changes in bleomycin hydrolase levels. *Biochem. Cell Biol.* **2002**, *80*, 789–796. [CrossRef] [PubMed]

172. Ma, Q.; Xu, Z.; Schroeder, B.R.; Sun, W.; Wei, F.; Hashimoto, S.; Konishi, K.; Leitheiser, C.J.; Hecht, S.M. Biochemical evaluation of a 108-member deglycobleomycin library: Viability of a selection strategy for identifying bleomycin analogues with altered properties. *J. Am. Chem. Soc.* **2007**, *129*, 12439–12452. [CrossRef] [PubMed]

173. Thomas, C.J.; Chizhov, A.O.; Leitheiser, C.J.; Rishel, M.J.; Konishi, K.; Tao, Z.F.; Hecht, S.M. Solid-phase synthesis of bleomycin A(5) and three monosaccharide analogues: Exploring the role of the carbohydrate moiety in RNA cleavage. *J. Am. Chem. Soc.* **2002**, *124*, 12926–12927. [CrossRef] [PubMed]

174. Chen, J.K.; Yang, D.; Shen, B.; Neilan, B.A.; Murray, V. Zorbamycin has a different DNA sequence selectivity compared with bleomycin and analogues. *Bioorg. Med. Chem.* **2016**, *24*, 6094–6101. [CrossRef] [PubMed]

175. Murray, V.; Chen, J.K.; Yang, D.; Shen, B. The genome-wide sequence specificity of DNA cleavage by bleomycin analogues in human cells. 2018; manuscript submitted for publication.

176. Yoshida, T.; Ogawa, M.; Ota, K.; Yoshida, Y.; Wakui, A.; Oguro, M.; Ariyoshi, Y.; Hirano, M.; Kimura, I.; Matsuda, T. Phase II study of NK313 in malignant lymphomas: An NK313 Malignant Lymphoma Study Group trial. *Cancer Chemother. Pharmacol.* **1993**, *31*, 445–448. [CrossRef] [PubMed]

177. Denny, W.A. DNA-Intercalating agents as antitumour drugs: Prospects for future design. *Anticancer Drug Des.* **1989**, *4*, 241–263. [PubMed]

178. Murray, V.; Chen, J.K.; Galea, A.M. The Potential of Acridine Carboxamide Platinum complexes as Anti-Cancer Agents: A Review. *Anti-Cancer Agents Med. Chem.* **2014**, *14*, 695–705. [CrossRef]

MDPI

St. Alban-Anlage 66

4052 Basel

Switzerland

Tel. +41 61 683 77 34

Fax +41 61 302 89 18

www.mdpi.com

International Journal of Molecular Sciences Editorial Office

E-mail: ijms@mdpi.com

www.mdpi.com/journal/ijms

www.ingramcontent.com/pod-product-compliance
Lightning Source LLC
Chambersburg PA
CBHW051726210326
41597CB00032B/5627